ANALYTICAL CHEMISTRY

(Problems and Solutions)

ANALYTICAL CHEMISTRY

(Problems and Solutions)

(SECOND EDITION)

S.M. Khopkar
Professor Emeritus
Department of Chemistry
Indian Institute of Technology, Bombay
Powai, Mumbai, India

Publishing Globally

NEW AGE INTERNATIONAL (P) LIMITED, PUBLISHERS
New Delhi • Bangalore • Chennai • Cochin • Guwahati
Hyderabad • Kolkata • Lucknow • Mumbai
Visit us at www.newagepublishers.com

Published by New Age International (P) Ltd., Publishers
First Edition: 2002
Second Edition: 2013
Reprint: 2014

BRANCHES

- **Bangalore** 37/10, 8th Cross (Near Hanuman Temple), Azad Nagar, Chamarajpet, Bangalore-560 018
Tel.: (080) 26756823, Telefax: 26756820, **E-mail: bangalore@newagepublishers.com**
- **Chennai** 26, Damodaran Street, T. Nagar, Chennai-600 017
Tel.: (044) 24353401, Telefax: 24351463, **E-mail: chennai@newagepublishers.com**
- **Cochin** CC-39/1016, Carrier Station Road, Ernakulam South, Cochin-682 016
Tel.: (0484) 2377004, Telefax: 4051303, **E-mail: cochin@newagepublishers.com**
- **Guwahati** Hemsen Complex, Mohd. Shah Road, Paltan Bazar, Near Starline Hotel, Guwahati-781 008
Tel.: (0361) 2513881, Telefax: 2543669, **E-mail: guwahati@newagepublishers.com**
- **Hyderabad** 105, 1st Floor, Madhiray Kaveri Tower, 3-2-19, Azam Jahi Road, Near Kumar Theater Nimboliadda, Kachiguda, Hyderabad-500 027, Tel.: (040) 24652456, Telefax: 24652457
E-mail: hyderabad@newagepublishers.com
- **Kolkata** RDB Chambers (Formerly Lotus Cinema) 106A, 1st Floor, S.N. Banerjee Road, Kolkata-700 014
Tel.: (033) 22273773, Telefax: 22275247, **E-mail: kolkata@newagepublishers.com**
- **Lucknow** 16-A, Jopling Road, Lucknow-226 001
Tel.: (0522) 2209578, 4045297, Telefax: 2204098, **E-mail:lucknow@newagepublishers.com**
- **Mumbai** 142C, Victor House, Ground Floor, N.M. Joshi Marg, Lower Parel, Mumbai-400 013
Tel.: (022) 24927869, Telefax: 24915415, **E-mail: mumbai@newagepublishers.com**
- **New Delhi** 22, Golden House, Daryaganj, New Delhi-110 002
Tel.: (011) 23262368, 23262370, Telefax: 43551305, **E-mail: sales@newagepublishers.com**

ISBN: 978-81-224-2746-2

₹ 275.00

C-13-09-7128

Printed in India at Print 'O' Pack, Delhi.
Typeset at Pagitek, Delhi.

PUBLISHING FOR ONE WORLD
NEW AGE INTERNATIONAL (P) LIMITED, PUBLISHERS
7/30 A, Daryaganj, New Delhi-110002
Visit us at **www.newagepublishers.com**

Dedicated to

Sucheta, Samir, Supriya

Preface to the Second Edition

The first edition of this book was published five years ago. Within this period there was phenomenal progress in analytical chemistry. Newer instrumental methods have been discovered or previous techniques which were less popular were now extensively used. In view of these changes it was felt worthwhile to revise the edition of this book. A mathematical aspect of newer methods were considered. In few instances where numerical data was not easily available a set of the multiple choice objective questions were introduced. Somehow there is growing trends in various examination to test knowledge through objective questions.

An endeavour was made to include new chapters on Electrophoresis, Raman Spectroscopy, Chemiluminescence inclusive of atomic fluorescence and ionisation spectroscopy, refractometry, X-ray spectroscopy, Mössbauer spectroscopy, Electron microscopy and chemical and biosensors. In addition the chapters on high performance liquid chromatography, super critical fluid chromatography and extraction, capillary electrophoresis, have been re-written. Every attempt was made to recheck some problems. All kind of printing, grammatical errors have been eliminated with due care.

I take this opportunity to thank Mr. Saumya Gupta, Managing Director and his efficient band of workers in New Age International Publishers (P) Ltd. for excellent cooperation and very good production. I also thank families of my son–Samir, Madhavi, Rohan and that of my daughter Dr. Supriya Ketan Ruchir for their cheering words and encouragement. Finally I thank my wife Dr. Sucheta for the wonderful cooperation and assistance during revision of this book.

I do hope readers would receive this book with same zeal and enthusiasm as was done for previous edition. I welcome constructive criticism and suggestions.

S.M. KHOPKAR

Professor Emeritus
Department of Chemistry
Indian Institute of Technology
Powa, Mumbai-400 076.

Preface to the First Edition

Analytical Chemistry is the science of measurement and instrumentation. This is the only branch of experimental science which involves a large number of measurement followed by intensive calculations. This is called quantitative chemical analysis. To procure reliable and accurate results, it is imperative to perform a series of experiments followed by intricate calculations before the final result is arrived at. Unfortunately in the curricula of our academic institutions more attention is paid to experimentation in the laboratory with virtually no emphasis on training in solving numerical problems. Generally there is apathy and reluctance on the part of students to do numerical. problems. During periodic evaluation of performance of candidates in examinations, it was observed that the candidates prefer writing long descriptive answers to solving simple problems. Precisely this was the reason for me to undertake writing this book on Analytical Chemistry through problems.

This book is divided into four major areas of analytical chemistry. The first portion comprising ten chapters covers statistical analysis, volumetry and gravimetry. The second part with ten chapters pertains to separation methods such as solvent extraction, ion exchange and various forms of the chromatography. The next part of twelve chapters relates to analytical spectroscopy involving absorption, emission and magnetic resonance spectroscopy. The surface analysis techniques like EŞCA or auger and also the photoacoustic spectroscopy are also included in this section. The last part of the book includes electroanalytical, thermal and radioanalytical methods. The sequence and order of chapters is identical to one given in my earlier textbook on *Basic Concepts of Analytical Chemistry*, 3rd edn. (2007). In fact this book should serve as complementary book to my original textbook on analytical chemistry.

To facilitate tackling of problems a thorough comprehension of theoretical aspects of each topic is essential. Hence an attempt is made to provide theory related to problems in the beginning of each chapter. This is followed by a large number of solved problems. Then requisite number of problems with answers are provided for practice. At the end of each chapter reference to well-known textbooks in chemistry which I have so freely used while compiling this book is appended to indicate source of similar problems. One needs clear understanding of theoretical principles, which the reader can procure by reading the initial portion of each chapter. However for getting greater depth of knowledge he may refer to my *Basic Concepts of Analytical Chemistry.* In newer areas since numerical data was not easily available, multiple choice questions to test the knowledge of students are provided. It is my cherished dream that after mastering the art of problem solving, faculty as well as students would realise how exciting is the subject of analytical chemistry.

I take this opportunity to thank New Age International (P) Ltd. Publishers for the excellent production and for publishing many more books on numerical problems in different disciplines. I desire to express my gratitude to the family of my son–Samir, Madhavi & Rohan as well as that of my daughter Dr. Supriya Ketan for rendering all assistance and encouragement during compilation of this book. I would be failing in my duty if I do not thank my wife Dr. Sucheta for her patience and for creating conducive atmosphere for writing this book. Constructive suggestions are welcome from the readers.

S.M. KHOPKAR

Acknowledgements

Thanks are due to following authors as well as publishers whose works have been freely used for formulating problems in this book.

1. *Analytical Chemistry*, D.A. Skoog and D.M. West, 3rd edn., Holt Reinhart Winston (1979).
2. *Analytical Chemistry: An Introduction,* D.J. Pretzyke and C.W. Frank, Academic Press (1974).
3. *Quantitative Analytical Chemistry,* J.S. Fritz, H. Schenk, Allyn & Bacon (1975).
4. *Chemical Analysis: An Advanced Text,* H.A. Laitinen and W.E. Harris, 2nd edn., McGraw-Hill & Co. (1975).
5. *Analytical Chemistry,* J.G. Dick, McGraw-Hill & Co. (1973).
6. *Quantitative Analytical Chemistry*, V. Alexcyeo, Mir Publishing House (1979).
7. *A Brief Introduction to Modern Chemical Analysis,* D.G. Peters, J.M. Hayes and G.M. Hieftze, Saunders & Co. (1980).
8. *A First Course in Quantitative Analysis*, R.D. Brumblay, Addison Wesley (1970).
9. *Quantitative Chemisry: Measurements & Equilibrium,* W.B. Guenther, Addison Wesley (1962).
10. *Quantitative Chemical Analysis,* G.H. Ayers, Harper and Row Brothers (1958).
11. *An Introductions to Separation Science,* B.L. Karger, L.R. Snyder and C. Horvath, Wiley-Interscience (1973).
12. *Chemical Separation Methods,* J.A. Dean, Van Nostrand Reinhold & Co. (1970).
13. *Chemical Separations & Measurements,* D.G. Peter, J.M. Hoyes and G.M. Hjeftje, Saunders Golden Series (1974).
14. *Solving Problems in Analytical Chemistry,* H. Brewster, John Wiley & Co. (1980).
15. *Calculations in Analytical Chemistry,* L.F. Hamilton, S.G. Simpson and D.W. Ellis, McGraw-Hill & Co. (1958).
16. *Problems in Quantitative Analysis*, A, Musakin Mir Publishing House (1984).
17. *Instrumental Methods of Analysis*, H.H. Willard, I.L. Merritt and J.A. Dean, D. Van Nostrand & Co. (1971).
18. *Instrumental Methods of Chemical Analysis,* 5th Edition, G.W. Ewing, McGraw-Hill & Co. (1985).
19. *Introduction to Instrumental Analysis*, R.D. Braun, McGraw-Hill & Co. (1987).
20. *Problems & Excercises in Analytical Chemistry,* A.A. Yaroslavasan, Mir Publishing House (1984).
21. *Modern Optical Methods of Analysis,* E.D. Olsen, McGraw-Hill & Co. (1975).

22. *Physico Chemical Analysis,* Y. Lyalikov, Mir Publishing (1964).
23. *Solving Problems with NMR Spectroscopy,* A.U. Rahman and M.I. Chandharey Academic Press (1996).
24. *Analytical Chemistry,* G.D. Christan, 5th edn., Wiley-Interscience (1997).
25. *Principles & Methods of Chemical Analysis,* H.F. Walton, Prentice Hall & Co. (1958).
26. *Physical & Chemical Methods of Separation,* E.W. Berg, McGraw-Hill & Co. (1963).
27. *Principles of Instrumental Analysis,* D.A. Skoog, D.M. West and J.J. Leory, Saunders, Holt & Co. (1980).
28. *Thermal Methods of Analysis: Principles Application & Problems,* P. Hains Blacki, Professional (1995).
29. *Principles of Instrumental Analysis,* D.A. Skoog, F.J. Holler and T.A. Neuman, 5th edn., Saunder Harcourt & Co. (1998).
30. *Spectroscopic Identification of Organic Compounds,* R.G. Silvestine, 5th edn., John-Wiley & Co. (1993).
31. *Spectroscopic Methods in Organic Chemistry,* D.H. Williams and I. Fleming, 4th edn., Tata McGraw-Hill & Co. (1987).
32. *Fundamentals of Molecular Spectroscopy*, C.N. Banwell and E.M. McCash, 4th edn., Tata McGraw-Hill & Co. (1994).
33. *Basic Concepts of Analytical Chemistry*, S.M. Khopkar, 3d edn., New Age International Publishers (P) Ltd. (2007).
34. *Environmental Pollution Analysis*, S.M. Khopkar, 2nd edn., New Age International Publishers (P) Ltd. (2011).
35. *Solvent Extraction of Metals*, A.K. De, S.M. Khopkar and R.A. Chalmers—Van Nostrand, Reinhold Co. (1970).
36. *Analytical Chemistry of Macrocyclic & Supermolecular Compounds*, S.M. Khopkar, 2nd edn., Springer-Verlag (2006).
37. *Solvent Extraction Separation of Elements with Liquid Ion Exchangers,* S.M. Khopkar, New Age International Publishers (P) Ltd. (2007).
38. *Environmental Pollution Monitoring and Control*, S.M. Khopkar, New Age International Publishers (P) Ltd. (2008).
39. *Instrumental Methods in Bioanalytical Chemistry*, S.M. Khopkar, New Age International Publishers (P) Ltd. (In press).

Contents

Chapter 1

Introduction

There is spectacular growth of knowledge in the field of analysis, making it almost impossible for us to keep track of the newer developments. An array of instruments with all sophistication with microprocessor and computers have arrived in the market to make the job of chemist very simple. A plethora of organic reagents are being synthesised to bring to limelight complexation, extraction and separation of elements at tracer concentration. The word tracer has lost its original significance as in last four decades anything at the level of 10^{-6} gm was considered as tracer concentration, but now we have instrumental methods permitting analysis up to 10^{-9}–10^{-15} gm concentration. The chemist has reached to the lowest or ultimate level of 10^{-15} gm of analysis termed as fematogram concentration. We have automated instruments needing no personal supervision, we have instruments reading out direct concentration of an analyt on the scale. The time factor has been so carefully controlled that we now have techniques which permit analysis within few seconds. We have also array of instrumental methods involving non-destructive methods like X-ray, neutron activation analysis or atomic emission spectroscopy facilitating rapid analysis. The science of analysis has so much advanced that it is very easy to know the level of constituent present in sample by few measurement of surfaces with electron spectroscopy for chemical analysis (ESCA) or aüger spectroscopy. The sound transmission is fruitfully used in photoacoustic spectroscopy for measurement. The components can be analysed at finite dilution by accomplishing clean separation at the lowest level with supercritical fluid chromatography or capillary electrophoresis. The old image of analytical chemistry involving qualitative or quantitative analysis with gravimetric or volumetric techniques is slowly disappearing inspite of the fact that for the basic standardisation of sample one still requires classical modes of analysis like volumetry or gravimetry.

The application of analytical chemistry has no boundaries as the horizon is limit for extending methods for the analysis of samples from the environment, space, biological products, material science and industrial samples where several physico-chemical or biological principles are used for such studies. Analytical chemistry is redesigned as analytical science. The methods for quantitative analysis are increasing by leaps and bounds. The criteria for selection of methods varies from sample to sample. Analytical chemistry not only encompasses chemical analysis but also covers boundaries of advances in physical, chemical and biological science. We can have either one step or multistep methods of analysis. The determination may involve one or more physical or chemical concepts. The role of instrumentation is of atmost importance in analytical science, it has all together changed the entire face of chemistry in last few decades.

In this chapter a large set of multiple choice questions are provided to test the readers knowledge of modern analytical chemistry. It tests the imagination to what extent analytical science has encircled other branches of knowledge. It provides exciting information on future scope of analytical chemistry. As you would proceed through different chapters of this work it will open a panorama of exciting inventions and challenges to chemist in general. You can also refer to *Basic Concepts of Analytical Chemistry* 3rd edn., New Age International Publishers (2007) p. 5, along with other excellent books listed at the end of this chapter.

OBJECTIVE QUESTIONS

Tick (√) mark the correct answer from the following multiple choices:

1.1 Analytical chemistry has to play vital role in
(*a*) elucidation and characterisation of structure (*b*) identification of the constituent
(*c*) estimation of components (*d*) separation of the constituents.

1.2 So far conventional analytical chemistry was confined to study of
(*a*) volumetric and gravimetric analysis (*b*) colorimetry
(*c*) chromatography (*d*) electroanalytical methods.

1.3 A level of 10^{-9} gm for analysis can be ascertained quantitatively by the technique of
(*a*) spectrophotometry (*b*) fluorimetry
(*c*) polarography (*d*) volumetry.

1.4 Instrumental methods have some limitation because
(*a*) the interference must be removed before analysis
(*b*) large volume of sample is required
(*c*) sample has to be in liquid state or solid form
(*d*) sample must be thermally stable.

1.5 The choice of particular method of analysis depends upon the
(*a*) speed of analysis (*b*) availability of instrument
(*c*) cost of analysis (*d*) level of analysis.

1.6 The method of analysis will vary depending upon the
(*a*) background of the sample or source (*b*) accuracy required
(*c*) selectivity and sensitivity (*d*) cost of analysis.

1.7 Chemical analysis and analytical chemistry are not similar as one
(*a*) involves qualitative while other involves quantitative analysis
(*b*) is adopted at milligram level while other is used at microgram concentration
(*c*) is focused over much broader and general aspect of analysis
(*d*) involves classical methods while other involves instrumental methods.

1.8 Excellent progress could be made in instrumental methods of analysis due to
(*a*) methods of separation (*b*) speed of analysis
(*c*) sensitivity of the method (*d*) accuracy of the analysis.

1.9 The analysis of metal at the 10^{-11} gram concentration is an example of
(*a*) macro analysis at milligram level (*b*) micro analysis at micro gram level
(*c*) micro analysis at nano gram level (*d*) trace analysis at picogram level.

1.10 The sample weighing less than 0.001 gm is termed as
(*a*) semi-micro sample (*b*) micro sample
(*c*) ultra or sub-micro sample (*d*) macro sample.

1.11 In sample if the constituents are present in less than 0.01% it is termed as
(*a*) minor constituent (*b*) major constituent
(*c*) trace concentration (*d*) ultra trace concentration.

1.12 The most difficult step in determination of sample is the
(*a*) procurement of sample (*b*) conversion to measurable form
(*c*) measurement of the constituent (*d*) calculation and interpretation.

1.13 The physiochemical methods of analysis involve measurement of
(*a*) physical property (*b*) chemical characteristics
(*c*) biological property (*d*) chemical reactions.

1.14 Physical methods of analysis differ from chemical methods as
(*a*) former are more dependable
(*b*) speed and accuracy of analysis is excellent
(*c*) physical property in former is examined
(*d*) latter are cheaper in cost.

1.15 A good analytical chemist has to consider
(*a*) limitation of method of analysis (*b*) cost of analysis
(*c*) steps involved in analysis (*d*) scope of the analysis.

1.16 The classical methods are as important as instrumental methods of analysis because
(*a*) they are rapid and selective
(*b*) it is necessary to standardise the analytical sample
(*c*) methods are more dependable
(*d*) methods are relatively cheap in cost.

1.17 Instrumental methods like polarography in environmental analysis facilitates
(*a*) rapid analyse sample
(*b*) to render method more sensitive
(*c*) analysis at microgram concentrations
(*d*) identifies the metal pollutant from knowledge of half wave potential.

1.18 The chief limitations of instrumental methods is that
(*a*) it is expensive due to cost of instruments
(*b*) it cannot be used for determination of large concentration of sample
(*c*) it is time consuming and needs removal of interferences
(*d*) it needs sophisticated instruments for accurate work.

1.19 The methods of analysis of iron irrespective of it being present in haemoglobin or steel is not always
(*a*) the same (*b*) slightly different
(*c*) totally different (*d*) exactly identical.

1.20 The statistical analysis of result in chemical analysis

(*a*) furnishes average value (*b*) indicate variation in result
(*c*) shows mean value of the results (*d*) gives median value.

1.21 The method which does not involve the wet chemical analysis is

(*a*) spectrophotometry (*b*) fluorimetry
(*c*) polarography (*d*) emission spectroscopy.

1.22 Amongst non-destructive methods best one for multi-element analysis is

(*a*) neutron activation analysis (*b*) X-ray fluorescence
(*c*) atomic emission spectroscopy (*d*) reflectance spectroscopy.

1.23 The complexometric titration method with EDTA is applicable at a level of

(*a*) miligram (*b*) macrogram
(*c*) microgram (*d*) submicrogram.

1.24 The titration of silver with ammonium thiocyanate is an example of

(*a*) redox titration (*b*) complexometric titration
(*c*) acid base titration (*d*) precipitation titration.

1.25 Metallochromic indicators are used in analysis of

(*a*) HCl against NaOH (*b*) $HgCl_2$ against NaSCN
(*c*) $FeCl_3$ against $K_2Cr_2O_7$ (*d*) $CaCl_2$ against EDTA

1.26 Analytical Chemistry is very much advanced now as it covers

(*a*) qualitative analysis (*b*) elemental analysis
(*c*) surfaces analysis (*d*) quantitative analysis.

1.27 A modern analytical technique always involves measurement of

(*a*) chemical characteristics (*b*) physical characteristics
(*c*) physico-chemical property (*d*) biological property.

1.28 Bio analytical Science is truly interdisciplinary as it has ramification over

(*a*) physics
(*b*) chemistry
(*c*) mathematics
(*d*) Biological Sciences.

ANSWERS

1.1 a, 1.2 a, 1.3 b, 1.4 a, 1.5 d, 1.6 a, 1.7 c, 1.8 c, 1.9 d, 1.10 b, 1.11 c, 1.12 a, 1.13 a, 1.14 b, 1.15 a, 1.16 b 1.17 d, 1.18 d, 1.19 c, 1.20 b, 1.21 d, 1.22 a, 1.23 a, 1.24 d, 1.25 d, 1.26 c, 1.27 c, 1.28 d.

REFERENCES

1.1 S.M. Khopkar, *Basic Concepts of Analytical Chemistry,* 3rd edn., New Age International Publishers (P) Ltd., p. 5 (2007).

1.2 I. M. Kolthoff and P. J. Elving (Ed.), *Treaties in Analytical Chemistry, Interscience,* New York, Part I-III all volumes (1959).

1.3 C.L.Wilson and D.W. Wilson, *Comprehensive Analytical Chemistry.* Elsevier Pub. Co. (1958).

1.4 Ju. Lurie, *Handbook of Analytical Chemistry,* Mir Publishing House, Moscow (1981).

1.5 G.D. Christen, *Analytical Chemistry,* 5th edn. Wiley-Inter-science Ltd. (1996).

1.6 D.A. Skoog and D.M. West 3rd edn., *Analytical Chemistry,* Holt Reinhart and Winston Ltd. (1979).

1.7 D.A. Skoog and L. Leory, *Principles of Instrumental Analysis,* 4th edn., Holt Reinhart and Winston Ltd. (1990).

1.8 D.A. Skoog, F.J. Holler, T.A. Nieman, *Principles of Instrumental Analysis,* 5th edn., Harcourt Butler and Co. (1998).

Chapter 2

Reliability of Analytical Data

INTRODUCTION

Analytical Chemistry is science where one collects large amount data from which one has to draw definite inferences. This is true of the data collected in environmental monitoring. If we carryout statistical analysis then one can draw definite conclusions on reliability and predict possibility of errors in measurements, as mere collection of data serves no purpose. In fact, good environmental chemist prepares from such data mathematical models and uses them for future forecasting of the environmental pollution in environmental impact assessment studies.

In chemical analysis, the measurements must be dependable. We must identify the determinate and indeterminate errors in analysis. Further we should check whether the readings are precise (*i.e.*, reproducible but may not be accurate) or readings are accurate (means they are both reproducible and accurate).

The various parameters which are calculated as shown in following:

1. We calculate **% of error** from following relationship, *viz.*,

$$\% \text{ error} = \left(\frac{E}{T} \times 100\right) = \% \qquad \text{if } E = \text{absolute error}$$

$$\text{Error per thousand } \left(\frac{E}{T} \times 1000\right) = \text{ppt} \qquad \text{if } T = \text{true value}$$

   One can experimentally test accuracy of method by independent experimental procedures and then decide as to how one can improve accuracy of analysis by selecting same blank, calibrating instrument. For all such work what one needs is the statistical analysis as it forms an integral part of the good analytical work. It comprises of evaluation of several parameters. Such parameters would be considered with appropriate examples in following sections.

2. **Absolute and relative difference:** If we have two values as 2.431 and 2.410 then absolute difference is (2.431 – 2.410) = 0.021 but relative difference is (0.021 × 1000) ÷ 2.40 = 8.75 ppt. This is relative difference in parts per thousand.

3. **Mean value and average deviation:** Suppose we have a set of five values as are 0.1010; 0.1020; 0.1005; 0.1030; 0.1015 the average value will be total of all values divided by 5 it is 0.1016.

4. **The average deviation:** The average deviation is calculated as

$$(0.0006 + 0.0004 + 0.0009 + 0.0014 + 0.0001) \text{ divided by } 5 = \frac{0.0034}{5} = 0.00068 = d.$$

The average deviation of mean is

$$D = \frac{d}{\sqrt{n}} = \frac{0.00068}{\sqrt{5}} = \frac{0.00068}{2.236} = 3.041 \times 10^{-4}$$

5. **Standard deviation and Standard deviation of mean:**

$$S = \sqrt{\frac{\Sigma\delta^2}{(n-1)}} \text{ as while SD of mean } S = \frac{S}{\sqrt{n}}. \text{ In above example}$$

$$\Sigma\delta^2 = (0.0006)^2 + (0.0004)^2 + (0.0009)^2 + (0.0014)^2 + (0.0001)^2 = 33 \times 10^{-7}$$

$$= 3.3 \times 10^{-6}$$

$$S = \sqrt{\frac{\Sigma\delta^2}{(n-1)}} = \sqrt{\frac{(3.30 \times 10^{-6})}{(5-1)}} = 0.0009$$ is standard deviation while the standard deviation of mean is $S = \frac{S}{\sqrt{n}} = \frac{0.0009}{\sqrt{5}} = \frac{0.0009}{2.236} = 0.00004025.$

6. **Median Value:** If we have 5 set of values in different order and if we select a middle value so that this value is having two reading higher and two reading lower than selected value this is called as median value. In present question 0.1051 is median value as 0.1030, 0.1020 are above while 0.1010 and 0.1005 are below this value.
7. **Range:** This is difference between the highest and smallest value, we have 0.1030 highest value and 0.1005 as lowest value so range is 0.0250.
8. **Relative Standard deviation** is also called coefficient of variation and is given as

$$\frac{\text{standard deviation}}{\text{mean}} \times 100 = \frac{S \times 100}{m} \; i.e., \; \frac{0.0009 \times 100}{0.1016} = 0.8858\%.$$

9. **Confidence limit:** If '*n*' is small than *S* or *s* will not be reliable and shall be uncertain. By calculating 90% confidence limit. For '*S*' upper and lower limits of confidence limits are obtained by dividing (n – 1) S^2 by two numbers taken from a Table of chi-square given in Appendix No. 7 *e.g.* If $n = 7$, $(n - 1) = 6$ then

$$\text{Upper confidence limit} = 3.67S^2 = \frac{(n-1)S^2}{\text{Chi-sq. value at 95\% confidence limit}}$$

$$\text{Lower confidence limit} = 0.47S^2 = \frac{(n-1)S^2}{\text{Chi-sq. value at 90\% confidence limit}} \text{ at 90\% limit.}$$

For a '*n*' value Chi-square value for 95% or 90% C.L. are available in Tables in the Appendix No. 7.

Based upon these confidence limits for 5% and 95% limit we can easily calculate the upper and lower confidence limits as

$$\text{Upper confidence limit} = \frac{(n-1)S^2}{3.940} = \frac{10\times(0.308)^2}{3.940} = 0.2407$$

$$\text{Lower confidence limit} = \frac{(n-1)S^2}{18.307} = \frac{10\times(0.308)^2}{18.307} = 0.0518$$

10. **The Q-test for rejection** of a results can be found out at 90 % level, by formula as shown below if M_1 is first reading, M_2 is second reading $Q_1 = \frac{(M_2 - M_1)}{(M_n - M_1)}$ is upper range

$$\text{Range } Q_n = \frac{(M_n - M_{n-1})}{(M_n - M_1)} \text{ is lower range.}$$

Now if Q_1, Q_n is larger than value in Tables in Appendix it must be rejected we calculate it from following examples. The readings are 6.593; 6.942; 7.098; 7.146; and 7.256 if $n = 5$ range is $(7.256 - 6.942) = 0.668$.

Now $$Q_1 = \frac{M_2 - M_1}{M_5 - M_1} = \frac{(6.942 - 6.593)}{(7.256 - 6.593)} = 0.527 < 0.64$$

Accepted if $(M_2 = 6.942, M_1 = 6.593)$ $(M_5 - M_1)$ range = 0.668

$$Q_2 = \frac{M_5 - M_4}{M_5 - M_1} = \frac{(7.256 - 7.146)}{(7.256 - 6.593)} = 0.17 < 0.64$$

Accepted if $(M_5 = 7.256, M_4 = 7.146)$ $(M_5 - M_1)$ range = 0.668

For 90% confidence limit $Q_{90} = 0.64$ from given Appendix-6. The value is lower than 0.64 hence acceptable.

11. **Student 't' test:** First standard deviation is determined and a value of 't' is obtained from 't' Tables in Appendix-4. The limits for given confidence level are given by equation $\left(\text{mean} \pm \frac{t.s}{\sqrt{n}}\right)$ supposing mean is 9.46 standard deviation is 0.17% and $n = 6$ then $t = 2.02$ for 90% confidence limit for six values from Table given in Appendix-8 $\left(9.46 \pm \frac{2.02 \times 0.17}{\sqrt{6}}\right)$ mean 9.46 ± 0.14%. This is acceptable value.

12. **Rejection of results:** Apart from criteria of rejection if the values from Q we can also reject by another criteria. If difference $(\delta) \geq 4d$ where d is given above the average deviation, then value is rejected, *e.g.*, $\delta \geq 2.5\ d$ is also allowed by some workers. If values 22.62 and if $d = 0.022$ then $\delta = 0.193 < 4d$, *i.e.* $(4 \times 0.022) = 0.088$ less than δ the value 22.62 is acceptable. With this knowledge we can solve several problems. The tables for 't' Q, F, Chi-square, etc., are provided at end of book in various Appendixes.

SOLVED PROBLEMS

2.1 Ten determination of lead gravimetrically as lead chromate gave a mean of 0.1462% with the standard deviation of 0.0074%. Calculate the 95% confidence interval of the mean and the standard deviation?

Ans: The 95% confidence interval of the mean is evaluated as per equation,

$\mu = x \pm 2.58 \frac{\sigma}{\sqrt{n}}$ with 95-99% confidence limit if $t = 2.58$. On substituting the values $x = 0.1462$, $n = 10$, $\sigma = 0.0074$.

$\mu = 0.1462 \pm 2.58 \times \frac{0.0074}{\sqrt{10}} = 0.1515$ to 0.1409 again for the (Appendix No. 4).

Standard deviation $q = \sqrt{\frac{s^2(n-1)}{n}} = \sqrt{\frac{\Sigma(x_1 \bar{x})^2}{n}} = \sqrt{\frac{0.0074 \times 9}{10}} = 0.00702.$

The 95% confidence level for S.D. = (0.0074) – (0.66 × 9) and + (1.77 × 9) = 0.0028 to 0.0198.

2.2 The following results were obtained for the gravimetric estimation of nickel as oxime 112.25, 112.36, 112.21, 112.30, 112.36, 112.32 mg. Is the mean of these values, *viz.*, 112.30 differ too much from accepted figure of 112.40?

Ans: We first calculate standard deviation as:

$S = \sqrt{\frac{\Sigma(x_1 - \bar{x})^2}{n-1}} = 0.060$. The quantity $t = [112.30 - 112.40] \sqrt{\frac{6}{0.006}} = 4.1$. From the table of confidence limit given in Appendix No. 8.

$V = 5$, $t = 4.032\%$ at 99% conf. and 4.773 at the 99.5% confidence level. The value has huge error.

2.3 Calculate average deviation, standard deviation, rejection of results, 90% confidence limit and *t*-distribution. The results for analyses of zinc (II) by complexometric titration with EDTA in micro burett are given below as 7.146, 7.098, 6.942, 7.256 and 6.593.

Ans: The average amount zinc = 7.007 ± 0.192 standard deviation of mean = 7.007 ± 0.256. Now rejection of the results is 4*d* or 3*s* values out of these mean and standard deviation values. Mean = 0.192 × 4 = 0.768. Limit of acceptance = 7.007 ± 0.768. Both values are acceptable standard deviation 0.256 × 3 = 0.765 limit of acceptance 7.007 ± 0.768 (also refer to Appendix No. 9).

The range is found out from highest (7.256) and lowest (6.593) values as 0.663. The *Q* value from Appendix No. 5 is

$$Q_1 = \frac{M_2 - M_1}{M_5 - M_1} = \frac{6.942 - 6.593}{0.663} = 0.53, \; Q_2 = \frac{M_5 - M_4}{M_5 - M_1} = \frac{7.256 - 7.146}{0.663} = 0.17.$$

Thus for 90% confidence limit Q_{90} is 0.64. Both values are acceptable.

Confidence level = mean $\pm \frac{ts}{\sqrt{n}}$ (*t* is obtained from Appendix No. 8 it is 22.132).

$7.007 \pm \frac{22.132(0.256)}{\sqrt{(6-1)}} = 7.007 \pm 0.244$. Thus there is a 90% probability that a distribution mean is within ± 0.244 of mean.

2.4 The crystalline lead sulphate had 14.70% of water of hydration from formula $PbSO_4.2H_2O$ calculate relative and absolute error.

Ans: The absolute error is $D = (14.70 - 14.75) = -0.05\%$ as from formula actual weight present is 14.75%. Now the relative error with reference to the value is

$$D = \frac{-0.05}{14.7} \times 100 = -0.34\%.$$

2.5 The following results were obtained for analysis of lead in sample of brass in % of metal as 1.12, 1.15, 1.11, 1.16, 1.12%. Calculate, the arithmetic mean; mean deviation; probable analytical error; confidence limit; standard deviation and standard deviation of mean.

Ans: The mean of these five readings is 1.13%. The deviation of mean is $\Sigma\delta^2 = (-0.1) + 0.02 + (-0.02) + 0.03 + (-0.01) = 0.018\%$ while average deviation of mean = 0.0001 + 0.0004 – 0.0004 + 0.0009

$$= \Sigma\delta^2 = 0.019, \text{ standard deviation is } (S) = \sqrt{\frac{\Sigma\delta^2}{n-1}} = \sqrt{\frac{0.0019}{4}} = \pm 0.022\%$$

The probable analytical error is $\Sigma = \pm \frac{t_\alpha^\sigma}{\sqrt{n}} = \frac{2.78 \times 0.022}{\sqrt{5}} = \pm 0.027\%.$

The confidence limit = 1.13 ± 0.027%.

$$\text{For 99\% confidence limit} = \frac{4.60 \times 0.022}{\sqrt{5}} = \pm 0.045\%.$$

So actual value = 1.13 ± 0.045.

2.6 In the analysis of calcium, following values were obtained: 15.42, 15.56, 15.68, 15.51, 15.52, 15.52, 15.53, 15.54, 15.53, 15.56. Indicate reading which is not acceptable due to presence of huge error?

Ans: The lowest value is 15.42 the highest value is 15.68, so range = 0.26

$$\left(\frac{15.51 - 15.42}{0.26}\right) = 0.35 \text{ is less than 0.41 so value is accepted.}$$

$$\left(\frac{15.68 - 15.56}{0.26}\right) = 0.46 \text{ is greater than 0.41 so the value is rejected.}$$

The average value (m) = 15.56. Now since difference in highest value (15.68) is greater than average difference of 0.41 the value is rejected. However, if we take 15.56 as highest value then lowest value being 15.42 it will be rejected as it involves huge error of 0.64 in such case lowest value will be 15.51 with average difference of 0.02 w.r.t. mean 15.56 and is therefore accepted.

2.7 A sample of pyrolusite on analysis showed % of manganese as : 22.62, 22.73, 22.75, 22.78, 22.79, 22.83, 22.84, 22.87, 22.92, 22.94. Calculate (*a*) mean and average deviation of mean, (*b*) standard deviation and standard deviation of mean, (*c*) median value and range, (*d*) relative difference in high and low value, (*e*) confidence limit.

Ans: (*a*) The mean value is average of eleven values it is 22.813.

$$d = \frac{\Sigma\delta^2}{n} = \frac{0.242}{11} = 0.022. \text{ This value is the average deviation of mean.}$$

(*b*) The Standard deviation $= \sqrt{\frac{\Sigma\delta^2}{n-1}} = 0.308$ while standard deviation of mean is

$\frac{S}{\sqrt{n}} = \frac{0.308}{\sqrt{10}} = 0.028$, The coeff. of variation $\left(\frac{S}{m} \times 100\right) = 1.22\%$ *i.e.*, Relative Standard deviation.

(*c*) Median value is middle of high and low value is = 22.83 while the range is (22.94 – 22.62) = 0.32.

(*d*) The relative difference between low and high value is 14 ppt.

The confidence limit lower and upper values are (as given in Appendix No. 8) is

$$\text{Upper level} = \frac{(n-1)S^2}{3.904} = \frac{10 \times (0.308)^2}{3.309} = 0.24$$

$$\text{Lower level} = \frac{(n-1)S^2}{18.307} = \frac{10 \times (0.308)^2}{18.307} = 0.0518.$$

Here 3.309 and 18.307 are obtained from tables for Chi-square given in Appendix No. 7 with 10-degress of freedom.

2.8 From following two set of values calculate the mean, standard deviation, confidence limit. rejection quotient (Q)
Set A = 107.861; 107.870; 107.881; 107.892
Set B = 107.777; 107.778; 107.779; 107.780 state which is best set of values?

Ans: The mean value in set (*A*) is 107.876 from set (*B*) is = 107.778
The deviation and average deviation for set A is 0.10 ppt and for set B = 0.01 ppt.

The standard deviations is $S = \sqrt{\frac{\Sigma d^2}{n-1}}$ for set A it is 0.12 ppt and for set B = 0.001 ppt.

The confidence limit is calculated as

Set (*A*) $= \frac{3.2 \times 0.013}{\sqrt{4}} = 0.021$ and for set (*B*) = 0.002 for 95% confidence level.

Set (*A*) $= \frac{5.8 \times 0.013}{\sqrt{4}} = 0.338$ and set (*B*) = 0.004 ppt for 99% confidence level.

$Q = a/b$ = different/range = 0.011/0.031 = 0.35 for set A
Q = 0.001/0.0031 = 0.33 for set B. Obviously set B is better.

2.9 Calculate the mean and median values in volumetric analysis in acid base titration. The readings are: 10.10; 10.06; 10.08; 10.20 ml.

Ans: Mean $(\bar{x}) = \frac{10.10 + 10.06 + 10.08 + 10.20}{4} = 10.11$ ml.

Median is average of 2nd and 3rd value it is greater $= \frac{10.08 + 10.10}{2} = 10.09$ ml.

2.10 What is value with 95% confidence limit when readings for absorbance of indicator in colorimeter are: 0.084; 0.089; 0.079; given t value for 3 readings is 4.30?

Ans: Mean $(\bar{x}) = \dfrac{0.084 + 0.089 + 0.079}{3} = 0.0840$

The standard deviation $(S) = \sqrt{\dfrac{\Sigma\delta^2}{n-1}} = 0.0050$ deviation.

Now Appendix 6 and 8 the value for 95% confidence is $0.084 \pm \dfrac{t \text{ value} \times 0.0050}{\sqrt{n}}$ if $n = 3$,

$t = 4.13$

We have $0.084 \pm \dfrac{4.3 \times 0.005}{\sqrt{3}} = 0.084 + 0.012.$

2.11 Calculate 50% and 95% confidence limit for results the values for analysis are 1.80, 1.58, and 1.64.

Ans: Mean $(\bar{x}) = \dfrac{1.80 + 1.58 + 1.64}{3} = 1.673.$

Now standard deviation $S = \sqrt{\dfrac{\Sigma\delta^2}{n-1}} = 0.10.$

From the Appendix No. 8 the value for 50% confidence limit $t = \pm 0.067$ and for 95% it is ± 1.96

So 50% confidence limit $= 1.67 \pm \dfrac{0.67 \pm 0.10}{\sqrt{3}} = 1.67 \pm 0.04$ as $\left[x + \dfrac{t+s}{n-1}\right]$

For 95% confidence limit $= 1.67 \pm \dfrac{1.96 \pm 0.10}{\sqrt{3}} = 1.67 \pm 0.11$ also $\left[x + \dfrac{t+s}{(n-1)}\right]$

We have to use formula as,

$$\text{Mean} \pm \frac{t \text{ value} + \text{standard deviation}}{\sqrt{\text{no. of readings}}} = \text{mean} \pm \frac{t \text{ value}}{\sqrt{n}}.$$

2.12 The analysis of hametite yielded % of iron as: 56.0; 55.95; 56.04 and 56.08; 56.23. Can the last value be rejected if it involves huge error?

Ans: The average $(\bar{x})$ is $\dfrac{56.0 + 55.95 + 56.04 + 56.23}{4} = \dfrac{29.32}{5} = 55.95$

The difference between mean and last value (56.23 – 55.95) = 0.28
The last and last but one value difference is (56.23 – 56.08) = 0.15
Q value for 5 determinant is 0.64. The difference between last value and the average value is (56.23 – 55.95) = 0.28 (Appendix No. 5)

$Q_{\text{exp}} = \dfrac{0.15}{0.28} = 0.54\, Q_s = 0.64 > 0.54$. Therefore reading is acceptable.

2.13 Calculate standard deviation and relative standard deviation for five values obtained in typical redox micro titrations of oxalic acid against potassium permanganate?
9.990; 9.985; 9.973; 9.980; 9.982 ml.

Ans: Mean $= \dfrac{9.990 + 9.985 + 9.973 + 9.980 + 9.982}{5} = \dfrac{49.910}{5} = 9.9820$ ml.

Now $\quad \text{S.D.} = \dfrac{\Sigma\delta^2}{\sqrt{n-1}} = \dfrac{\Sigma\delta^2}{\sqrt{4}} = \dfrac{158.0 \times 10^{-6}}{\sqrt{4}} = 6.3 \times 10^{-3}$

Relative standard deviation $\dfrac{6.3 \times 10^{-3} \times 100}{9.9820} = 0.06311\%$

$\left(i.e., \dfrac{\text{Standard deviation}}{\text{Mean value}} \times 100 = \text{R.S.D.} \right)$

2.14 Calculate the mean, average deviation and standard deviation of mean and median value for set of results given as:
9.990; 9.982; 9.985; 9.980; 9.973

Ans: Mean is 9.9820. median = 9.982, average deviation of mean is $D = \dfrac{d}{\sqrt{n}}$

$$= \frac{158 \times 10^{-6}}{5} = 31.6 \times 10^{-6} = 0.00000316$$

standard deviation = 6.3 × 10^{-3} ; $\dfrac{(6.3 \times 10^{-3} \times 100)}{9.982} = 0.0631$ relative standard deviation.

2.15 Calculate the (*i*) average deviation of 9.8, 10.6, 10.8 and 10.4 litre values, (*ii*) standard deviation of 10.1, 10.5, 9.9; 9.5, 10.6, 9.4, 11.5, 9.5, 9.5 10.0, 9.5, (*iii*) Evaluate standard deviation and estimate range for the values as 19.13, 19.25, 19.30 in typical precipitation titration.

Ans: (*i*) The average deviation is sum of values divided by number of values, *viz.*, 4 so average = 10.65, sum of deviation = 1.2 average 1.2/4 = 0.30.

(*ii*) The standard deviation is $(S)\sqrt{\dfrac{\Sigma\delta^2}{n-1}}$. Here $n = 11$, $\Sigma S^2 = 4.224$, $S = \sqrt{\dfrac{4.24}{10}} = 0.65$.

The standard deviation of mean $= \dfrac{0.65}{\sqrt{10}} = 6.5$ ppt. $= \dfrac{S}{\sqrt{n-1}}$

(*iii*) Standard deviation can be calculated from range (Appendix No. 9)
(19.30 – 19.13) = 0.17 is the range for three ⁝ 0.59 *n* deviation for range
Standard (*d*) = 0.17 × 0.59 = 0.10
Standard deviation = $(|\vec{\delta}| \times \text{Range})$

Actual calculation of (*S*) will be also 0.10.

2.16 If for set of ten results in gravementric analysis the mean is 56.06, with the standard deviation as 0.2% with use of t-values (t – 1.833) calculate confidence limit for 90% limit.

Ans: $\frac{\text{Mean} \pm \text{Confidence } t - \text{value} \times \text{standard deviation}}{\sqrt{\text{No. of reading}}} = \frac{56.06 \pm 1.833 \times 0.21}{\sqrt{10}} = 5.6 \pm 0.12\%$

(Appendix-4).

2.17 In polargraphic analysis results obtained for zinc were: 30.13, 30.2, 31.23 which value for zinc be rejected?

Ans: Q test be considered from Tables in Appendix No. 6

$$Q = \text{Smallest Value} = \frac{X_3 - X_2}{X_3 - X_1} = \frac{31.23 - 30.2}{31.23 - 30.13} = \frac{1.03}{1.10} = 0.94 \text{ if } n = 3$$

(Please refer to Appendix No. 6 also calculated as)

$$\text{Largest value} = \frac{X_n - X_{n-1}}{X_n - X_1} \text{ since } 0.94 = Q_{90} = 0.94$$

The value may be rejected.

2.18 Which value should be rejected for % recovery of uranium in mineral pitchblend. The values are: 40.12, 40.15, 40.55 on the basis of Q = test?

Ans: $Q = \frac{X_3 - X_2}{X_3 - X_1} = \frac{40.55 - 40.15}{40.55 - 40.12} = \frac{0.40}{0.43} = 0.93 \text{ for } Q_{90}^{3} = 0.94$

On referring to Appendix No. 5

The value cannot be rejected as Q for 90% limit is 0.94.

2.19 From following seven result in titrimetry by using Q test indicate which result is rejected? The values are 6.22. 6.32, 6.02, 6.32, 6.12, 6.82 and 5.12, smallest value = 5.12, largest value 6.82.

Ans: $Q = \frac{X_2 - X_1}{X_n - X_1} = \frac{6.02 - 5.12}{6.82 - 5.12} = 0.53 \text{ for } Q_{90} \text{ for } n = 7 \text{ is } 0.51$

For large value,

$$Q = \frac{X_n - X_{n-1}}{X_n - X_1} = \frac{6.82 - 6.32}{6.82 - 6.07} = 0.625 \text{ if } n = 7,\ Q_{90} = 0.51$$

As 0.625 > 0.56 the last value (*viz.*, 6.82) is rejected.

On referring to Appendix No. 5 we have for value 5.12

The value cannot be rejected as Q for 90% limit is 0.94.

2.20 From following results in titrimetry by using Q-test indicate which result is rejected if 6.82 and 5.12 are largest and smallest value.

Ans: On referring to Table in Appendix No. 5

So, small result be rejected (*viz* 5.12) for large value as 0.625 > 0.56 last result is rejected.

2.21 The relative standard deviation = 5 ppt. The result first obtained for the gravimetric analysis of magnesium were 40.12, 40.15, 40.55. New additional results are 40.20, 40.39. Calculate standard deviation for first three and all five values from range.

Ans: For first set range = (40.55 – 40.12) = 0.43

The range of second set (40.55 – 40.12) = 0.43 so we have

On refering to Table in Appendix No. 9

S = range × K if K = 0.59 for 3 values is K = 0.43 for five values

$$S_3 = 0.59 \times 0.43 = 0.25; \text{ relative value } \frac{0.25}{\sqrt{3}} = 6.33 \text{ ppt. } S = x \pm \frac{ts}{\sqrt{n}}$$

$$S_5 = 0.43 \times 0.43 = 0.18; \text{ relative value} = \frac{0.18}{\sqrt{5}} = 4.6 \text{ ppt.}$$

PROBLEMS FOR PRACTICE

2.22 From the three values in titration of ferrous (*II*) ion with permanganate values obtained as 6.40, 6.49, 6.40 ml. What is the mean median and difference between two?

[**Ans:** mean 6.43; median 6.40]

2.23 The following results were obtained in volumetric analysis of calcium with mureoxide as an indicator against EDTA

The result for the volumetric analysis of calcium in complexometry are:

3.6052, 3.6049, 3.6063, 3.6053, 3.6052

3.6047, 3.6060, 3,6081, 3.6048, 3.6051

Calculate (*i*) mean, (*ii*) median, (*iii*) average deviation, (*iv*) standard deviation, (*v*) Is any result rejectable? (n = 10)?

[**Ans:** (*i*) 3.06056, (*ii*) 3.6052, (*iii*) 30.00076, (*iv*) 0.0010 (*v*) No]

2.24 On the basis of criteria of rejection of result, state whether any one of the following values is rejected from knowledge of Q-test, *viz*., 31.34, 31.50, 31.68 and 31.82?

[**Ans:** Q = 0.44, 0.76 No rejection]

2.25 For the following sets of results for gravimetric determination of manganese as the pyrophosphate following results were obtained:

136.4, 136.9, 135.8, 137.1, 136.3, 136.5,136.3, 135.9, 135.7 and 136.5 mg

What is (*a*) mean, (*b*) median, (*c*) range, (*d*) average deviation, (*e*) standard deviation?

[**Ans:** (*a*) 136.3 (*b*) 135.8 (*c*) 1.1 (*d*) ±0.3]

2.26 What is the mean and median of following results in volumetric analysis 10.10, 10.08, 10.20, 10.06 ml? [**Ans:** mean = 10.11, median = 10.09]

2.27 What is the standard deviation for the following five results in titration : 9.982, 9.980, 9.973, 9.985, 9.990? [**Ans:** mean = 9.9820, standard deviation = ±0.006]

2.28 For the following set of values in spectrophotometric analysis of Chromium (*VI*) with s-diphenylcarbazide calculate 95% and 99% confidence limit of the mean. The value for absorbances are 0.770, 0.850, 0.860, 0.750.

2.29 Two girls gave following set of results.

Girl (*A*) : 1.73, 1.70, 1.75, standard value 1.76

Girl (*B*) : 7.41, 7.44, 7.46 standard value 7.49

What are absolute and relative deviations of two sets of values and absolute and relative errors with means of two sets of values?

[**Ans:** Girl (*A*) :	0.018,	10 ppt	–	0.03%	–	19 ppt
Girl (*B*) :	0.018,	2 ppt	–	0.005%	–	7 ppt
	abs. devi.	rel. devi.		ab. error		error]

2.30 Calculate average deviation of following set of values in a typical titration experiment: 9.8, 10.6, 10.8 and 10.4 ml. [**Ans:** 0.3 average deviation]

2.31 What is the relative standard deviation of following eleven results in acid base titration: 10.1, 10.5, 9.9, 9.5, 10.6, 9.4, 11.5, 9.5, 10.0, 9,5 ml? [**Ans:** 6.5 ppt relative standard deviation]

2.32 What is the standard deviation of following results in precipitation titration 19.13, 19.25, 19.30 ml.? Give also the range. [**Ans:** 0.17, S = 0.10]

2.33 Which of the following values be rejected in volumetric analysis on statistical performance? The values are 30.13, 30.20, 31.23 ml. [**Ans:** *Q* 0.094 Third result is bad]

2.34 Should any of the following values be rejected on the basis of *Q* test as given in Appendix-5? The values are 22.64, 22.64, 23.01. [**Ans:** Last value 23.01]

2.35 From the following results, calculate indeterminate error of analysis and confidence limit for these results which are 8.25, 8.15, 8.08, 8.20. [**Ans:** ± 0.10% m = 8.14]

2.36 From following results for % of silica in feldspar 1.12, 1.15, 1.16 and 1.12%. Calculate mean, deviation from mean, standard deviation. [**Ans:** m = 5.66, d = 0.018, SD = ±0.022%]

2.37 Apply *Q*-test and find out whether these results should be retained or rejected.

Set I	24.26,	24.50,	24.73,	24.63
Set II	26,49,	26.56,	26.84	

[**Ans:** (*a*) 0.51, 0.76 retain (*b*) 0.80, 0.94 retain]

2.38 How many significant figures are there in following values: (*a*) 0.00607 (*b*) 0.0037. [**Ans:** (*a*) 3 (*b*) 2]

2.39 The amount of magnesium in hard water was 2.86, 2.95, 2.84 meq/Litre. What is 95% confidence interval for the mean of data if no information and precision on accuracy is known? [**Ans:** S = 0.059 or 95% CL = 2.88 ± 0.15 (*b*) 2.88 ± 0.07]

2.40 The three readings for % of tin in bronze are: 6.38, 5.99, and 6.25. Calculate the 90% confidence level for mean of three values for percentage of tin in the bronze. [**Ans:** 6.21 ± 0.34, 6.21 ± 0.26]

2.41 Evaluate the absolute and relative values of the standard deviation of the following titration values of potassium permanganate against calcium oxalate.
40.02, 40.11, 40.16, 40.18, 40.18, 40.19 ml.
Can value 40.02 be rejected on the basis of *Q*-test? [**Ans:** mean = 40.14, absolute s = 0.06]

REFERENCES

2.1 S.M. Khopkar, *Basic Concepts of Analytical Chemistry*, 3rd edn. New Age International Publishers (P) Ltd., (2007).

2.2 H.A. Latinen and W.E. Harris, *Chemical Analysis –An Advanced Textbook*, McGraw-Hill Co. 2nd edn., p. 541-3 (1975).

2.3 D.J. Pietryzyke and C.W. Frank, *Analytical Chemistry: An Introduction*, Academic Press, pp. 73-7 (1974).

2.4 V. Alexyen, *Quantitative Analysis*, Mir Publishing House, Moscow, 2nd edn., pp. 51-4, 57, 63 (1979).

2.5 D.A. Skoog and D.M. West 3rd edn. Holt Rinehart Winston Publishers, pp. 47, 71-3, 64, 80-3 (1979).

2.6 J.S. Fritz and G.H. Schenk, *Quantitative Analytical Chemistry*, Allyn and Bacon Ltd., pp. 31-39, 41-7 (1979).

2.7 G. Dick, *Analytical Chemistry*, McGraw-Hill Co. pp. 78 (1973).

2.8 G.D. Christen, *Analytical Chemistry*, 5th edn., Harcourt Butler and Co. (1997).

Chapter 3

Sampling in Analysis

INTRODUCTION

In chemical analysis, the clear cut separation is most essential for having reliable results specially if subsequent determination is to be carried out by any instrumental technique. Similarly the kind or nature of sample taken for the analysis is of great significance to get reproducible results. As a rule in quantitative chemical analysis, much attention is paid on isolation and determination but not much attention is paid on procedure of sampling. It is the traditional practice to accept whatever sample is provided and to proceed directly with the analysis. In fact, one never bothers to know as to what was sampling procedure? In fact, many times source of error in the analysis originates from the sample collected rather than due to faulty handling of instruments. Therefore in this context, one has to define what is sample?

A small portion or segment taken from bulk of the sample should have same chemical characteristics as it is for the bulk material. A small variation in sample should cause serious error. Therefore we always insist that the sample should be representative in all its chemical as well as physical properties with the sample of original lot of material.

This leads to consideration of mode of sampling. A random sample can be taken by device called thief from bulk of material for quick analysis, however, for getting representative sample we will have to collect large set of increments (*i.e.*, fraction of small samples). It may be even necessary to get sub samples (*i.e.*, part of one increment subdivided further into small portions). After analysis the quality of increment sample or subsample must match with the original bulk of material. 'Gross sampling' technique consists of analysis of small portions or increments to know its quality. The structural analysis plays a significant role in deciding number and size of sampling unit. As opposed to random sampling, statistical sampling gives best results. The stratified sample consists of several portions representing original bulk material. If an estimate of the population means to be reliable with least errors then number of each stratum is given by an expression as:

$\dfrac{\eta_\gamma}{\eta} = \dfrac{W_\gamma (\sigma_0)^r}{\Sigma W_\gamma (\sigma_0)^r}$, η_γ = number of samples from 'γ' stratum, while η = total number of samples, W_γ is wt. of stratum, $(\sigma_0)_\gamma$ = standard deviation of stratum. The stratum is to be weighted and is evaluated by an expression as: $X_\gamma = \dfrac{\Sigma W_\gamma \bar{X}_\gamma}{\Sigma W_\gamma}$ if $\bar{X}_\gamma$ = mean of all 'γ' stratum, $\bar{X}$ is estimation of population mean if size difference is there but not in variance σ_λ we have $\dfrac{\eta_\gamma}{\eta} = \dfrac{W_\gamma}{\Sigma W_\gamma}$ as $(\sigma_0)_\gamma$ will be fixed.

We can minimize variance in the stratified sampling also.

The final figure for number of unit is now considered. A number of units which would vary around expected value (ρ_n) with standard deviation of (σ_n) is evaluated from following formula:

$\sigma_n = \sqrt{n\rho(1-\rho)}$ for fixed value of σ^2 the value of n_1, n_2, n_3 are given as

$n_1 = \left(\sqrt{\dfrac{\sigma_1^2 C_1}{\sigma_2}}\right)\left(\sqrt{\sigma_1^2 C_1} + \sqrt{\sigma_2^2 C_2} + \sqrt{\sigma_3^2 C_3}\right)$ if n_1 n_2 n_3 represent number of samples analysed

$n_2 = \sqrt{\dfrac{\sigma_2^2 C_1}{\sigma_2^2 C_2}}$ and $n_3 = \sqrt{\dfrac{\sigma_3^2 C_2}{\sigma_\gamma^2 C_3}}$

The relative standard deviation is calculated in following manner:

$$\sigma_n/n_1 = \sigma_n/n_p \text{ and } \sigma_n/n_2 = \sigma_n/(n-p) \text{ if } \sigma = \text{expected value.}$$

Finally in calculation involving sampling with least standard deviation we use (*f*), *i.e.*, fraction of sample it is calculated from the expression $\dfrac{f}{(1-f)} \doteq \dfrac{d_2(f-f_2)}{d_1(f_1-f)} = 1.903$ so $f = 0.656$ and the average density is given by equation. $fd_1 + (1-f)\,d_2 = 4.34$ and for σ is calculated. In order to get value of '*n*' we must use expression

$$n = f(1-f)\left[\frac{d_1 d_2}{d_2}\right]^2\left[\frac{100(f_1 - f_2)}{\sigma_s\, f_{ar}}\right]^2$$

We can use sets of such expression to arrive at sampling unit and sampling size with least value for standard deviation.

SOLVED PROBLEMS

3.1 The occurrence of defective sample in the chemical unit is 2% and if random sample of 104 unit is taken, what number of defective samples you would encounter and what is the standard deviation?

Ans: A number of units which would vary around expected value (σ_n) with standard deviation of (σ_n) is evaluated from formula $\sigma_n = \sqrt{np(1-p)}$, therefore, we can forecast the number of defective samples as 0.02 × 10,000 if n = 10,000 p = 0.02 we get standard deviation as $\sigma_n = \sqrt{[10{,}000](0.02)(1-0.02)} = 14$ we calculate the relative standard deviation as on $\frac{\sigma_n}{n_1} = \frac{\sigma_n}{n_p} = 100\sqrt{\frac{(1-p)}{n_p}} = \sigma_s$ substituting values of relative standard deviation

$$\frac{\sigma_n}{n_2} = \frac{\sigma_n}{(n-p)} = 100\sqrt{\frac{l_p}{n(1-p)}} \text{ if } n_1 = n_p,\ n_2 = (1-p).$$

3.2 A magnetite sample has density of 5.2 (d_1) with gangues density = 2.7 (d_2). If percentage of the iron in mineral is 75% and the percentage of the gangues is 5% with an average percentage (P_{av} = 60%) indicate how many particles of mineral must be taken so as to have relative standard deviation of < 1 ppt (0.1%) due to sampling error?

Ans: The portion 'p' of mineral is evaluated as follows:

$\frac{p}{(1-p)} = \frac{d_2(P-P_2)}{d_1(P_1-P)} = 1.903$ so $p = 0.656$ since average density = 5.2

$5.2 = pd + (1-p)d_2$ 4.34. As the value of P_{av} = 60 so 60 × 0.1 = 6 the number of samples is (1.7×10^5), *i.e.*, for spherical particles it corresponds to sample weight of 0.3 g for particles of 0.01 cm. diameter or 0.386 for particle of 10 mm diameter.

3.3 If standard deviation in 0.07 strats is 0.1 and of single analysis as 0.21, if expenses are in ratio of 4:2:1. Sampling strata : sampling st.: single analysis. Calculate sampling pattern and (0.07) least expenses with standard deviation of 0.08.

Ans: Now $n_3 = \sqrt{\frac{\sigma_1^2 \times C_1}{\sigma_2}}\left(\sqrt{\sigma_1^2 C_1} + \sqrt{\sigma_2^2 C_2} + \sqrt{\sigma_3^2 C_3}\right),\ n_2 = \sqrt{\sigma_2^2 C_2}\ n_3 = \sqrt{\sigma_1^2 C_2}$ and if C_1, C_2, C_3 sampling strats, σ_1, σ_2, σ_3 is standard deviation.

Substituting these values

$$n_1 = \sqrt{(0.07)^2/4(0.07)^2 \times 4 + \sqrt{(0.10)^2 \times 2 + \sqrt{\left[(0.21)^2 \times 1\right]/(0.08)^2}}} = 3$$

$$n_2 = \sqrt{(0.10)^2 \times (4)/(0.07)^2 \times 2} = 2.02$$

$$n_3 = \sqrt{(0.21)^2 \times (2)/(0.1)^2 \times 1} = 2.97$$

Taking n_1 = 3, n_2 = 2, n_3 = 3 by approximation we get σ_0 = 0.076 and C = (3 × 4) + (6 × 2) + (18 × 1) = 42 on relative standard deviation.

3.4 With identical values of the standard deviation and relative expenditure as in above example find out minimum standard deviation for thirty expenditure of one unique analysis.

Ans: $n_1 = 30\sqrt{(0.07)^2/4(\sqrt{(0.07)^2 \times 4} + \sqrt{(0.10)^2 \times 2} + \sqrt{(0.21)^2 \times 1}} = 2.14$

$n_2 = 2,\ n_3 = 2.97$ By approximation if we assume;

if n_1 = 2, n_2 = 2, n_3 = 3 the expenditure is C = 28

$\sigma_0^2 = (0.07)^2/2 + (0.01)^2/4 + (0.21)^2/12 = 0.00862;\ \sigma_0 = 0.093$

3.5 What is the size of sample necessary to have 10 M strength with relative standard deviation to be smaller than 0.1%?

Ans: This involves sampling from two types of analyst, *viz.*, (A) and (B) now in one liter of the solution $N_1 = 10^{-8} \times 6 \times 10^{23} = 6 \times 10^{15}$ (if 6×10^{23} is Avagadro's number)

$N_2 = 55.6 \times 6 \times 10^{23} = 3.3 \times 10^{25}$

Now if $p = \dfrac{N_1}{(N_1 + N_2)} = 1.8 \times 10^{-10}$ and $(1-p)$ (from equation)

$$\frac{\sigma_n}{n_1} = \frac{\sigma_n}{n_p} = 100\sqrt{\frac{(1-p)}{n_p}} = \sigma_s.$$ This is relative standard deviation,

$$\text{So, } N = 10^6/p = 5.5 \times 10^{15} = \frac{5.5 \times 10^{15}}{55.5 \times 6 \times 10^{23}} = 1.6 \times 10^{-10} \text{ liter.}$$

3.6 A sample is made of uniform grains of Devadra's alloy of density 5.2 and a mixture of alloy of density 2.7. If % aluminium in alloy is 75% and in mixture 5%. As sample contains average 60% aluminium how many grains must be taken of Devadra's alloy so that the relative standard deviation is < 1 ppt in average % due to sampling error?

Ans: The fraction (f) of an Devadra's alloy grain is calculated from data as

$$\frac{f}{(1-f)} = \frac{d_2(f - f_2)}{d_1(f_1 - f)} = 1.093 \text{ if } f = 0.656$$

The average density of Devadra's alloy is given by an equation:

$$fd_1 + (1-f)d_2 = 4.34$$

The value of f_{av} $\sigma = 60 \times 0.1 = 6$ d n calculated from expression

$$n = f(1-f)\left[\frac{d_1 d_2}{d_2}\right]^2 \left[\frac{100(f_1 - f_2)}{\sigma_s f_{ar}}\right]^2 = 1.7 \times 10^5 \text{ grams.}$$

3.7 From the following sets of five samples, evaluate mean, average and standard deviation with Q and t-test also to verify the results. The values for analysis are 7.146, 7.698, 6.942, 7.256 and 6.393.

Ans: The mean is $\dfrac{7.146 + 7.698 + 6.942 + 7.256 + 6.393}{5} = 7.087$

Average deviation = ±0.912 ($\bar{\delta}$) and standard deviation = ±0.256.

Rejection of results if any = 4 × $\bar{\delta}$ = 4 × 0.192 = 0.768

Limit of acceptance = 7.007 ± 0.768

Standard deviation = 3S = 3 (0.256) = 0.768

Same limit of acceptance as above on referring to tables in Appendix 6 we have values of Q as

Q-test the value 6.593, 6.942, 7.098, 7.146, 7.256 median = 7.098

Range = $(M_5 - M_1) = (7.256 - 6.593) = 0.663$

$$Q_1 = \frac{M_2 - M_1}{M_5 - M_1} = \frac{6.942 - 6.593}{(0.663)} = 0.53$$

$$Q_2 = \frac{M_5 - M_4}{M_5 - M_1} = \frac{7.256 - 7.146}{(0.663)} = 0.17$$

For 90% confidence limit $Q_{0.90} = 0.64$ (as readings are, $Q = 0.64$ from tables)

For t-distribution mean $\pm \frac{ts}{\sqrt{n}} = \frac{7.007 \pm t \times 0.256}{\sqrt{5}}$ if $t_{90} = 2.13$

With reference to tables in Appendix No. 10

We have $\frac{7.007 \pm 2.13 \times 0.256}{\sqrt{5}} = 7.007 \pm 0.244\%$ as $t_{90} = 2.13$

3.8 From the following two sets of values for analysis of silver in mineral Horn Silver is Set A: 107.861; 107.870; 107.881 and 107.892 gm
Set B: 107.777; 107.778; 107.779; and 107.780 gm
Calculate the mean, standard deviation, confidence limit (95%) rejection quotient Q in both set of values?

Ans: The mean for first set A is 107.876 and for second set B is 107.778. The deviation and average deviation in first set A is 0.10 ppt for set B 0.01 ppt.

The standard deviation for set A is 0.12 ppt and set B is 0.001 ppt.

The confidence limit is calculated at 95% level for four values is

The chi-square value for 95% limit for four reading is 3.2 (From Tables in Appendix)

The chi-square value for 99% limit for four reading is 58

So for first Set A $= \frac{3.2 \times 0.013}{\sqrt{4}} = 0.021$ Set B $= \frac{3.2 \times 0.001}{\sqrt{4}} = 0.002\%$ as 95%

So for first Set A $= \frac{5.8 \times 0.013}{\sqrt{4}} = 0.038$ Set B $= \frac{5.8 \times 0.001}{\sqrt{4}} = 0.004\%$ as 95%.

The $Q = a/b$ = difference = 0.011/0.031 = 0.35 Set A (Refer table in Appendix 6)

$Q = 0.001/0.003 = 0.33$ for Set B.

3.9 If two analyst have standard deviation of 0.055 for six readings and 0.022 for four readings state whether 0.055 is greater than 0.022 from values of F-test at 95% confidence limit?

Ans: For first $n = 5$ is for second set $n = 3$ and from Table for F-values given in Appendix No. 7 F = 9.01. Now variance ratio is calculated as

$$\frac{V_1}{V_2} = \frac{(S_1)^2}{(S_2)^2} = \frac{(0.055)^2}{(0.022)^2} = \frac{0.0030}{0.00048} = 6.25.$$ Since F = 9.01 greater than the variance ratio the variation is good hence null hypothesis is true.

3.10 If for a given set values the mean value is 1.67 with the standard deviation of 0.10 for n-determination being 3. What is the value at 50% and 95% confidence limit?

Ans: Confidence limit $\mu = \bar{x} + \frac{Z\sigma}{\sqrt{N}}$ if $\bar{x}$ is mean, n = nos. *of* readings, σ-deviation, Z = value for confidence limit. Referring to Appendix No. 9, we have

So, 50% confidence level $1.67 \pm \frac{0.67 \pm 0.10}{\sqrt{3}} = 1.67 \pm 0.04$

For 95% confidence limit = $1.67 \pm \frac{1.96 \pm 0.10}{\sqrt{3}} = 1.67 \pm 0.11.$

3.11 For analysis of uranium (VI) spectrophotometically as its oxinate complex the mean value for uranium was 1.80 μg, with standard deviation of 0.10. Calculate the confidence level at 50% and 95% limit?

Ans: $\bar{x}$ = 1.80, S = 0.10, σ = 0.10, n = 3

We calculate from expression $CL(\mu) = \bar{x} \pm \frac{Z\sigma}{\sqrt{n}}$

For 50%, Z = 0.07 for 90% Z = 1.64, n = 3, σ = 0.1, S = 0.10

50% CL =1.80 ± 0.67 × 0.010 =1.80 ± 0.07 Here $\mu = (x \pm Z\sigma)$

95% CL = 1.8 ± 1.96 × 0.1 = 180 ± 0.20 (Z - from Appendix No. 9)

3.12 For set of values for gravimetric analysis of manganese as its pyrophosphate following values were obtained: 0.1010, 0.1020, 0.1005, 0.1030, 0.1015 mgs. Calculate the (*i*) mean, (*ii*) median, (*iii*) average deviation, (*iv*) standard deviation.

Ans: (*i*) Mean = $\frac{0.1010 + 0.1020 + 0.1030 + 0.1005 + 0.105}{5} = 0.1016$

(*ii*) Median = 0.1015.

(*iii*) Average deviation = $\frac{0.006 + 0.0004 + 0.0009 + 0.0014 + 0.0001}{5} = 0.0006$

(*iv*) Standard deviation =

$$\sqrt{\frac{(0.006)^2 + (0.0004)^2 + (0.0009)^2 + (0.0014)^2 + (0.0001)^2}{(5-1)}} = \frac{3.3 \times 10^{-6}}{4} = 0.009.$$

3.13 For volumetric titration of oxalic acid with permanganate three values were obtained as 24.39; 24.19; 24.36 ml. Calculate mean, median, deviation average, derivation standard, deviation standard, deviation of the mean?

Ans: The median = 24.31; deviation = 0.08, mean = 24.31, n = 3

Average deviation = $\frac{0.077 + 0.123 + 0.0047}{3} = \frac{0.247}{3} = 0.082.$

$$\text{Standard deviation} = \sqrt{\frac{(0.077)^2 + (0.123)^2 + (0.0047)^2}{3-1}} = 0.077$$

$$\text{The relative deviation from mean} = \frac{0.077 \times 100}{24.31} = 0.32, \textit{ i.e.}, 0.3\%.$$

$$\text{The relative average deviation of median} = \frac{0.067 \times 1000}{24.36} = 2.8, \textit{ i.e.}, 3 \text{ ppt}.$$

3.14 In analysis of silicon from lime stone following results were obtained: 1.12, 1.15, 1.11, 1.16 and 1.12%. Calculate mean, median, average deviation, standard deviation, confidence limit for 99.7% and 95% level.

Ans: Median = 1.12, Mean = 1.13

$$\text{Average deviation} = 0.018, \text{ Standard deviation} = \sqrt{\frac{\Sigma\delta^2}{n-1}} = \sqrt{\frac{0.0019}{5-1}} = 0.022.$$

The value for 95% and 99.7% confidence limit (are obtained from Appendix No.9)

$$\text{Value} = \frac{\text{Confidence limit} \times \text{Standard deviation}}{\sqrt{(\text{Nos. of readings} - 1)}}$$

For	95% C.L. = $M \pm 2\sigma$	*i.e.,*	1.13 + 2 × 0.022 = (1.086 – 1.174%)
	99% C.L. = $M \pm 3\sigma$	*i.e.,*	1.13 + 3 × 0.022 = (1.064 – 1.1960%)
	50% C.L. = $M \pm 2/3\sigma$	*i.e.,*	1.13 + 2 × 0.022/3 = (1.115 – 1.45%)

PROBLEMS FOR PRACTICE

3.15 How large a sample of equimolar mixture of oxygen gas is measured at 0°C and 10^{-6} mm pressure in order to have the sampling error not more than 0.1%.

[**Ans:** 2.8×10^{-5} ml]

3.16 From a sample of feldspar with density of 4.5 and gangue density 2.2 containing 60 and 15% of metal, if 0.20 gm sample is taken having approximately 40% metal, and if particles are in homogeneous mixture with average diameter of 0.01 mm what is the expected standard deviation? [**Ans:** 0.0020%]

3.17 If repeation of the faulty sample from lot is 2%, and if 10,000 units are taken. Calculate the expected number of faulty units and the relative standard deviation?

[**Ans:** $\sigma = 14$, RSD $= 100\left(\frac{1-p}{np}\right)^{1/2}$]

3.18 If standard deviation of sampling strata is 0.07 and for sampling 0.10 and for one measurement is 0.21, if relative expenses are in ratio of 4:2:1, calculate best sampling scheme and lowest expenses with standard deviation of 0.08?

[**Ans:** $n_1 = 2.14$, $n_2 = 2.02$, $n_3 = 2.97$, $C = 42$]

3.19 If standard deviation and relative expenditure is same as in above example no. 3.18, calculate the minimum standard deviation so as to have expenditure 30 times that of the single measurement. [**Ans:** $n_1 = 2.14$, $n_2 = 2.02$, $n_3 = 2.97$, $C = 28$ $\sigma_0 = 0.093$]

3.20 Indicate the sample size required from uniform mixture of halides to be at concentration of 10^{-8} M with standard deviation > 0.1% relative.

[Ans: 1.6×10^{-10}]

REFERENCES

3.1 S.M. Khopkar, *Basic Concepts of Analytical Chemistry,* 3rd edn; New Age International Publishers (P) Ltd. (2007).

3.2 H.A. Latinen and W.E. Harris, *Chemical Analysis : An Advanced Text,* McGraw-Hill and Co. 2nd edn., pp. 570-2, 576-8, 580-2 (1975).

3.3 D.J. Pietrzyke and C.W. Frank, *Analytical Chemistry: An Introduction,* Academic Press, p.72 (1979).

3.4 D.A. Skoog and D.M. West, *Analytical Chemistry,* 3rd edn., Holt Reinhart Winston, p. 48 (1979).

3.5 G.D. Christian, *Analytical Chemistry,* 3rd edn., Harcourt Butler and Co. (1997).

3.6 D.A. Skoog and J.J. Leory, *Principles of Instrumental Analysis,* Holt Reinhart Winson (1996).

3.7 G.W. Ewing, *Instrumental Methods for Chemical Analysis*, McGraw-Hill and Co. (1980).

3.8 H.F. Walton, *Principles and Methods of Chemical Analysis,* Prentice Hall and Co. (1958).

Chapter 4

Gravimetric Analysis

INTRODUCTION

The gravimetric analysis is very useful for the estimation of the elements either by precipitation using inorganic reagents or by the use of sensitive specific and selective organic reagents. Most of the weights are measured in milligram, where estimation at 10^{-3} gram level is only possible. The most important criteria in these methods is the weighing must be accurate and it should be recorded to the fourth place of decimal on sensitive balance.

There are several precipitation methods, *e.g.*, one can use organic precipitant because we get large weight of complex due to high weight of reagent in spite of having small weight of metal to be analysed. In such exercises one must carefully guard against phenomena of coprecipitation. One should also strictly observe all the conditions of optimum precipitation. In very few cases, precipitation from the homogeneous solution is attempted. Knowing the correct weight of the precipitate and ignition temperature are most important steps in analysis.

There are large array of organic precipitants used in the inorganic gravimetric analysis. They include, oxines. oximes, azonapthols, dithiozone, cupferron, dithol, and several kind of the organic complexing ligands. A best ligand to be used for precipitation should be one which is very selective for particular metal when under the same conditions no other metal gets precipitated by it. Dimethyglyoxime is specific for nicked; 1-nitro 2-napthol is specific for cobalt. The reagent should be sensitive. It must permit us to determine smallest concentration of particular metal ion under specified experimental conditions.

The problems in gravimetery involves calculation of solubility product from intrinsic solubility or vice versa. Many of the problems pertain to calculation of percentage composition of the mixture from the weight of the precipitate. Few problems deal with calculation of metal contents from their precipitate by using conversion factor. Such of the expressions or formulas which are used for computing these various parameters are given in the following section of this chapter. The evaluation of the solubility product (K_{sp}) in different experimental conditions is important.

The solubility product is of importance in the gravimetric analysis. If M is the metal R^- is the ligand to form complex, as MR we have $MR = M^+ + R^-$. So $K_d = \dfrac{[M^+][R^-]\gamma i^2}{[MR]\gamma_0}$ is dissociation equilibria. If MR is saturated MR (solid) = MR (soln.) = $M + R^-$. Then we have, If γ = activity coefficient then = K_d

$[MR]\gamma_0$ so molar solubility (S^0) which is related to the solubility product (K_{sp}) as

$$K_{sp} = [M^+][R]\gamma \pm^2 = K_d \times S\gamma_0 \text{ is assumed to be unity.}$$

For calculating solubility of precipitate in pure solution, we consider following equation. If $\gamma = 1$ and if $S^0 = 0$ then we have
$[M^+] = [R^-] = S$ (molar solubility) so

$S = \sqrt{K_{sp}}$ for $M_m R_a$ so $M_m R_a\,(\text{solid}) M_m R_a\,(\text{liquid}) = mM^{n+} + aR^{b-}$ and if 'm' and 'a' are valancies.

$$K_{sp} = [M^{n+}]^m [A^{b-}]^a = [MS]^m [aS]^a \text{ molar solubility.}$$

The effect of $[H^+]$ on the magnitude of K_{sp} will be:

$MR_n = M^{n+} + nR^-$ and $R^- + H^+ = HR$ if C_R = concentration of R so $C_R = [R^-] + [HR]$
If α_1 fraction of total (R) in ionized form

$$K_{sp} = [M^{n+}][R^-]^n = [M^{n+}]\alpha_1^n C_R^n \text{ if } \alpha_1 = \frac{[R^-]}{[R^-]+[HR]} = \frac{K_a}{[H^+]+K_a}$$

The conditional solubility product is $K_{sp} = [M^{n+}]C_R^n = \dfrac{K_{sp}}{\alpha}$.

The effect of foreign complexing agent such as (HL) on solubility product is

$$HL \rightleftharpoons H^+ + L^-,\ M^+ + L^- = ML \text{ so } K_1 = \frac{[ML]}{[M][L]}[M^+][L^-]$$

$$ML + L = ML_2,\ K_2 = \frac{[ML_2]}{[ML][L]} = \frac{[ML_2]}{K_1[M][L]^2}$$

If $K_1, K_2 \ldots K_n$ are successive formation constants also called as the stepwise formation constant. overall constant (β) = $K_1, K_2 \ldots K_n$ total metal concentration is

$$C_M = [M]/\alpha_M \text{ as } K_{sp} = [M][L] = \alpha_M C_M [L] \text{ so solubility } (S) \text{ is}$$

$$S = C_M = \sqrt{\frac{K_{sp}}{\alpha_M}} = [L]$$

The effect of complex formation with precipitating ion on solubility of precipitate is given as

$$S = [M] + [MR_1] + [MR_2] + [MR_3] + \ldots [MR_n]$$

$$S = \frac{K_{sp}}{[R]} + K_{sp} \sum_{R-1}^{n} \beta_1 [R]^{1-1} \text{ and if nos. higher complex is formed then we have}$$

$$S = \frac{K_{sp}}{[R]} + S^0 + \beta_2 K_{sp} [R]$$

while evaluating solubility product, we must know intrinsic solubility and from equation of dissociation K_{sp}, it can be easily evaluated:

$$[M^+]=[R^-]=S \text{ so } K_{sp}=[M^+][R^-] \text{ if } S= \text{ solubility then}$$

$$[M^+]=[R^-]=[M^+][M^+]=[M^+]^2=S \text{ so } K_s=\sqrt{S^2}$$

Finally in problems involving gravimetric estimation one must know the conversion factor, *e.g.*, if Al is precipitated as $Al(OH)_3$, then we have

$Al^{3+}+3OH^-=Al(OH)_3$, $Al=Al(OH)_3$ so conversion factor is

$2Al=2Al(OH)_3=Al_2O_3+3H_2O$ so $2Al=Al_2O_3$

$x=\dfrac{Al_2O_3}{2Al}$ so we multiply weight of hydroxide by (x) to get weight of $Al(OH)_3$.

Now if weight of sample is known, it is quite easy to calculate percentage of Al.

Another example is precipitation of Mg as $Mg_2P_2O_7$ with microcosmic salt here the reaction proceeds as $2Mg=Mg_2P_2O_7$ so conversion factor is $\dfrac{Mg_2P_2O_7}{2Mg}=y$ so we multiply by (y).

SOLVED PROBLEMS

4.1 In a sample of unsoluble mineral (0.5 gm) the total weight of chlorides of sodium and potassium is 0.1180 gm. On treatment with silver nitrate it gave 0.2451 gm of insoluble chloride of silver. Calculate the percentage oxides of sodium and potassium in feldspar.

Ans: Let us assume x = wt. of KCl and y = wt. of NaCl; $x+y=0.1180$.

Now amount of AgCl obtained from x gm of KCl $=x\left(\dfrac{AgCl}{KCl}\right)=1.92\,x$.

Similarly amount of Ag derived from y gm of NaCl $=y\left(\dfrac{AgCl}{NaCl}\right)=2.45y$.

Since $1.92x+2.45y=0.2451$. Solving simultaneous equation $x=0.0834$ KCl and $y=0.0346$ gm of NaCl (as mol wt. KCl = 74.56 and mol wt. of NaCl = 58.5)

(*a*) $AgNO_3+NaCl \rightleftharpoons AgCl+NaNO_3$

(*b*) $AgNO_3+KCl \rightleftharpoons AgCl+KNO_3$

Silverchloride from (*a*) + (*b*) = 0.2451gm

$$\%K_2O=\left(\frac{K_2O}{2KCl}\right)=\frac{0.0834\times 100}{0.500}=10.5\%\ K_2O$$

$$\%Na_2O=\left(\frac{Na_2O}{2NaCl}\right)=\frac{0.0346\times 100}{0.500}=3.67\%\ Na_2O.$$

4.2 A mineral for alkali metals weighs 0.4150 gm with weight of alkali chloride such as sodium and potassium is 0.0715 gram. On precipitating it with potassium hexachloroplatinate the weight is 0.1548 gm. Calculate the percentage of oxides of sodium and potassium.

Ans: Let x be wt. of NaCl, in combined chloride then $(0.0715 - x)$ will be wt. of KCl.

$2KCl = K_2PtCl_6$ so $(0.0715 - x)$ gm, weight of KCl = 0.1548 gm of K_2PtCl_6

By substituting the values in equation for molecular weight, we have

$$(0.0715 - x)\frac{K_2PtCl_6}{2KCl} = (0.0715 - x) \times \frac{485.95}{2 \times 74.56} = 0.1548. \text{ So } x = 0.0240 \text{ gm as}$$

$2NaCl = Na_2O$ and $2KCl = K_2O$. We substitute

$$\frac{0.0240 \times (Na_2O/2NaCl)}{0.4150} \times 100 = 3.07\% \; Na_2O$$

So oxide of potassium is

$$\frac{(0.0715 - 0.024) \times K_2O}{2KCl \times 0.4150} \times 100 = \frac{0.0475 \times 84.20 \times 100}{2 \times 0.4150 \times 74.56} = 55.44\%.$$

This is oxide of potassium.

Hence in given sample Na_2O = 3.07% and K_2O = 55.44%.

4.3 How is the solubility of strontium in solution of pH 3.0 containing excess of oxalate of 0.01 M is determined?

Ans: $\alpha_2 = \dfrac{K_1 K_2}{[H^+]^2 + K_1[H^+] + K_1 K_2}$ with usual notations

α_2 is fraction of anion in the solution

$$\alpha_2 = \frac{[A^{2-}]}{C_{H^2}A}$$

Now $\alpha_2 = 0.0056\, S = (Sr^{2+})C_{ox} = S + 0.001,\; S(S + 0.01) = \dfrac{K_{sp}}{\alpha_2}$

i.e., 3.6×10^{-8}; $S = 3.6 \times 10^{-6}$ moles per liter.

4.4 The solubility product of $CaF_2 = 4 \times 10^{-11}$. Calculate its actual solubility (S) ignoring hydrolysis of salt with S° = 0 (*i.e.*, molar solubility)?

Ans: $(Ca^{2+}) = S\,(F) = 2s$, $4s^{3-} = 4 \times 10^{-11}$, $S = 2.2 \times 10^{-9}$ moles/ litre.

4.5 Evaluate the solubility of mercuric carbonate in water assuming carbonate anion undergoes hydrolysis. Given ($K_{sp} = 8 \times 10^{-12}$, $K_1 = 4 \times 10^7$, $K_2 = 3 \times 10^{-11}$).

Ans: $Hg_2\,CO_3 + H_2O = 2Hg^+ + HCO_3 + OH$ so $K = \dfrac{K_{sp}Kw}{K_2}$

Since $(HCO_3) = (OH) = S$; $(Hg^+) = 2S$,

$$4S^4 = \frac{8 \times 10^{-12} \times 10^{-14}}{5} \times 10^{-11} \text{ so } S = 1.4 \times 10^{-4} \text{ mols/L}$$

$$(Hg+) = 2(CO_3^2) + (HCO_3) + (H_2CO_3) = \frac{2[CO_3^2]}{\alpha^2}$$

$$(OH) = 1.4 \times 10^{-4},\; \alpha_2 = 0.41,\; 4S_3 = \frac{K_{sp}}{\alpha} = \frac{8 \times 10^{-12}}{\alpha_2} = 1.7 \times 10^{-4} \text{ mol/L}.$$

4.6 What is solubility of HgI in 1 × 10^{-2} M ammonia if $K_{sp} = 9\times10^{-7}$ for HgI? If $\log K_1 = 3.2$ while $\log K_2 = 3.8$?

Ans: $C_M = (M) + (MR_1) + (MR_2) .... + (MR_n).$

$$C_M = (M) + (K_1)(R) + K_1K_2(R^2) + ... K_1K_2K^n .... [Rn]^n$$

$$C = \frac{[M]}{\alpha M} \text{ so } \frac{1}{\alpha M} = 1\times10^{3.2}\times10^{-2} + 10^{7.0}\times10^{-4} = 10^3. \text{ Now since}$$

$$S = C_M = (A) = \sqrt{\frac{K_s}{\alpha M}} \text{ substituting the values we have}$$

$$S = \sqrt{9\times10^{-17}\times10^3} = 3\times10^{-7} \text{ moles/litre if Hg(NH}_3\text{), is negligible.}$$

4.7 What is the solubility of halide ion concentration of 10^{-2} M for metal fluoride with $\log K_1 = 7.8$, $\log K_2 = 6.1$, $\log K_3 = 4.7$ and $K_{sp} = 4\times10^{-28}$?

Ans: $$S = \frac{10^{-27.4}}{10^{-8}} + \frac{10^{7.8}\times10^{-27.4}}{10^{-6}} + \frac{10^{13.9}\times10^{-27.4}}{10^{-4}} + \frac{10^{18.6}\times10^{-27.4}}{10^{-2}},$$

$= 10^{-6.8}$, *i.e.*; 1.6 × 10^{-7} M due to halide.

4.8 What is the solubility of AgBr given $K_{sp} = 1.8\times10^{-10}$?

Ans: AgCl = Ag^+ + Br^-, $K_{sp} = (Ag^+)(Br^-) = 1.8\times10^{-10} = S\times S = S^2$
Since $(Ag^+) = (Br^-) = S$ (Solubility)

i.e., $$S = 1.34\times10^{-20} \text{ M.}$$

4.9 If lead sulphate has solubility of 1.14 × 10^{-5} M. What would be its solubility product?

Ans; $PbSO_4 = Pb^{++} + SO_4^{2-}$, K_{sp} = (Pb) (SO_4) Given (Pb) = 1.14 × 10^{-5}

$$= (Pb^{2+}) = (SO_4^{2-}) = (1.14\times10^{-5})$$

$$K_{sp} = (Pb^{2+})(SO_4^{2-}) = (1.14\times10^{-5})\times(1.14\times10^{-5}) = 1.30\times10^{-10}.$$

4.10 What is solubility of Hg_2CrO_4 if its $K_{sp} = 1.1\times10^{-12}$ for mercury chromate?

Ans: $$Hg_2CrO_4 = CrO_4^{2-} + 2Hg^+$$

$$K_{sp} = (Hg^+)^2(CrO_4^{2-}) = 1.1\times10^{-12}$$

$$[Hg^+] = 2S, [CrO_4^{-2}] = S \text{ So } [2S]^2[S] = [Hg^+]^2[CrO_4^{2-}]$$

i.e., $$1.1\times10^{-12} = [2S^2][S] = [Hg^+]^2[CrO_4^{-2}] = 6.5\times10^{-5}\text{M.}$$

4.11 Calculate the percentage of iron in sample, when 0.210 gm of iron sample was dissolved and iron was precipitated as hydroxide which on further ignition and weighing gave 0.11069 as ferric oxide?

Ans: $Fe^{3+} + OH^- = Fe(OH)_3 + xH_2O; 2Fe(OH)_3.xH_2O \xrightarrow{\Delta} Fe_2O_3 + (x+3)H_2O$

$$\% \text{ Fe} = \frac{\text{wt. of } Fe_2O_3 \times \text{conversion factor of Fe to } Fe_2O_3 \times 100}{\text{wt. of sample}}$$

$$\% \text{ E} = \frac{0.1106 \times \dfrac{2Fe}{2Fe_2O_3}}{0.2010} = \frac{0.1106 \times \dfrac{2 \times 55.85}{159.7}}{0.2010} \times 100 = 38.48\%$$

4.12 Calculate the amount of chloride in AgCl weighting 0.1290, the amount of magnesium in magnesium phosphate weighing 0.3515 and the weight of magnesium sulphate heptahydrate in pyrophosphate weight as 0.35159.

Ans: (*a*) $\dfrac{AgCl}{Cl} = \dfrac{143.5}{35.45} = \dfrac{0.1290}{x}$ so $x = \dfrac{0.1290 \times 35.45}{143.5} = 0.03186$ of chloride in AgCl

(*b*) $\dfrac{Mg_2P_2O_7}{2\,Mg} = \dfrac{226.2}{24.31} = 0.3515x = \dfrac{0.3515 \times 2 \times 24.31}{226.2} = 0.07555$ of magnesium

(*c*) Since $2MgSO_4 \cdot 7H_2O = Mg_2P_2O_7$ as $7H_2O = 246.5$, $MgSO_4 = 226.2$ and $Mg_2P_2O_7 = 0.3515$ we have as the following expressions.

$222.6 \times 246.5 = 0.3515 \times (\overline{x})$ so $x = \dfrac{0.3515 \times 2 \times 24.31}{226.2} = 0.7784$ of magnesium sulphate hexahydrate.

4.13 Calculate weight of Galena if it contains 66% of Lead in analysis of lead as insoluble sulphate.

Ans: Pb = 207, $PbSO_4$ = 303 If x = wt. of Lead then we have

$207\,Pb = 303\,PbSO_4$; wt. of $PbSO_4 = 0.5$. So $x = \left(\dfrac{207 \times 0.5}{303}\right) = x$

Now 66 gm Pb = 100 gm galena

So, Amt. of Pb $= \dfrac{100 \times 207 \times 0.5}{66 \times 303} = 0.5$ gm. of galena mineral.

4.14 What is the weight of alum $\left(K_2SO_4 \cdot Al_2(SO_4)_3 \cdot 12H_2O\right)$ which should be taken for gravimetric analyse of aluminum as its oxide?

Ans: $K_2SO_4 \cdot Al_2(SO_4)_3 \cdot 12H_2O = 2Al(OH)_3 = Al_2O_3$.

$$2 \times 474 = 102\ Al_2O_3 \text{ so } x = \frac{0.1 \times 94.6}{102} = 0.92 \text{ of alum.}$$

4.15 Derive an expression for conversion of Mg to $Mg_2P_2O_7$.

Ans: $\dfrac{Mg}{Mg_2P_2O_7} = \dfrac{2Mg\ (\text{Formal})}{Mg_2P_2O_7\ (\text{Formal})}$

So, $Mg = \dfrac{(Mg_2P_2O_7) \times 2\,Mg(\text{Form})\,\text{wt.}}{Mg_2P_2O_7\,(\text{Form})\,\text{wt.}}$

PROBLEMS FOR PRACTICE

4.16 What is the volume of solution containing 40mg/ml of silver nitrate needed to precipitate the sulphate from one gram of sodium chloride? **[Ans:** 57ml $AgNO_3$]

4.17 The phosphate in one gram of sample of manure was precipitated and weighted as magnesium pyrophosphate. The weight was 0.2550 gram. Calculate the concentration of the phosphorus in the sample of manure? **[Ans:** 7.10% of P]

4.18 Dolomite weighing 0.9876 gm is composed of carbonates of magnesium and calcium. The weight of combined oxide, *viz.*, 0.3123 gm. What is the weight of calcium carbonate in the dolomite? **[Ans:** $CaCO_3$ = 0.4887]

4.19 How much of ferrous oxide (Fe_2O_3) is obtained from 1.63 gm of Fe_3O_4? Calculate the conversion factor. **[Ans:** 1.69 gm; 1.035]

4.20 Potassium oxalate on heating gives potassium carbonate. If 1.3906 gm of $K_2C_2O_4$ gives residue of 1.1436 gm. What is the purity of sample? **[Ans:** 84.97%]

4.21 What is weight of copper sulphate obtained from 0.676 gm of Na $[Cu(CN)_2]$ complex? **[Ans:** 0.576 grams]

4.22 Calculate the volume of 0.5% dimethylglyoxime which is needed to precipitate bismuth from 0.080 gm of bismuth chloride if 4% excess of reagent is added to ensure complete precipitation. **[Ans:** 30ml of DMG]

4.23 If mixture of oxides of gallium and selenium weighs 0.5 gm while their oxinates weigh 1.86gm. What should be percentage of oxides of each in the sample? **[Ans:** 17.6% SeO_2, 82.4% Ga_2O_3]

4.24 When zinc is determined as $Zn_2P_2O_2$, 1000 mg of sample gave 6611 mg pyrophosphate? Calculate the concentration of zinc in the given sample. **[Ans:** 28.37%]

4.25 If 50ml of potassium chloride solution of sea weed calchie when titrated gave 53.62 ml of 0.63 M of silver nitrate. Calculate the concentration of potassium chloride in the sea weed? **[Ans:** 0]

4.26 If 20 ml of sea water largly containing sodium chloride gave 0.2311 gm of precipitate of silver chloride. Calculate the concentration of sodium chloride in the sea water? **[Ans:** 0.0806M]

4.27 A combined precipitate of oxalates of calcium and magnesium were analysed by thermo-gravimetric analysis. On heating its weighs was 0.4123 gm as oxide. The carbonate of calcium on further heating gave oxides which weighed 0.2943 gm. Calculate the weight of calcium oxide in the sample? **[Ans:** 0.1504 gm]

4.28 A mixture of magnesium and calcium carbonates in limestone weighs 0.9876 gram. Calculate the weight of residue if it is ignited to oxide of magnesium assuming carbonate of calcium remains unaffected? **[Ans:** 0.7272 gm]

4.29 Zinc and cadmium sulphide weight 0.4987 gm. If on dissolution they were precipitated with dihydrogen ammonium phosphate and weighed as pyrophosphate which had weight of 0.6987 gm. What was weight of zinc sulphide? **[Ans:** 0.0567 gm]

4.30 What is the percentage weight of uranium in the mineral pitchblende? **[Ans:** 84.80%]

4.31 A 0.5 gram of iron ore magnetite changed to ferric oxide with weight as 0.411 gram. What is the amount of Fe_3O_4 in the magnetite in percentage? **[Ans:** 79.46%]

4.32 Calculate the weight of ferrosoferric oxide (Fe_3O_4) which will furnish 0.5630 gram of ferric oxide. **[Ans:** 0.5249 gm Fe_3O_4]

4.33 A sample of calcium mineral gave following results in percentages: CaO = 45.18, MgO = 8.10, Fe_2O_3 = 4.0. SiO_2 = 6.02, CO_2 = 34.7, H_2O = 2.03. On ignition no water was detected and carbon dioxide was 3.30%. Calculate percentage of calcium oxide and ferric oxide in the burnt sample? **[Ans:** 68.54%, CaO 6.74%. Fe_2O_3]

4.34 A platinum crucible weighs 25.6142 gram in air with brass weights. Calculate the weight of crucible in vacuum if the density of platinum is 21.4 while that of brass is 8.0? The weight of one ml of air is 0.0012 gram. **[Ans:** 25.6118 gram]

4.35 If 340 mg of sample sodium and potassium chlorides furnish 706.2 mg of silver chloride. Calculate the amount of each compound in percentage? **[Ans:** 28.9% NaCl; 71.7%, KCl]

REFERENCES

4.1 S.M. Khopkar, *Basic Concepts of Analytical Chemistry,* 3rd edn., New Age International Publishers (P) Ltd. (2007).

4.2 H.A. Latinen & W.E. Harris, *Chemical Analysis – An Advanced Textbook,* McGraw-Hill & Co., 2nd edn., pp. 129-130. 133, 136 (1975).

4.3 D.J. Pietrzyke & C.W. Frank, *Analytical Chemistry: An Introduction,* Academic Press pp. 127-130, 207, 216-7, 245-6 (1979).

4.4 J.S. Fritz & G.H. Schenk, *Quantitative Analytical Chemistry,* Allyn & Bacon Inc., pp. 58, 66-7, 85 (1979).

4.5 D.A. Skoog & D.M. West, *Analytical Chemistry,* 3rd edn., Holt Rienhart Winston, pp. 29, 119, 121, 139, 142 (1979).

4.6 J.G. Dick, *Analytical Chemistry,* McGraw-Hill & Co., p. 188 (1973).

4.7 L. Erdey, *Gravimetric Analysis,* Pergamon Press (1963).

4.8 H.F. Walton, *Principles and Methods of Chemical Analysis,* 2nd edn., Prentice Hall & Co. (1966).

4.9 J.F. Flagg, *Organic Reagents in Gravimetry and Volumertic Analysis,* Interscience (1948).

4.10 F.J. Welcher, *Organic Analytical Reagents,* Van Nostrand (1959).

Chapter 5

Volumetric Analysis —Acid Base Titrations

INTRODUCTION

In spite of greater strides in the development of the modern instrumental methods of analysis, volumetric analysis has still retained its importance in quantitative chemical analysis; because in instrumental methods of analysis one needs standard solution for calibration. Such standard solutions can be obtained by carrying out either gravimetric or volumetric analysis. In fact, some of the volumetric methods like complexometric titration are so good that many times they compete with instrumental techniques if proper metallochromic indicators are available. Many environmental analysis like analysis of dissolved oxygen from water can be done by Winkler's method involving redox titration. The dissolved chloride from polluted water can easily be analysed by Volhard's method involving precipitation titrations, or analysis of moisture from air by Karl Fischer titration.

The acid base titrations have special role to play in analytical chemistry. From such titrations one can calculate pH of the solution at equivalence point. One can also ascertain magnitude of K_a or K_b *i.e.*, the acid dissociation constant of an acid or base respectively or K_W *i.e.*, the ionic product of water. The magnitude of K_a or K_b depends upon the nature of the acid or base. *viz.*, whether it is strong acid or strong base or weak base or acid or any such combination in titration. The pH is related to hydronium (H_3O^+) ion concentration as

$$\text{pH} = \frac{1}{\log[\text{H}_3\text{O}]^+} \text{ or pH} = -\log[\text{H}_3\text{O}]^+ \text{ similarly [OH]}^- \text{ is related to pH}$$

$$\text{as pOH} = \frac{1}{\log[\text{OH}^-]} = -\log[\text{OH}^-] \text{ assuming in all cases the ionic product is}$$

$$K_w = 1.0 \times 10^{14} \text{ or pH} + \text{pOH} = 14 = pK_w \text{ (ionic product)}$$

Since $pK_w = \dfrac{1}{\log K_w}$ or $-\log K_w$. In above expression if HA is acid, B = base

Since we have

$$\text{HA} + \text{H}_2\text{O} \rightleftharpoons \text{H}_3^+\text{O} + \text{A}^-$$

$$HB^+ + OH^- \rightleftharpoons H_2O + B^-$$

$$H_2O \rightleftharpoons H^+ + OH^-$$

$$K_a = \frac{[H_3O^+][A^-]}{[HA][H_2O]} \text{ and } K_b = \frac{[HB^+][OH^-]}{[B^-][H_2O]} \text{ also } K_w = \frac{[H^+][OH^-]}{[H_2O]} \text{ as reaction is:}$$

$HA + H_2O = H_3O^+ + {}^-A^-$ (conjugate base) HA is acid or

$B + H_2O = BH^+ + OH^-$ (conjugate acid) BH^+ is base

Normally it is very easy to evaluate magnitude of (H^+) and hence pH of solution in different combination of acids and bases in titrations.

We can consider different equations for [H^+] in combinations of acid and base.

(*a*) For instance for titration of strong acid and strong base (*e.g.*, NaOH and H_2SO_4)

$$(H^+) = \sqrt{K_w} \text{ (assuming } K_H \geq [OH^-])$$

(*b*) For strong base and weak acid (*e.g.*, CH_3COOH and NaOH)

$$(H^+) = \sqrt{\frac{K_a K_w}{K_a}} \text{ (assuming } [HA] = [OH])$$

(*c*) For strong acid and weak base it is (*e.g.*, H_2SO_4 against weak base like K_2CO_3)

$$(H^+) = \sqrt{\frac{K_w K_a}{K_a}} \text{ and (assuming } [B] = [H^+]) \text{ and finally for}$$

(*d*) Weak acid and weak base we have (*e.g.*, formic acid against sodium formate)

$$(H^+) = \sqrt{\frac{K_a K_w}{K_b}} \text{ (assuming } [B] = [HA])$$

Thus the indicator pH is evaluated from the relationship.

Indicator pH range = $pK_a \pm 1$.

Apart from above expression other formulae which are used for the exact calculation of pH and titration curves are given in following equations.

$$(H^+{}_3O) = \frac{V_R M_R - V_T M_T}{(V_R + V_T)} = [H_3^+O] = [R]V \text{ if } V_R V_T \text{ ..., are Volumes of reactant and titrant respectively.}$$

(Before equivalence point) and $M_R M_T$ is moles of reactant and titrant resp.

While after equivalence point we have;

$$(OH^-) = \frac{V_T M_T - V_R M_R}{(V_R + V_T)} = [R_T]V_T \text{ with same notations } [R_T]V_T = [OH^-] \text{ and } [R_T]V_T = [H_3O^+].$$

These two equations are invariably used to ascertain (H_3O^+) and (OH^-) during titrations. From these equations one can therefore find out pH = –log $(H_3O)^-$ or pOH –log(OH). Now the normality is expressed in terms of gms equivalent per liter. For simple calculations we use $N_1V_1 = N_2V_2$ where N_1 and N_2 are normalities of soln. No. 1 and No. 2 while V_1 or V_2 are volumes of solution number one and two. The kind of indicators used are different. The derivatives of the triphenylbasic dyes are used as indicator, *e.g.*, Methyl orange. The primary standard selected are such which maintains constant weight on exposure to atmosphere. Therefore compounds with water of crystallisation are not preferred as standards. The mixed indicators are generally used for ascertaining pH of the solution instead of detecting end point in titration.

SOLVED PROBLEMS

5.1 If 0.59 gm of mixture of carbonates of lithium and potassium required 30 ml of 0.25 N acid solution for neutralisation, what was the percentage composition of the mixture?

Ans: Let $x = K_2CO_3$, $y = Li_2CO_3$, $(x + y) = 0.59$ gm.

$$\text{No. of meq of } K_2CO_3 = \frac{x}{K_2CO_3/2000} = \frac{x}{0.06}$$

$$\text{No. of meq of } Li_2CO_3 = \frac{y}{Li_2CO_3/2000} = \frac{y}{0.0375}$$

Acid consumed (30 × 0.25) ml on solving simultaneous equation
$x = 0.247$ gm $y = 0.253$ gm. Total amount = 0.59 gm Hence percentage compositions.

$$\% K_2CO_3 = \frac{0.247 \times 100}{0.59} = 49.4\% K_2CO_3 \qquad \% Li_2CO_3 = \frac{0.253 \times 100}{0.59} = 50.6\% Li_2CO_3$$

5.2 How much of 0.6 N potassium hydroxide must be mixed with 750 ml of potassium hydroxide (0.2N) to prepare new solution of 0.3N?

Ans: Let x = ml of 0.6 N of KOH is added, volume after dilution = $(750 + x)$ ml. Now nos. of gm equivalent present = $[(750 \times 0.2) + (x \times 0.6)]$

$$\text{estimated volume} = \frac{(750 \times 0.2) + (x \times 0.6)}{(750 + x)} = 250 \text{ ml.}$$

5.3 A 1.2 gm sample of a mixture of sodium carbonate and sodium bicarbonate with inert impurities was dissolved and then titrated with 0.5N hydrochloric acid. With phenolpthalein as the indicator the end point was 15 ml while on further titration with methyl orange as indicator end point was 22 ml. Calculate percentage composition of the mixture?

Ans: (15 + 15) ml = 30 ml of acid was required to neutralise Na_2CO_3 in mixture while total neutralisation needed (15 + 22) = 37 ml. That showed
(37 – 30) = 7 ml. was required to neutralise $NaHCO_3$.

$$\frac{30 \times 0.5 \times Na_2CO_3/2000}{1.2} \times 100 = 66.25\% \; Na_2CO_3$$

$$\frac{7.0 \times 0.5 \times NaHCO_3/1000}{1.2} \times 100 = 24.3\% \; NaHCO_3$$

5.4 Calculate the pH of the solution having 0.4 M in formic acid and having 1M sodium formate and 0.28 M in ammonium chloride and 0.07M in ammonia?

Ans: The reactions occurring are as follows:

$$H_2O + HCOOH = H_3O^+ + HCOO^- \quad ...(i)$$

$$NH_3 + H_2O = |NH_4^+| + OH^- \quad ...(ii)$$

We calculate K_a for both reactions

$$K_a = \frac{[H_3O][HCOO^-]}{[HCOOH]} = 1.77 \times 10^{-4} \text{ as } [HCOO^-] = 1.0, [HCOOH] = 0.4 \text{ given}$$

$$K_a = \frac{[NH_4^+][OH^-]}{[H_2O][NH_3]} = 1.76 \times 10^{-5} \text{ as } [NH_4^+] = 0.28, [NH_3] = 0.7. \text{ Now}$$

$$[H_3O^+] = K_a \frac{(HA)}{(NaA)} = \frac{K_a \cdot HCOOH}{HCOONa} = \frac{1.77 \times 10^{-4} \times 0.4}{1.0} = 7.08 \times 10^{-5} \text{ also}$$

$$[OH^-] = \frac{1.76 \times 10^{-4} \times 0.07}{0.28} \text{ so } p[OH] = 4.40 \times 10^{-5}$$

So $[H_3O^+] = 7.08 \times 10^{-5}$, pH = $-\log [7.08 \times 10^{-5}] = 4.15$

So also $[OH^-] = 4.04 \times 10^{-5}$, pOH = $-\log[4.04 \times 10^{-5}] = 5.36$

pH = (14.0 – 5.36) = 8.64.

5.5 Calculate the hydronium concentration (H_3O^+) of 0.1 M (*a*) Sodium bicarbonate and (*b*) 0.001 M of disodium hydrogen phosphate, and (*c*) 0.01 M dihydrogen sodium phosphate solution?

Ans: The formula used is $[H_3O^+] = \sqrt{K_1K_2}$ first part and

$$[H_3O^+] = \sqrt{\frac{K_1K_2M(Na)HA + K_1K_2}{M(Na)H|A| + K_1}} \text{ so we substitute the given values,}$$

(*a*) $K_1 = 4.45 \times 10^{-7}$; $K_2 = 4.7 \times 10^{-11}$ (Given), K_1K_2 M NaHA

So $\quad H_3O^+ = \sqrt{K_1K_2} = \sqrt{4.45 \times 10^{-7} \times 4.7 \times 10^{-11}} = 4.6 \times 10^{-9}$

(*b*) If K_2 for HPO_4^{2-} is 6.34×10^{-8} and for H_3PO_4 $K_3 = 6.4 \times 10^{-13}$

$$[H_3O^+] = \sqrt{\frac{6.34 \times 10^{-8} \times 4.2 \times 10^{-13} \times 0.001 + 6.34 \times 10^{-22}}{1.0 \times 10^{-3}}} = 8.1 \times 10^{-10}$$

(*c*) K_1 for $HPO_4^{2-} = 7.11 \times 10^{-3}$ and K_2 for $H_3PO_4 = 6.3 \times 10^{-8}$

$$[H_3O^+] = \frac{7.11 \times 10^{-3} \times 6.34 \times 10^{-8} \times 1 \times 10^{-2}}{(0.01 + 7.11 \times 10^{-3})} = 1.6 \times 10^{-3}$$

5.6 A 0.1196 g of sodium carbonate was titrated with hydrochloric acid with methyl orange as an indicator which gave titer value of 42.45 ml. Now if acid was standardised with pure sodium carbonate weighing 0.1425 g against 24.30 ml of hydrochloric acid.
Calculate the concentration of carbon dioxide present in the sample?

Ans: The reaction was $Na_2CO_3 + 2HCl = 2\ NaCl + H_2O + CO_2\uparrow$ (reaction ratio is 1:0.5) wt. of Na_2CO_3 = (ml of HCl × ratio × Na_2CO_3)

$$145.2 = \left(24.30 \times M_{HCl} \times \frac{1}{2} \times 106.0\right) \text{So } M_{HCl} = 0.1107\ M$$

$$\%CO_2 = \frac{\text{ml HCl} \times M_{HCl} \times CO_2 \times 100}{\text{wt. of sample}} = \frac{42.45 \times 0.117 \times 0.5 \times 44.01 \times 100}{119.6\ \text{mg}} = 86.46\%\ CO_2$$

5.7 A 0.2744 gram of sodium carbonate and sodium hydroxide with inert impurities was titrated with 0.104 M hydrochloric acid. If 34.11 ml of acid was required with phenolpthalein and further 7.12 ml was required to attain an end point with methyl orange. What is the percentage composition of mixture?

Ans: The reactions were

(*a*) $NaOH + HCl \rightarrow NaCl + H_2O$ (with phenolphthalein)

(*b*) $Na_2CO_3 + HCl \rightarrow NaHCO_3 + H_2O + NaCl$ (phenolphthalein indicator)

(*c*) $NaHCO_3 + HCl \rightarrow NaCl + CO_2 + H_2O$ (Methyl orange indicator)

The volume of acid from one indicator to other indicator (*viz.*, 7.12 ml) was amount consumed by bicarbonate. Since $Na_2CO_3 = NaHCO_3$ the reaction ratio was 1:1 as shown in equation (*b*) above

$$\%\ Na_2CO_3 = \frac{ml_{HCl} \times M_{HCl} \times \text{ratio} \times Na_2CO_3 \times 100}{\text{wt. of sample}}$$

$$\%\ Na_2CO_3 = \frac{7.12 \times 0.104 \times 1 \times 106 \times 100}{0.2744} = 28.66\%\ Na_2CO_3.$$

As 7.12 ml was required for reaction (*c*) same volume was also required for reaction (*b*) So (34.11 − 7.12) = 26.99 ml acid required to neutralised NaOH only

$$\%NaOH = \frac{ml_{HCl} \times M_{HCl} \times \text{ratio} \times NaOH \times 100}{\text{wt. of sample}}$$

$$\%NaOH = \frac{26.99 \times 0.104 \times 1 \times 40 \times 100}{0.2744} = 41\%\ NaOH.$$

5.8 Calculate the normality of 20% sulphuric acid having specific gravity = 1.14?

Ans: $V = \frac{P}{d} = \frac{1000}{1.14} = 87.7$ ml. For 100 gm of sulphuric acid

(if d = sp. gravity, P = total volume)

If 87.7 ml had 20 gram of H_2SO_4 how much was acid in 1000 ml of acid?

$$x = \frac{1000 \times 20}{87.7} = 228 \text{ gm of acid eq. wt. } H_2SO_4 = \frac{1}{2} \text{ Gram mol} = \frac{98.08}{2} = 49.04$$

So $\frac{228}{49.04}$ = 4.65 N. The mol wt. = 1/2 eq. wt. So $\frac{4.65}{2}$ = 2.32M is H_2SO_4.

5.9 Predict the total volume of uninormal solution containing equal amounts of the dissolved substance as much as 30 ml of 0.2 N solution?

Ans: Amount of substance was same in both solutions, we had

$N_1V_1 = N_2V_2$, *i.e.*, $1 \times x = 0.2 \times 30$ so $x = 6$ ml.

Gm equivalent = 6 × 1.0 = 6.0.

5.10 What is the permanent hardness of water if excess of 20 ml sodium carbonate with water is taken and was back titrated with 0.1023 N acid showed volume of 16.48 ml? The direct titration of 20 ml sodium carbonate took 19.16 ml of acid.

Ans: Amount of acid consumed due to hardness = (19.16 – 16.48) = 2.68 of 0.1023 N acid. Therefore permanent hardness = normality × volume of acid assumed = 0.1023 × 26.8 = 2.7416 mg/equivalent/Liter = 0.00274 g/eq/L.

5.11 What is the carbonate hardness if 200 ml of hard water if titrated with 0.120 N hydrochloric acid to have titre values of 10.13 ml with methyl orange as indicator?

Ans: $Ca(HCO_3)_2 + 2HCl = CaCl_2 + 2H_2O + 2CO_2\uparrow$

If 200 ml needed 10.13 ml then one liter needed 10.13 × 5 = 50.65 ml of 0.1023N acid. Since 1 liter acid had 0.1023 gm eq of HCl so ml would contain 0.1023 milligram equivalent. So 0.103 × 50.65 = 3.181mg/eq/ liter is carbonate hardness of water.

5.12 Define the indicator error in titration of 0.1 N hydrochloric acid against 0.1N sodium hydroxide with methyl orange as an indicator with end point at pH = 4.0?

Ans: Eq. point at pH = 7.0. During titration volumes were same also with same normality so

$V_2 = 2V_1$. The H^+ error was given as $\left(\frac{10^{-4} \times V_2}{NV_1} \times 100\right)$ so by substituting H^+ error

$$= \frac{10^{-4} \times 2V_1}{10^{-1} \times V_1} \times 100 = 0.3\% \text{ was negative error.}$$

5.13 If 50 ml of 0.1 M hydrochloric acid was titrated against 0.1 M sodium hydroxide, Calculate the pH at the beginning of titration and after addition of 10.50, and 60 ml of alkali hydroxide.

Ans: (*i*) Initial pH, HCl was dissociable as $\{H_3O\}^+$ = 0.1 log (0.1) = pH. It was pH = 1.0.

(*ii*) The pH after addition of 10 ml of base we use expression as

$[R]V_T = \frac{V_RM_R - V_TM_T}{V_R + V_T}$ if V_R, V_T is volume of reactant and titrant and M_R, M_T was molarity of reactant and titrant.

$$\left[H_3O^+\right] = \frac{(50 \times 0.1) - (10 \times 0.1)}{(50 + 10)} = 6.67 \times 10^{-2} \text{ M.pH} = (2 - \log 6.67) = 1.18$$

(*iii*) pH at eq point when 50 ml of base was added. So $\{H_3O\}^+ = [OH^-] = 1\times10^{-7}$. So

pH = 7.0 (Because $\sqrt{Kw} = 1\times10^{-7}$)

(*iv*) After addition of 60 ml of NaOH the value was

$$R_T[V_R] = \frac{V_T M_T - V_R M_R}{V_T + V_R} = \frac{(60\times0.1)-(50\times0.1)}{(50+60)} = 9.1\times10^{-3}\text{ M of }[OH^-]$$

$$pH = (3-\log 9.1) = 2.4 \quad pH = (14-2.04) = 11.96.$$

5.14 When 50 ml of 0.1 M solution of weak acid (K_a = 1 × 10^{-5}) was titrated with 0.1N sodium hydroxide, what is the pH of initial and after addition of 10 and 50 ml of sodium hydroxide?

Ans: Since *Ka* × *Kb* = *Kw* and *Kt* = 1/*Kb*. If *B*-anion of weak acid

(*i*) Initial pH of reaction $[H_3O^+] = [A^-]$ so $[HA] = 0.1 - [H_3^+O] = 0.1$

$$\text{So } Ka = \frac{[H_3O^+][A^-]}{[HA]} = 1\times10^{-5} \text{ if } [HB] = 0.1 \text{ then } Ka = 1\times10^{-5}$$

(*ii*) pH after addition of 10 ml of base if K_1 = 1 × 10^9

$$[R]V_R = \frac{V_R M_R - V_T M_T}{(V_R + V_T)} = \frac{50\times0.1-(10\times0.1)}{(50+10)} = \frac{4}{60}\times[B] = \frac{1}{60}$$

$$Ka = 1\times10^{-5} = \frac{[H_3O^+][B^-]}{[HB]} = \frac{[H_3O^+][1/60]}{[4/60]}$$

Since $[H_3O^+] = 4\times10^{-5}$, $pH = (5-\log 4) = 4.40$

$$\left(pH = pKa - \log\frac{[HB]}{[B^-]}\right)$$

(*iii*) pH at equivalence point 50 ml of sodium hydroxide was added, ...

$[B^-] = 0.05$ M, as $[HB] = [OH^-]$ so

$$Kp = 1\times10^{-9} = \frac{[HB][OH^-]}{[B^-]} = \frac{[OH^-]^2}{[B^-]} = \frac{[OH^{-2}]^2}{0.03} = 1\times10^{-9}.$$

So $[OH]^2 = 5\times10^{-11}$; $2\times pOH = (11-\log 5)$ so $pOH = 5.15 pH = 8.85$.

PROBLEMS FOR PRACTICE

5.15 What volume of 6N and 3N sulphuric acid should be mixed to prepare a litre of 5N sulphuric acid? [**Ans:** 667 ml of 6N + 333 ml of 5N]

5.16 A mixture containing carbonates of lithium and barium weighing 1 gram required 15 ml of hydrochloric acid for neutralisation. What is the percentage of the carbonates of the lithium and barium? [**Ans:** $LiCO_3$ = 71.26% $BaCO_3$ = 5.40%]

5.17 How many ml of 0.1421 N Lithium hydroxide was needed to neutralise 13.72 ml of 0.06860N of the phosphric acid (H_3PO_4)?

[**Ans:** 6.6 ml of LiOH]

5.18 Calculate the pH of a solution which was 0.4 M in formic acid and 1M of sodium formate?

[**Ans:** pH = 4.15]

5.19 Indicate the weight of ammonium chloride required to be added to 1000 ml of 0.125 M ammonia in order to have buffer of pH = 10, if there was change in volume.

[**Ans:** 0.816 grams.]

5.20 Calculate the normalilty of barium hydroxide if it was titrated with 1.50 ml of perchloric acid?

[**Ans:** 0.015N]

5.21 The carbonate and bicarbonates of sodium were dissolved in water and were titrated with 0.1 M hydrochloric acid. The end points with phenolpthalein was 12 ml but with methyl orange it was 34 ml. Calculate percentage of carbonate and bicarbonates?

[**Ans:** 1.2 m moles CO_3^{-2}; 1.0 m moles of HCO_3^-]

5.22 A one gram of mixture sodium carbonate and bicarbonates when titrated with 0.1 M hydrochloric acid gave end point of 15.5 ml with phenolpthelin and 40.1 ml with methyl orange as indicators. What is the composition of the mixture?

[**Ans:** Na_2CO_3 = 16.42 % $NaHCO_3$ = 7.64%]

5.23 A mixture of carbonate and bicarbonate of sodium and inert material weighed 0.2905 gm. It was titrated with 0.1141 M hydrochloric acid with end point with the phenolptheline at 22.15ml and end point with methyl orange as 46.34 ml, what is the percentage composition of the mixture ? [**Ans:** Na_2CO_3 = 99.2% $NaHCO_3$ = 6.7%]

5.24 If 50 ml of 0.1 M acid ($Ka = 10^{-5}$) was titrated with 0.1 M potassium hydroxide. Calculate the pH after addition of 60 ml of base? [**Ans:** pOH = 2.04 and pH = 11.96]

5.25 Major amount of sodium hydroxide with traces of sodium carbonate as impurity was present. If 50 ml was titrated with 0.50 M hydrochloric acid then end point with phenopthalein was 30 ml. After evolution of CO_2; 30 ml required same amount of acid with methyl orange as indicator. What is its composition? [**Ans:** NaOH = 0.808 + Na_2CO_3 = 0.0137]

5.26 A glycine (0.36 gm) was dissolved in strong acetic acid and titrated with 0.0163 M percholeric acid. Calculate the volume of acid consumed for total neutralisation? [**Ans:** 45.57ml]

5.27 If a mixture contains neutral salt and calcium oxide and if one gram of it was dissolved in 50 ml of 0.2560 M hydrochloric acid and if excess of acid was neutralised with 0.3311 N sodium hydroxide had titration volume of 9.54 ml. Calculate the percentage composition of calcium oxide in the mixture? [**Ans:** CaO = 27%]

5.28 A sample of complan was analysed by Kjeldahl method. When 0.9865 gm was dissolved in 50 ml of 0.1085 M hydrochloric acid it needed 39.65 ml of sodium hydroxide of strength 0.1108N. Calculate percentage of nitrogen in the Complan? [**Ans:** 2.72 % N]

5.29 When mixture of calcium carbonate and inactive matter weighed 1.0825 gm if treated with 50 ml of 0.1 N hydrochloric acid needed 36.8 ml of sodium hydroxide for end point. What is the composition of mixture ? [**Ans:** $CaCO_3$ = 5.27%]

5.30 Oxalic acid dihydrate weighing 0.1950 gram needed 25.42 ml of potassium hydroxide for neutralisation,what was normality of titratant if known oxalate (0.2915 gm) needed 21.50 ml of sodium hydroxide for neutralisation? What is the amount of oxalic acid in the mixture?

[**Ans:** 40.41%]

5.31 How many grams of potassium tetraoxalate must be dissolved in water and made up to 780 ml to make 0.03 N solution?

5.32 What is the volume of 0.0545 N hydrochloric acid required to titrate 30 ml of solution that was 0.0225 M in sodium phosphate and 0.0361 M in sodium hydroxide with thymolpthalein as indicator? **[Ans:** 32.1 ml]

5.33 Sodium azide (0.2484 gm) was reacted with sodium nitrite and 50 ml of 0.1 N perchloric acid to liberate nitrogen, N_2O gas and water and excess acid was titrated with 13 ml of 0.08 N sodium hydroxide. What is the amount of sodium azide in the sample? **[Ans:** 45.3%]

5.34 What is the equivalent weight of an acid if 1.26 gm sample required for titration 31.7 ml with 0.112 N sodium hydroxide with phenolpthalein as indicator? **[Ans:** 3.55g/eq]

5.35 A gas from polluted atmosphere liberated by Green house effect was 4000 ml at 22°C and one atmospheric pressure when passed through solution containing 25 ml of 0.0207 N barium hydroxide when barium carbonate was precipitated. The resulting base was back titrated with phenolphthalein as an indicator with end point volume was 17.2 ml with 0.0135 N hydrochloric acid. What is the gas and amount present in the atmosphere? **[Ans:** 863 ppm of CO_2]

REFERENCES

5.1 S.M. Khopkar, *Basic Concepts of Analytical Chemistry,* 3rd edn., New Age International Publishers (P) Ltd. (2007).

5.2 D. A. Skoog and D. M. West, *Analytical Chemistry,* 3rd edn. Holt, Reinhart and Winston pp. 196-7, 212, 243 (1979).

5.3 D.J. Pietrzyke and C.W. Frank, *Analytical Chemistry; An Introduction,* Academic Press pp. 300-1, 307 (1979).

5.4 V. Alexeyev, *Quantitative Analysis,* 2nd edn., Mir Publishing House, Moscow, pp. 218-20; 270; 300-1 (1979).

5.5 J.S. Fritz and G.H. Schenk, *Quantitative Analytical Chemistry,* Allyn and Bacon Inc. pp. 185, 191 (1979).

5.6 H.A. Latinen and W.E. Harris, *Chemical Analysis: An Advanced Text,* 2nd edn., McGraw-Hill and Co., pp. 124 (1995).

5.7 J.G. Dick, *Analytical Chemistry,* McGraw-Hill and Co., pp. 307-9 (1973).

5.8 H.F. Walton, *Principles & Methods of Chemical Analysis*, 2nd edn. Prentice Hall & Co. (1966).

Chapter 6

Redox Titrations

INTRODUCTION

The redox reaction plays a significant role in the chemical analysis. A redox process is accompanied by both process of the oxidation as well as reduction. An oxidation process is one which involves increase in the oxidation number while reduction is the process which involves decrease in oxidation number. This is because in the oxidation process is one which involves loss of electrons while reduction is the process which involves the gain of electrons. While calculating the equivalent weight of an oxidising/reducing agent one must examine the reaction and note by how many units oxidation number of the reactant has changed *e.g.*, the reaction of $K_2Cr_2O_7$ in acid media proceeds as

$Cr_2O_7^{2-} + 14H^+ + 6e^- \Leftrightarrow 2Cr^{3+} + 7H_2O$; $E^0_{Cr} = 1.33$. The oxidation number of chromium has changed from +6 to +3, *i.e.*, by 3 units, so for dichromate ion change is 6 units. Hence Eq. Wt = 1/6 mol. wt. of $K_2Cr_2O_7$ = 49.03. Again in reaction *e.g.*, $Na_2S_2O_3.5H_2O$ undergoes transformation in the following manner $2S_2O_3^{2-} = S_4O_6^{2-}$. The oxidation number of sulphur has changed by 1/2 unit, for 2S atom the change is one unit hence the eq wt = mol wt. for $Na_2S_2O_3$ = 248.2.

The redox titration involves two process one involving oxidation while another involving simultaneously reduction, *e.g.*, $Fe^2 + Ce^{4+} \Leftrightarrow Fe^{3+} + Ce^{3+}$. Here Fe (*II*) gets oxidised to Fe (*III*) while Ce (*IV*) is reduced to Ce (*III*) state. In linear titration one can trace course of titration by plotting graph of change of potential against volume of titrant added. Actually one can easily calculate potential of the system by making use of

$$\text{Nernst Equation: } E = E^0 - \frac{RT}{nF}\log\frac{[\text{reductant}]}{[\text{oxidant}]} = E^0 - \frac{0.0591}{n}\log\frac{[\text{reductant}]}{[\text{oxidant}]}$$

In this equation E = final potential, E^0 = standard electrode potential against hydrogen (see Appendix) R = Universal gas constant T = absolute temp, F = Faradays of electricity and n = nos. of electrons transferred.

For all practical purposes at room temp. expression $\frac{RT}{nF} = \left(\frac{0.0591}{n}\right)$ is used.

After each addition of an increment of the titrant one can easily calculate the resulting potential of a mixture of titrant and reactant. What is required is to ascertain molar concentration of the reductant as well as that of oxidant followed by evaluation of cell potential provided if standard electrode potential is known; using Nernst, equation *e.g.* if 50 ml of 0.05 N Fe(*III*) is titrated with 0.1 N Ce (*IV*) solution in H_2SO_4, with $E^0_{Ce(IV)} = 1.44\,V$ and $E^0_{Fe(III)} = 0.68\,V$. Then after 5 ml of addition titrant is:

$$[Ce^{3+}] = \frac{5\times 0.1}{55} - [Ce^{4+}] = \frac{0.5}{55} \text{ and, for iron (III)}$$

$$[Fe^{3+}] = \frac{5\times 0.1}{55} - [Ce^{4+}] \text{ so}$$

$$[Fe^{3+}] = \left(\frac{50\times 0.05 - 5\times 0.1}{55}\right) + Ce^{4+} = \frac{2}{55}. \text{ Now potential shall be}$$

$$E = E^0_{Ce^{3+}} - 0.0591\log\left[\frac{Ce^{3+}}{Ce^{4+}}\right] = E^0_{Fe} - 0.0591\log\left[\frac{Fe^{2+}}{Fe^{3+}}\right]$$

By substituting value of Fe (II) and Fe (III) concentration, we have

$$E = 1.44 - 0.0591\log\left[\frac{Ce(III)}{Ce(IV)}\right] = 0.68 - 0.0591\log\left[\frac{Fe(II)}{Fe(III)}\right] \text{ as } E^0_{Fe} = 0.68\,V,\ E^0_{Ce} = 1.44\,V$$

$$E = 0.68 - 0.0591\log\left(\frac{2/55}{0.5/55}\right) = +0.64\,V \text{ which can be plotted on graph.}$$

During course of redox titrations, several common oxidising agents like $Ce(SO_4)_2$, $K_2Cr_2O_7$, $KMnO_4$ $KBrO_3$ or also reducing agents like oxalic acid, sodium thiosulphate, ferrous sulphate are used. As told before their equivalent weights can be evaluated by knowing the units for change of oxidation number, *e.g.*, $KMnO_4$ change is +5 units, $K_2Cr_2O_7$ change is 3 units, for oxalic acid change is 2 units, for thiosulphate change is by +1 unit.

The classic example of redox reaction in chemistry is Winklers method and Karl Fischer titration. Former is used to determine dissolved oxygen in water while latter is used for determination of the moisture. The reaction in Winklers method proceeds as:

$$2Mn(II) = 4OH^- + O_2 \leftrightarrow 2MnO_2 + 2H_2O \text{ (at pH} = 9.0)$$

$$MnO_2 + 4H^+ + 2I^- \rightarrow I_2 + Mn(II) + 2H_2O$$

While in Karl Fischer, titration reaction is as follows:

$$C_5H_5N\cdot I_2 + C_2H_5N\cdot SO_2 + C_5H_5N + H_2O \rightarrow C_5H_5N\cdot SO_3$$

$$C_5H_5NSO_3 + CH_3OH \rightarrow C_5H_5\ HNOSO_2CH_3 \text{ and finally pyridine-methyl sulphate}$$

$$C_5H_5NSO_3 + H_2O \rightarrow C_5H_5NHO\cdot SO_2OH \text{(pyridine-hydrogen sulphate)}$$

Thus by making use of redox titration one can easily evaluate not only cell potential using Nernst equation but also one can find out equivalent weight of oxidising or reducing agent or determination of the solubility product of salt or dissociation constant or analysis of alkaline earths, transition metals, or sketch a titration curve also. The redox titrations are one which are popularly used by chemist as good number of redox reagents are easily available. The most important component of these experiments is one must know exactly what kind of chemical transformations are progressing during the course of redox reactions. This will enable one to calculate not only molar concentration of oxidant and reductant but also find out cell potential.

In this chapter several examples dealing with calculation of cell potential; E_{cell} for the system, molar concentrations of oxidant/reductant; solubility product and dissociation constant, concentration of an ion or percentage composition of mixture are calculated.

SOLVED PROBLEMS

6.1 Calculate the equivalent weight of following reducing agents (*i*) $FeSO_4 \cdot 7H_2O$, (*ii*) $H_2C_2O_4$ $2H_2O$, (*iii*) $Na_2S_2O_3 \cdot 5H_2O$, (*iv*) H_2O_2 (acid media).

Ans: (*i*) First ascertain from chemical reactions what is the redox potential and oxidation number (O.N) : $Fe^{2+} \rightarrow Fe^{3+} + e^-$ as O.N changes by one unit so mol. wt. = Eq. wt. = 278.

(*ii*) $C_2O_4^{2-} \rightarrow 2CO_2 + 2e^-$; O.N of Carbon changes by $+3 \rightarrow 4$ for $C_2O_4^{2-}$ by 2 units, Therefore eq. wt. = 1/2 mol wt., *i.e.*, 126.6/2 = 63.3.

(*iii*) $2S_2O_3^{2-} \rightarrow S_4O_6^{2-} + 2e^-$; O.N of (S) changes from +2 to 5/2, *i.e.*, 1/2 unit for 2S atom for one atom is unit so eq. wt = mol.wt = 248.2.

(*iv*) $5H_2O_2 + 2MnO_4^- + 6H^+ \rightarrow 5O_2 + 2Mn^{2+} + 8H_2O$; so O.N of Mn change from –1 to 0, *i.e.*, 2 unit so eq. wt. = 1/2 mol wt. is 17.01.

6.2 Calculate equivalent weight of following oxidising agents:
(*i*) $KMnO_4$, (*ii*) $K_2Cr_2O_7$, (*iii*) I_2, (*iv*) H_2O_2 .

Ans: (*i*) $MnO_4^- + 8H^+ + 5e^- \Leftrightarrow Mn^{2+} + 4H_2O$; O.N of Mn changes from $+7 \rightarrow +2 = 5$ units so eq. wt = 1/5 mol wt = 31.6.

(*ii*) $Cr_2O_7^{2-} + 14H^+ + 6e^- \rightarrow 7H_2O + 2Cr^{3+}$; O.N of Cr changes from $+6 \rightarrow +3$ by 3 units for one dichromate ion and for two dichromate units so
Eq wt. = 1/6 mol wt = 49.03.

(*iii*) $H_2O_2 + 2H^+ + 2e^- \rightarrow 2H_2O$; O.N of oxygen changes; from –1 to –2 per mol two unit
Eq. wt = 1/2 mol wt = 17.01.

6.3 What was the amount of potassium dichromate present in solution if excess of alkali iodide was added to it and liberated iodine was titrated with 48.8 ml of 0.1 N sodium thiosulphate to get end point ?

Ans: The reaction was $Cr_2O_7^{2-} + 6I^- + 14H^+ \rightarrow 2Cr^{3+} + 3I_2 + 7H_2O$.

$2S_2O_3^{2-} + I_2 \rightarrow S_4O_6^{2-} + 2I$. So eq. wt of $K_2Cr_2O_7$ = 1/6 mol wt (*i.e.*, 49.03)

So amount of $K_2Cr_2O_7$ = vol. of Na_2SO_3 × normality of $Na_2S_2O_3$ × eq.wt of $K_2Cr_2O_7$

So 48.8 × 0.1 × 49.03 = 0.2393 gm.

6.4 What is the potential of a half cell consisting of cadmium electrode in a solution of 0.01 M in cadmium salt (Given $E^0_{cd} = -0.403\,V$) ?

Ans: The redox reaction was $Cd^{2+} + 2e^- = Cd$; $E^0_{cd} = -0.403$ V

By Nernst equation $E = E^0_{cd} - \frac{RT}{nF}\log\frac{1}{[Cd^{+2}]}$ if $\frac{RT}{F} = 0.059$ and $n = 2$, $E^0_{cd} = 0.403$

$E = E^0_{cd} - \frac{0.0591}{2}\log\frac{1}{[Cd^{+2}]}$ substituting values as given above we have

$$E = 0.403 - \frac{0.0591}{2}\log\frac{1}{[0.100]} = -0.403 - \frac{0.0591}{2}(+2) = -0.462\text{V}.$$

6.5 What is the potential of platinum electrode if immersed in solution of (*i*) 0.0100 M of potassium bromide, (*ii*) 1×10^{-3} M of bromine (Given $E^0_{Br} = 1.065$ V)?

Ans: (*i*) $Br_2 + 2e^- = 2Br^-$; $Br_2O \neq Br_2$ (satd.) by Nernst equation

$$E = E^0_{Br} - \frac{RT}{nF}\log\frac{[Br^-]^2}{[0.01]} = 1.065 - \frac{0.0591}{2}\log\frac{[Br^-]^2}{[1.0]}$$

$$E = 1.065 - \frac{0.059}{2}\log(0.0100)^2 = 1.065 - \frac{0.0591}{2}(-4) = 1.183\text{V}$$

(*ii*) $$E^0_{Br} = 1.087 - \frac{RT}{nF}\log\frac{[Br^{-2}]}{[Br_2]} = 1.087 - \frac{0.0591}{2}\log\frac{(1\times10^{-2})^2}{(1\times10^{-3})} = 1.117\text{V}.$$

6.6 Evaluate the potential of the half cell shown below:

$$Pt|H_2(0.800\text{ atm})|HI(0.20\text{ M})\ AgI(\text{Satd.})|Ag\left(\text{Given } E^0_H = 0,\ E^0_{Ag} = -0.151\right)$$

Ans: $2H^+ + 2e^- \Leftrightarrow H_2(g)$ and $Ag(s) + e^- \Leftrightarrow Ag(s) + I^-$ for hydrogen electrode

$$E = E^0_H + \frac{RT}{nF}\log\frac{(0.8)}{(0.2)^2} \text{ so } E = 0 - \frac{0.0591}{1}\log 0.2 = -0.038 \text{ for Ag electrode}$$

$$E = E^0_{Ag} + \frac{RT}{nF}\log\frac{(0.8)}{(0.2)^2} \text{ so } E = -0.110\text{ V} - \frac{0.0591}{1}\log 0.2 = -0.110\text{ V as } n = 1.$$

$$E_{cell} = (E_{Ag} - E_H) = -0.110\text{ V} - (-0.038) = -0.072\text{ V}.$$

6.7 Evaluate the potential of the half cell given below.

$\left(\text{Given } E^0_{Cu} = 0.337,\ E^0_{Zn} = -0.763\text{ V}\right)$.

cell reaction is represented as:

(*a*) $Zn^0 | Zn^{+2}(1M) || Cu^{2+}(1M) | + Cu^+$ or also

(*b*) $Cu^0 | Cu^{+2}(1M) | Zn^{2+}(1M) + Zn^0$.

Ans: Since (*a*) $E_{cell} = (E_R - E_L) = (E_{red} - E_{ox}) = (0.337 + 0.703) = +1.100\text{V}$

$E_R^0 - E_L^0 = E_{cell}^0$. $0.337 - (-0.763) = +1.100\text{V}$ (for first cell)

If direction of current was reversed then

(*b*) $\left(E_R^0 - E_L^0\right) = E_{cell}^0 - 0.763\text{ V} + (-0.337) = -1.100\text{V}$ using reduction potential

$$E_R^0 - E_L^0 = E_{cell}^0\ (-0.763) - (+0.337) = -1.100\text{ V}$$

6.8 What is the potential at platinum rod if half cell contains 1×10^{-4} M ferric (*III*) and 1×10^{-2} M of ferrous (*II*) ion in hydrochloric acid? (Given $E_{Fe} = 0.700\text{V}$)

Ans: $Fe^{3+} + e^- \rightarrow Fe^{2+}$; $E = E_{Fe}^0 - \dfrac{RT}{nF}\log\dfrac{(\text{oxi})}{(\text{red})}$ By Nernst Equation.

$$E = E_{Fe}^0 + \frac{0.0591}{1}\log\frac{[Fe^{3+}]}{[Fe^{+2}]} = +0.700 + 0.0591\log\frac{[0.0001]}{[0.01]} \text{ we get}$$

$$E = 0.700 - 0.118 = +0.582\text{ V}.$$

6.9 Evaluate the half potential for cell combination given below:

$$\text{Pt}\left|Fe^{2+}\ (0.01\text{ M})\ Fe^{3+}\ (0.0001)\right|\left|\ Ce^{4+}\ (0.01\text{M})Ce^{3+}\ (0.001\text{M})1\text{M HCL}\right|\text{Pt}$$

$\left(\text{Given } E_{Fe}^0 = 0.700,\ E_{Ce}^0 = 1.28\text{ V}\right)$.

Ans: The reaction in the cell was

$$Fe^{2+} + Ce^{4+} = Fe^{3+} + Ce^{3+}$$

$$E_{cell} = E_{Ce}^0 + \frac{0.0591}{1}\log\frac{(Ce^{4+})}{(Ce^{3+})} - E_{Fe}^0 - \frac{0.591}{1}\log\frac{(Fe^{3+})}{(Fe^{2+})}\quad (n = 1)$$

$$E_{cell} = 1.28 + 0.0591\log\frac{[0.01]}{[0.001]} - 0.700 - \frac{0.0591}{1}\log\frac{[0.0001]}{[0.01]} \text{ we get}$$

$$E_{cell} = 1.28 + 0.0591 - 0.700 - 0.118 = 0.521\text{ V}.$$

6.10 Calculate the solubility product of silver chloride if

$$E_{Ag}^0 = +0.799 E_{AgCl}^0 = 0.220\text{ V}.$$

Ans: $Ag^+ + e^- = Ag$; $E_{Ag}^0 = +0.799\text{V}$

$$AgCl + e^- = Ag^+ + Cl- \ E_{AgCl}^0 = +0.222;\text{ as } AgCl + e^- = Ag(s) + Cl^-$$

$$AgCl(s) = Ag^+ + Cl^-.\text{ So }\log K_{sp} = \frac{1[0.222 - 0.799]}{0.059} = \frac{E_{AgCl}^0 - E_{Ag}^0}{RT/F}$$

$$K_{sp} = 1 \times 10^{-10}.$$

6.11 What is the magnitude of an equilibrium constant for interaction between Ferrous and Cerric solution $\left(E_{Ce^{4+}}^0 = 1.70,\ E_{Fe^{3+}}^0 = 0.771\right)$?

Ans: The reaction were $Ce^{4+} + Fe^{2+} = Ce^{3+} + Fe^{3+}$

$$Ce^{4+} + e^- = Ce^{3+} \qquad E^0_{Ce} = 1.70 \text{ V}$$

$$Fe^{3+} + e^- = Fe^{2+} \qquad E^0_{Fe} = 0.771 \text{V}$$

So $$K = \frac{[Ce^{3+}][Fe^{3+}]}{[Ce^{4+}][Fe^{2+}]} \text{ so } \log K = 16.9(E_R - E_L)$$

$$\log K = 16.9 \times (1)(1.70 - 0.77) = 15.76,\ K = 5.75 \times 10^{15}.$$

6.12 The amount of calcium was determined by redox titration by precipitating it with oxalic acid and then on dissolving it in acid it was titrated with permanganate solultion. If 0.08554 M permanganate required 27.5 ml for completion of reaction. What is the amount of calcium in the sample ?

Ans: $2MnO_4^- + 8H^+ + 5e^- = Mn^{2+} + 4H_2O,\ 5C_2O_4^{2-} = 2CO_2 + 2e^-$

$2MnO_4^- + 5C_2O_4^{2-} + 16H^+ \rightarrow 2Mn^{2+} + 10CO_2 + 8H_2O$ ratio = 5/2

wt. of Ca = vol of $KMnO_4$ × [$KMnO_4$] × reaction ratio × [Ca] if eq wt. of Ca = 40.08 amount of Ca = 27.50 × 0.08554 × 5/2 × 40.08 = 235.7 gm.

6.13 A 0.115 gm of Devadra's alloy and 0.2129 gm of copper pyrite containing copper was taken and in iodometric titration volume of thiosulphate required was 39.42 ml for alloy and 28.42 ml for copper pyrite. What is the amount of copper in alloy and ore?

Ans: The reaction was $Cu^{2+} + 4I^- = 2CuI + I_2$ total and every reactions were

$Cu^{2+} + 2I^- = 2Cu^{2+} + I_2$ and $I_2 + 2S_2O_3^{2-} = 2I^- + S_4O_6^-$ reaction ratio was 2/2 so wt. of Cu = ml of $Na_2S_2O_3$ × [$Na_2S_2O_3$] × Reaction ratio [Cu]

$$110.5 = 39.42 \times \left[S_2O_3^{-2}\right] \times \frac{2}{2}[63.54] \text{ as } \left[S_2O_3^{2-}\right] = 0.0441 \text{ M}$$

$$\text{amt. of Cu} = \frac{\text{ml of } Na_2S_2O_3 \times \left[S_2O_3^{-2}\right] \times \text{ratio} \times \text{Cu} \times 100}{\text{wt. of sample}}$$

$$\%\ \text{Cu (ore)} = \frac{28.42 \times 0.0441 \times 1 \times 63.54 \times 100}{212.9} = 37.41\% \text{ in ore.}$$

$$\%\ \text{Cu (alloy)} = \frac{39.42 \times 0.0441 \times 63.54 \times 100}{115 \text{ mg}} = 96.05\% \text{ in Devadra's alloy.}$$

6.14 For a sample (0.2801 gm) containing chromium 75 ml of 0.1 M ferrous sulphate was added and the excess of iron solution was titrated with 0.02507 M, of dichromate to give 16.85 ml titre value. What is the percentage of chromium ?

Ans: The overall reaction was $Cr_2O_7^{2-} + 14H^+ + 6Fe^{2+} = 2Cr^{3+} + 6Fe^{3+} + 7H_2O$ ratio = 1/6

$$\%\ \text{Cr} = \frac{(\text{mill moles taken} - \text{mill moles found}) \times \text{ratio} \times \text{Cr} \times 100}{\text{wt. of sample in mgs.}}$$

$$\% \text{ Cr} = \frac{\text{ml Fe} \times [FeSO_4] - \text{ml } K_2Cr_2O_7 \times [K_2Cr_2O_7] \times \frac{6}{1} \times \frac{1}{6} \times [Fe] \times 100}{280.1}$$

$$\% \text{ Cr} = \frac{75 \times (0.1) - 16.85 \times 0.0257 \times \frac{6}{1} \times \frac{1}{6} \times [Fe] \times 100}{280.1} = 15.60\% \text{ Cr}$$

6.15 Depict a titration curve for the titration of 50 ml of 0.05 N iron (*III*) with 0.10 N cerium (*IV*) in 1M sulphuric acid. $\left(\text{Given } E^0_{Fe^{2+}} = 0.68,\ E^0_{Fe^{3+}} = 1.44\text{ V}\right)$.

Ans: The reaction was $Fe^{2+} + Ce^{4+} = Fe^{3+} + Ce^{+3}$

(*i*) at start of titration $E = E^0_{Fe} - 0.0591 \log \frac{[Fe^{2+}]}{[Fe^{3+}]} = 0.68 - 0.0591 \log 1000 = 0.50\text{ V}$

(*ii*) If the addition of 50 ml of Ce^{4+} soln. to 50ml of Fe^{3+}, concentration of Ce^{4+} was

$[Ce^{3+}] = \frac{5 \times 0.1}{55} - [Ce^{4+}] = \frac{0.5}{55}$, and concentration of Fe^{3+} was

$$[Fe^{3+}] = \frac{5 \times 0.1}{55}[Ce^{4+}] = \frac{0.5}{55} = \frac{(50 \times 0.05) - (5 \times 0.1)}{55} + [Ce^{4+}]\frac{2}{55}$$

Now potential $E = E^0_{ce} - 0.0591 \log \left[\frac{Ce^{3+}}{Ce^{4+}}\right] = E^0_{Ce} - 0.0591 \log \frac{[Fe^{2+}]}{[Fe^{3+}]}$

On substituting values $E = 0.68 - 0.0591 \log \left[\frac{2/55}{0.5/55}\right] = +0.64\text{ V}$

(*iii*) The potential is $= \frac{E^0_{Ce^{4+}} + E^0_{Fe^{3+}}}{2} = \frac{0.68 + 1.44}{2} = +1.06\text{ V}$

(*iv*) 25 ml of Ce^{4+} soln. to 50ml. of 0.005 N Fe^{3+}, excess of Ce^{4+} was present

$$[Fe^{3+}] = \frac{25 \times 0.1}{75.1} - [Fe^{2+}] = \frac{2.5}{73.10},\ [Ce^{3+}] = \frac{25 \times 0.1}{75.1} - [Fe^{2+}] = \frac{2.5}{75.10}$$

$[Ce^{4+}] = \frac{(25.1 \times 0.1) - (0.05 \times 50)}{75.10} + [Fe^{2+}] = \frac{0.01}{75.10}$. The cell potential was

$$E_{cell} = 1.44 - 0.0591 \log \frac{(Ce^{3+})}{(Ce^{4+})} = 1.44 - 0.0591 \log \left[\frac{2.5/75.10}{0.01 \times 75.1}\right] = 1.30\text{ V.}$$

PROBLEMS FOR PRACTICE

6.16 Calculate the potential of half cell containing cadmium electrode dipped in solution of 1.0×10^{-3} M of cadmium solution? **[Ans: –0.462V]**

6.17 What is the potential of typical platinum electrode dipped in the solution of 0.0100 M of potassium bromide also containing bromine in solution of 0.1 M potassium bromide with 0.001 M bromine? **[Ans:** (*i*) 1.183V (*ii*) 1.117V]

6.18 What is the concentration of the various components as well as end products at equivalance point in titration of 0.1 N ferrous sulphate with 0.1 N of Cerric sulphate from 1M sulphuric, acid media? **[Ans:** $Ce^{3+} = Fe^{3+} = 0.050$ $Ce^{4+} = Fe^{e+} = 2 \times 10^{-8}$]

6.19 Evaluate the solubility product of silver oxalate from following reaction:

$$AgCr_2O_4 + 2e^- = 2Ag + C_2O_4^{2-} \left(\text{Given } E^0_{Ag} = 0.4901\right).$$ **[Ans:** 3.5×10^{-11}]

6.20 Calculate the weight of the iron ore containing 55.8% iron if 30 ml of 0.1 N dichromate is required for end point titration. **[Ans:** 37.34%]

6.21 Calculate the amount of copper in one gram of malachite mineral which required 12.1 ml of 0.1 M Sodium thiosulphate in excess of potassium iodide. **[Ans:** 7.69%]

6.22 What is the amount of arsenic oxide in impure sample if 0.5201 gm of sample was titrated with 0.0946 M equivalence iodine solution at pH 9.0 with equivalence point at 24.9 ml? **[Ans:** 44.8%]

6.23 A standard steel (0.800 gm) was dissolved in acid when manganese from the sample was titrated in the presence of fluoride with 0.005M permanganate with 20 ml. The resulting solution was further titrated with 0.04M ferrous sulphate to 30 ml. Calculate the composition of chromium and manganese in the steel. **[Ans:** Cr = 1.52 % Mn = 2.75%]

6.24 Galena (1.234 gm) with lead in form of oxides was reacted with 20 ml of 0.25 M oxalic acid to reduce oxides to lead and then lead was precipitated as oxalate. The filtrate free from lead was titrated with 10 ml of 0.04 M permanganate to have 30 ml as end point. Find out the composition of the galena? **[Ans:** PbO = 36.18, PbO_2 = 19.38%]

6.25 What is the amount of iron in a sample of hametite weighing 0.71 gram after reduction if it needed 48.08 ml of 0.1 M potassium permanganate ? **[Ans:** 0.00310 gm]

6.26 When 200 mg of oxalic acid needed 30.12 ml of potassium hydroxide for neutralisation, find out the strength of alkali solution. **[Ans:** 0.153 N]

6.27 If magnetite (0.3gm) was titrated with mercury chloride what is the volume of 0.1 M mercury chloride if sample was dissolved in acid with CO_2 and excess of nitric acid? **[Ans:** 14.4; 21.6 ml]

6.28 A mineral Carnotite containing iron and vanadium was dissolved and made up to 250 ml. A 25 ml solution on reduction with John's reductor required 43.21 ml of 0.1667 N of potassium dichromate for end point. Another fraction of 25 ml on reduction in silver reductor required 23.45 ml of dichromate. Calculate the concentration of iron and vanadium in the mineral? **[Ans:** V = 5035 mg; Fe = 7576 mg]

6.29 When ferrous salt was titrated with ceric solution in sulphuric acid media the end point was at 0.008 ml with the corresponding formal potential of ferric ferrous and ceric cerous was 0.68V and 1.44 V respectively? **[Ans:** 0.04%]

6.30 An alloy with presence of manganese, chromium and vanadium was oxidised to give potassium permanganate, potassium dichromate, and vanadium. When titrated it gave 40 ml of 0.1 M ferrous sulphate. The resulting vanadyl ion requires 2.5 ml of 0.02 M permanganate. After addition of (microcosmic salt) the manganese requires 4 ml of 0.02 M permanganate. What is the concentration of different metals in the alloy? **[Ans:** V = 12.7 mg, Cr = 41.6 mg, Mn = 14.8 mg]

6.31 What was the purity of oxalic acid dihydrate if 0.2003 gm/cm^3 needed 29.30 ml of permanganate (given if one ml of the solution was equivalent to 6.02×10^{-3} gm of Fe)?

6.32 How many ml of potassium dichromate solution containing 23 gm of material per liter should react with 3.402 gm of Mohr's salt in the acid solution ?

6.33 About 2.010 gm of strontium sample was dissolved in acid and was precipitated as strontium oxalate which on further dissolution in sulphuric acid was titrated with 18.5 ml of 0.01 M permanganate. Calculate the amount of strontium oxide in the given sample. **[Ans:** 1.294% SrO]

6.34 What is the potential of the system against saturated calomal electrode if 60 ml of 0.1 M ceric (*i*) solution was mixed with 20 ml of 0.2 M ferrous (*ii*) ion in 1M sulphuric acid?

[Ans: 1.64 v]

6.35 What is the potential of half cell obtained by mixing 40 ml of 0.1 M ferric ion with 25 ml of 0.1 M ferrous ion in homogeneous acid concentration of 1M sulphuric acid?

[Ans: +0.771 V]

REFERENCES

6.1 S.M. Khopkar, *Basic Concepts of Analytical Chemistry,* 3rd edn., New Age International Publishers (P) Ltd. (2007).

6.2 D.A. Skoog and D.M. West, *Analytical Chemistry,* 3rd edn. Holt Reinhart Winston, pp. 283-4; 299; 311 (1979).

6.3 D.J. Pietryzke and C.W. Frank, *Analytical Chemistry: An Introduction,* Academic Press, pp. 334; 339-41; 380; 390; 400 (1979).

6.4 J.S. Fritz and G.H Schenk, *Quantitative Analytical Chemistry,* Allyn and Bacon Inc., pp. 270. 280. (1979).

6.5 J. G. Dick, *Analytical Chemistry,* McGraw-Hill and Co., p. 411 (1973).

6.6 H.A Latinen and W.E. Harris, *Chemical Analysis,* 2nd edn., McGraw-Hill and Co., pp. 305, 323, 336 (1973).

6.7 H.F. Walton, *Principles and Methods of Chemical Analysis,* 2nd edn., Prentice Hall Co. (1960).

6.8 D.A. Skoog and J.J. Leoroy, *Principles of Instrumental Analysis,* Harcourt Butler and Co. (1996).

Chapter 7

Precipitation Titrations

INTRODUCTION

The phenomena of precipitation is very important in the gravimetric analysis. The precipitant may be inorganic or organic reagent. We have already seen in chapter on gravimetry the significance of terms, coprecipitation, precipitating agents, the precipitation from homogeneous solutions and the term solubility product.

In precipitation titrations such as $AgNO_3$ against KSCN or of $BaCl_2$ against H_2SO_4 we get precipitate in direct proportion to the amount of cation (*viz.* Ag or Ba resp) present in the solution. This forms basis of quantitative analysis. One can easily calculate molar solubility from the knowledge of solubility product (K_{sp}) *e.g.*, AB = $[A^+] + [B^-] = [A^+] = [B^-]$ = S (solubility) conversely we can easily calculate the solubility of A^+ if K_{sp} of AB is known.

The presence of common ion effect or diverse salt effect substantially affects solubility product. The well known example of precipitation titration is Volhard's or Mohr's method. In the former $AgNO_3$ is titrated against Na_2CrO_4. The end point is typical colouration while indicators like fluoroscein or methyl violet or p-ethoxychrysodine a (cationic or anionic) indicator in nature are used. The colour change is very sharp and the reaction by absorption is rapid.

In the precipitation titration, one can calculate the titration curve, from given set of values of reactant [*R*] and titrant [*T*]. If V_R is volume of reactant, V_T = volume of titrant and M_R = molarity of reactant and M_T is molarity of titrant then the course of titration can be followed. The concentration of anion (*e.g.*, B^-) can be calculated before end point as:

$$[B^-] = \frac{V_R M_R - V_T M_T}{(V_R + V_T)}$$ while the concentration of the cation can be easily computed after equivalence point as:

$$[A^+] = \frac{V_T M_T - V_R M_R}{(V_T + V_R)}$$ where A^+ is cation B^+ is anion when

K_{sp} of AB = $[A^+][B^-]$.

It will be thus observed that from these two equations, during the titration of the reactant with titrant one can easily calculate the concentration of anion reactant and concentration of cation of titrant, *e.g.*, in titration of $AgNO_3$ with NaCl, one can evaluate the magnitude of $[Cl^-]$ as well as that of $[Ag^+]$ during course of addition of each increments of titrant.

The number of factors such as temperature, nature of solvent, common ion effect or diverse ion effect, pH, hydrolysis, complexation invariably affects the solubility of the precipitate. One uses the adsorption indicators like sodium rhodizonate, fluoroscein, methylviolet; p-ethxoychrysodine, eosin, dichlorofluoroscein. During course of reaction, these adsorption indicators react with either cation (*e.g.*, Fluoroscein with Ag^+ or *e.g.*, methylviolet with Cl^-) anion to form a coloured precipitate by process of adsorption. Since this is surface phenomena those of the precipitates with large surface area give better results. The coagulation of precipitate adversely affect visibility of adsorpted precipitate. In such case a protective colloid (*e.g.*, Nitrobenzene) is used however it renders reaction very slow.

In the number of numerical problems we use expression given for calculation of anion (reactant) and cation (titrant). The other examples are based upon calculation of solubility product or vice versa. Some problems pertaining to evaluation of percentage composition of mixture are also given.

SOLVED PROBLEMS

7.1 A sample of barium chloride weighed 0.500 gram. About 50 ml of 0.21N silver nitrate was added to precipitate it as silver chloride, the excess of silver nitrate was titrated against 0.28 N potassium thiocyanate to give 25.5 ml titre value. What is the amount of barium chloride in the sample?

Ans: meq of $AgNO_3$ added = 50 × 0.21 = 10.50

meq of KSCN added = 25.5 × 0.28 = 7.14

Total meq needed (10.5 – 7.14) = 3.36 ml consumed

$$\text{amount of } BaCl_2 = \frac{3.36\times\left(\frac{BaCl_2}{2000}\right)\times 100}{0.50} = \frac{3.36\times 0.104\times 100}{0.50} = 69.88\%\ BaCl_2$$

7.2 A sample of feldspar (1.50 gram) was dissolved to give 0.801 gram mixture of chlorides of sodium and potassium. The residue was dissolved in water and 50 ml of 0.08333 N of silver nitrate was added to precipitate as halide. The excess was titrated with 0.1 N ammonium thiocyanate with 1.47 ml at end point. What is the amount of sodium oxide in the sample?

Ans: x = wt. of KCl while (1.801 – x) = wt. of NaCl. If wt. of mixture is 0.1801 g

meq of $AgNO_3$ = 50 × 0.0833

meq of NH_4SCN = 16.47 × 0.1

Total meq needed = [(50 × 0.0833) – (16.47 × 0.1)] = (4.16 – 1.64) = 2.52 ml consumed.

Alternatively we can solve it as total halide =

$$\frac{x}{KCl/1000} = \frac{(0.1801-x)}{NaCl/1000} = \frac{x}{0.0746} + \left(\frac{0.1801-x}{0.0585}\right)$$

i.e. [(50 × 0.08333) – (16.47 × 0.10)]. Solving we get x = 0.152 KCl, (0.1801 – 0.152) = 0.0281 NaCl

2KCl = K_2O is 0.528 2NaCl = Na_2O is 0.53 are conversion factors

$$\%\text{ amt. of oxide} = \frac{0.0281\times 0.530\times 100}{1.5} = 9.88\%\ Na_2O$$

$$\%\text{ amt. of oxide} = \frac{0.152\times 0.528\times 100}{1.5} = 5.3\%\ K_2O.$$

7.3 In Mohr's method 50 ml of 0.1 M sodium chloride was titrated with 0.1 M silver nitrate. Calculate the chloride ion concentration during titrations, initially and after addition of 10 ml; 49.9 ml; is at eq. point and after addition of 60 ml (K_{sp} of AgCl = 1.56 × 10^{-10}).

Ans: (*i*) At start $[Cl^-] = 0.1$ so, $-\log [Cl] = pCl = 1.0$

(*ii*) After adding 10 ml (V_T) in 50 ml (V_R) we have

$$[Cl^-] = \frac{V_R M_R - V_T M_T}{(V_R + V_T)} = \frac{(50 \times 0.1) - (10 \times 0.1)}{(50 + 10)} = 0.067, \text{ So pCl} = 1.17$$

(*iii*) After addition of 49.9 ml of silver nitrate

$$[Cl^-] = \frac{(50 \times 0.1) - (49.9 \times 0.1)}{(50 + 49.9)} = 1 \times 10^{-4} \text{ so pCl} = 4.0$$

(*iv*) At end point of titration

$$[Ag^+][Cl^-] = [Cl^-][Cl^-] = [Cl^-]^2 \text{ if}$$

$$[Cl^-]^2 = [1 \times 10^{-10}]^2 \text{ so } [Cl^-] = 1 \times 10^{-5} \text{ so pCl} = 5.0$$

$$K_{sp} = [Ag^+] = [Cl^-] = [Ag^+][Cl^-] = \sqrt{K_{sp}} = \sqrt{1.56 \times 10^{-10}}$$

(*v*) After addition of 60 ml of silver nitrate

$$[A^+] = \frac{V_T M_T - V_R M_R}{(V_T + V_R)} = \frac{(60 \times 0.1) - (50 \times 0.1)}{(60 + 50)} = 9.1 \times 10^{-3} \text{ so pAg} = 2.04$$

However pAg + pCl = 10, pCl + 2.04 = 10, so pCl = 7.96.

7.4 A silver coin (0.1755 gram) was dissolved and titrated with 0.0710 M potassium thiocyanate to have titre value of 20.92. Calculate the amount of silver in the coin?

Ans: $Ag^+ + SCN^- = AgSCN$ reaction ratio 1 : 1

$$\% \text{ Ag} = \frac{\text{ml KSCN} \times [\text{KSCN}] \times \text{reaction ratio} \times [\text{Ag}] \times 100}{\text{wt. of silver coin}}$$

$$\% \text{ Ag} = \frac{20.92 \times 0.0710 \times 1/1 \times 107.9 \times 100}{175.5 \text{ mg}} = 91.33\%.$$

7.5 Depict the titration curve if 50 ml of 0.005 M sodium bromide when titrated with 0.01 M silver nitrate at (*a*) initial point, (*b*) after adding 5 ml, (*c*) equivalence point, and (*d*) after addition of 25.1 ml?

Ans: (*a*) Initial point 0.005 M Br^- and 0.0001M Ag^+

$$pBr^- = -\log (0.005) = -\log (5 \times 10^{-3}) = 2.30 \text{ pAg}$$

(*b*) After adding 5 ml we have as $V_R = 50$, $M_R = 0.005$, $V_T = 5$, $M_T = 0.01$ M

$$\frac{V_R M_R - V_T M_T}{(V_R + V_T)} = \frac{(50 \times 0.005) - (5 \times 0.010)}{(50 + 5)} = 3.64 \times 10^{-3}$$

$$\left[Br^-\right]=3.64\times10^{-3}+\left[Ag^+\right]\left[Ag\right]=3.64\times10^{-3}\ [Br]=3.64\times10^{-3}$$

$$pBr=-\log\left(3.64\times10^{-3}\right)=2.439$$

(*c*) At equivalance point NaBr or $AgNO_3$ was in equal proportion

$$[Ag]=[Br]=\sqrt{5.2\times10^{-13}}=7.2\times10^{-7}$$

$$pAg=pBr-\log\left(7.2\times10^{-7}\right)=6.14$$

(*d*) After addition of 25.10 ml of $AgNO_3$

$$=\frac{(25.10\times0.010)-(50\times0.0050)}{(25.1+50)}=1.33\times10^{-5}$$

$$[Ag]=\left(1.33\times10^{-5}\right)+\left[Br^{-1}\right]=1.33\times10^{-5}$$

(Since V_T = 25.10, M_T = 0.010, V_R = 50, M_R = 0.005)

$$pAg=-\log\left(1.33\times10^{-5}\right)=4.88$$

$$pBr=(12.28-4.88)=7.40.$$

7.6 If a solution containing 0.08 M iodide and 0.100 M chloride ion was titrated with 0.2 M silver nitrate. Predict nature of the titration curve after addition of (*a*) initial, (*b*) equivalence point, (*c*) 25 ml of solution.

Ans: (*a*) $\left[Ag^+\right]\left[I^-\right]=8.3\times10^{-17}:\left[Ag^+\right]\left[Cl^-\right]=1.82\times10^{-10}$; Ratio $=4.56\times10^{-7}$

$$\frac{\left[Ag^+\right]\left[I^-\right]}{\left[Ag^+\right]\left[Cl^-\right]}=4.56\times10^{-7}\text{ so }[I]=4.56\times10^{-7}\left[Cl^-\right]$$

(*b*) Now $\left[Cl^-\right]=\dfrac{50\times0.10}{(50+20)}=0.0714, [I]=4.56\times10^{-7}\times0.0714=3.26\times10^{-8}$

$$\left[Cl^-\right]=0.074,\left[Ag^+\right]=\frac{1.82\times10^{-10}}{0.0714}=2.55\times10^{-9},\ pAg=-\log\left(2.55\times10^{-9}\right)=8.59$$

(*c*) At 25 ml addition

$$\left[Cl^-\right]=\frac{V_TM_T-V_RM_R}{\left(V_R+V_T\right)}=\frac{(50\times0.1)+(50\times0.08)-(25\times0.2)}{50+25}=0.0533$$

$$\left[Ag^+\right]=\frac{1.82\times10^{-10}}{0.0533}=3.41\times10^{-9};\ pAg=8.47.$$

7.7 Draw the titration curve for titration of 100 ml of 0.1 N sodium chloride with 0.1 N silver nitrate (*a*) initial, (*b*) after addition of 90 ml, (*c*) 100 ml, (*d*) 100.1 ml of silver nitrate.

Ans: (*a*) Initially pCl = –log [Cl^-] = –log 10^{-1} = 1.0

(*b*) After 90 ml addition, [Ag^+] [Cl^-] = 10^{-10} *i.e.*, solubility product so

$$\left[Ag^+\right] = \frac{10^{-10}}{\left[Cl^-\right]} = \frac{10^{-10}}{10^{-2}} = 10^{-8} \text{ gm/liter.}$$

(*c*) After addition of 100 ml [Cl] = [Ag^+] = $\sqrt{K_{sp}} = \sqrt{10^{-10}} = 10^{-5}$

pCl = pAg = 5. This is equivalence point

(*d*) After addition of 100 ml of solution

[Ag^+] = 10^{-4} pAg = 4. [Cl^-] = 10^{-7} pCl = 6 with 100.1 ml of $AgNO_3$

[Ag^+] = 10^{-3} pAg = 3, [Cl^-] = 10^{-7} pCl = 7.0 so values for [Ag^+]

7.8 If barium sulphate is 1.14 × 10^{-5} M. Calculate its solubility product?

Ans: $BaSO_4 = Ba^{2+} + SO_4^{2-}$

So $K_{sp} = \left[Ba^{+2}\right]\left[SO_4^{2-}\right]$

Since [Ba^{2+}] = 1.14 × 10^{-5} given so also

$SO_4^{2-} = \left[1.14 \times 10^{-5}\right]$ so $K_{sp} \approx 1.30 \times 10^{-10}$ M.

7.9 Find out molar solubility of silver chromate if K_{sp} = 1.1 × 10^{-12}? When 40 ml of 0.01 M NaCl was titrated against 0.01 M silver nitrate and if K_{sp} of AgCl is 1 × 10^{10}. What is the pAg after addition of 20, 40 ml of silver nitrate?

Ans: (*a*) $AgCrO_4$ = $2Ag^+$ + CrO_4^{2-}

K_{sp} = [Ag^+]^2 [CrO_4^{2-}] = 1.1 × 10^{-12}

Since [Ag^+] = 25 and [CrO_4^{2-}] = 5 from above reaction

1.1 × 10^{-12} = [25]^2 [S] = [Ag^+]^2 [CrO_4^{2-}]

S = 6.5 × 10^{-5} M.

(*b*) Cl^- not consumed = 20 × 0.01 = 0.20 M so

The concentration of Cl^- was

$$= \frac{0.20}{60} = 0.0033\text{M as } \frac{V_R M_R - V_T M_T}{(V_R + V_T)} = \frac{(40 \times 0.1) - (20 \times 0.01)}{(40 + 20)} = 0.0033 \text{ M.}$$

$$K_{sp} = \left[Ag^+\right]\left[Cl^-\right] \text{ so } 1 \times 10^{-10} = \left[Ag^+\right][0.0033] \text{ so } [Ag] = 3 \times 10^{-8}$$

$$\log\left[Ag^+\right] = (0.5 - 8) = -7.5 \text{ so pAg} = 7.5 \text{ after first addition.}$$

(*c*) Now after addition of 40 ml $AgNO_3$ to 40 ml NaCl soiution so [Ag^+] = [Cl^-]

$$K_{sp} = \left[Ag^+\right]\left[Cl^-\right] = \left[Ag^+\right]\left[Ag^+\right] = \left[Ag^+\right]^2$$

So $\left[Ag^+\right] = \sqrt{10^{-10}} = 10^{-5} = \log K_{Ag}$; pAg = 5.0.

7.10 With K_{sp} given AgCl gives pAg = 5.0 and K_{sp} of $AgCrO_4$ gives pAg as 1.1 × 10^{-12} . Find out the concentration of chromate present to precipitate as silver chromate at pAg = 5.0?

Ans: Since $Ag_2CrO_4 = 2Ag^+ + CrO_4^{2-}$

$$K_{sp} = [Ag^+] = [CrO_4^{2-}] = 1.1 \times 10^{-12} \text{ so } [Ag^+] = 10^{-5} \text{ so pAg} = 5.0.$$

7.11 An iron sample (0.201 gm) was dissolved and if iron was precipitated as its hydroxide to weigh latter as oxides as 0.116 gm. Calculate the amount of iron in the sample?

Ans: $Fe^{3+} + 3OH^- \xrightarrow{H_2O} Fe(OH)_3\, x\, H_2O \xrightarrow{\Delta} Fe_2O_3 + (x+3)H_2$

$$\% \text{ Fe} = \frac{Fe_2O_3 \times (2Fe/Fe_2O_3) \times 100}{\text{wt. of sample}} = \text{Since } Fe_2O_3 = 0.116, Fe_2O_3 = 55.80/159.7$$

$$\% = 0.116 \times \left[\frac{2 \times 55.80 \times 100}{159.7}\right] = 39.89\% \text{ in the sample.}$$

7.12 What is the solubility of silver chloride if its solubility products is 1.8×10^{-10}?

Ans: $AgCl = Ag^+ + Cl^-$ so $K_{sp} = [Ag^+][Cl^-] = 1.8 \times 10^{-10}$

Since 1.8×10^{-10} = solubility of salt, $K_{sp} = [1.8 \times 10^{-10}]^2$

$$S = 1.34 \times 10^{-5}.$$

7.14 Calculate the weight of sodium chloride required for the titration of 0.1 N solution of silver nitrate with 25 ml from burette?

Ans: The sample weight must be such that the titration uses up to 20 ml of standard solution. So the sample should be weighing as 0.1 × 20 × 58 = 120 mg = 0.120 gm.

When the titre was determined by a method of back titration, to the sample was added 25 ml of silver titrate with excess of 5 ml. The answer is 120 mg = 0.12 gram.

7.15 Calculate the concentration of potassium cyanide in milligram if 26.05 ml of the solution was consumed with 0.1015 N of silver nitrate solution?

Ans: The titration reaction was

$2CN^- + Ag^+ \rightarrow [Ag(CN)_2]^-$ and equivalent weight of KCN twice molecular weight

Hence $x = 26.05 \times 0.1015 \times 2\text{KCN} = 344$ mg.

7.16 When 30 ml of silver nitrate was treated with 0.1173 gram sodium chloride and the excess of Ag^+ was back titrated with 3.20 ml ammonium thiocyanate and 20 ml of silver nitrate equal to 21 ml of ammonium thiocyanate solution, what is the normality of thiocyanate solution?

Ans: The volume ratio of $AgNO_3$ to NH_4SCN was $\dfrac{20.0}{21.0} = 0.952$

$$\text{Normality } AgNO_3 = \frac{117.3}{NaCl(30 - 3.2 \times 0.952)} = 0.0745 \text{ N}$$

$$\text{The titre value of } AgNO_3 \text{ was } AgNO_3 \times \frac{Cl}{1000} = \frac{0.75 \times 35.5}{1000} = 0.002640$$

Therefore, normality of (NH_4SCN) = normality $AgNO_3 \times 0.952$

= 0.0745 × 0.952

= 0.070 N.

7.17 Calculate volume of 0.05 M potassium ferrocyanide was required for titration of 25 ml of 0.1020 M Zinc sulphate solution?

Ans: The overall reaction during titration was

$$3Zn^{2+} + 2K_4Fe(CN)_6 = K_2Zn_3[(FeCN)_6]_2 + 6K^+$$

Thus $3Zn^{2+}$ ions were equivalent to two molecules of $K_4Fe(CN)_6$

Hence if x = ml of $K_4Fe(CN)_6$ required for titration, then

$$x = \frac{2}{3} \times 25 \times \frac{0.1020}{0.050} = 34.0 \text{ ml.}$$

7.18 Calculate the weight of the substance containing 60% of sodium chloride and 37% potassium chloride required for analysis so that by action of 25 ml of 0.1 N silver nitrate the excess of silver was back titrated with 5 ml of thiocyanate solution.

Ans: Since NaCl = 60% KCl = 37%

25 ml of $AgNO_3$ of 0.1 N was added in excess and back titration was carried out with 5 ml of NH_4SCN solution. Let x – be wt. of substance of mixture of (NaCl + KCl) so

$$\frac{x \times 0.60}{\text{NaCl}} + \frac{x \times 0.37}{\text{KCl}} = (25 - 5 \times 1.10)0.10 = 128 \text{ mg.}$$

(as NH_4SCN / $AgNO_3$ = 1.0 /1.10 because 1 ml of NH_4SCN = 1.10 ml of NO_3)

PROBLEMS FOR PRACTICE

7.19 A silver wire (0.1755 gm) required 20.92 ml of 0.07101 M potassium thiocyanate in Volhard's titration. Calculate amount of silver in wire? **[Ans: 91.33%]**

7.20 Calculate pPb^{2+} and pIO^-_3 if solution were obtained from 25 ml of 0.1 M lead nitrate alongwith 25 ml of 0.1 M sodium iodate? **[Ans:** pPb = 1.6, pIO_3 = 5.70**]**

7.21 One gram sample of type metal when analysed gravimetrically for zinc gave 0.6611 gram of zinc pyrophosphate. Calculate amount of Zinc in type metal? **[Ans: 28.37%]**

7.22 If sodium chloride (15 ml) was titrated with 53.62 ml of 0.63 M silver nitrate. Calculate the concentration of sodium chloride? **[Ans: 131.6 g/ L]**

7.23 If a sodium chloride (0.01 M, 40ml) was titrated with 0.01 M silver nitrate (Given $K_{sp}^{AgCl} = 1 \times 10^{-10}$). Calculate pAg or (–log Ag) when 20 ml or 40 ml of the titrant was consumed? **[Ans: 7.5, and 5.0]**

7.24 Sodium salt (0.30 gram) was analysed by Volhard method when 40 ml of excess of 0.1234 M silver nitrate showed volume of 0.093 M thiocyanate on back titration as 12.20 ml. Calculate the chloride amount? **[Ans: 44.47%]**

7.25 Calculate the normality of silver nitrate solution if 24 ml was required to react with 18.5 ml of potassium thiocyanate? **[Ans: 0.0663 N]**

7.26 When zinc (0.317 gm) was titrated with 45.1 ml of sodium sulphide. Calculate the normality of sodium sulphite solution? **[Ans: 0.215 M]**

7.27 Calculate the molarity of sodium bromide if 36.45 ml solution required 40.18 ml of 0.0995 M of silver nitrate at equivalence point? **[Ans: 0.1097 M]**

7.28 A water (50 ml) sample when analysed by Mohr's method containing (20 g/gallon of chloride ion) with 0.15 M silver nitrate required 25.80 ml of titrant for end point. Calculate the conce tration of sodium chloride required? **[Ans: 12.5 g]**

7.29 What is the titration error for analysis of 0.1 M sodium bromide in presence of chloride iodide of sodium with no co-precipitation if solubility product of AgCl = 2×10^{-10} ?

[Ans: 0.005%]

7.30 If silver image (0.59) contained 90% silver, what was the least normality of thiocyanate solution which might require 50 ml in the titration? **[Ans:** 0.08343 N]

7.31 Calculate weight of sample required for titration of chloride so as to have volume of 0.1 M silver nitrate equivalent to percentage of chloride and sodium chloride in titration?

[Ans: 99.44% ; 70.5 ml]

7.32 A sample weighing 340 mgs containing a mixture of chlorides of sodium and potassium required 48.2 ml of 0.10 M of silver nitrate in Mohr's method. Calculate composition of the mixture. **[Ans:** NaCl = 27.8% rest = 72.2 KCl]

7.33 Calculate the error per thousand in Mohr's method of 4 meq of Cl^- with 0.1 M silver nitrate to a final volume of 100 ml in the presence of 1×10^{-3} M of potassium dichromate at end point?

[Ans: 0.08 ml or 2 ppt]

7.34 An ore (150 mg) containing 7.58% sodium oxide and 9.93% potassium oxide when titrated with silver nitrate gave precipitate as chloride. What is concentration of titrant?

[Ans: 003008 N]

7.35 Calculate the strength of silver nitrate solution if 24 ml were required to react with (*a*) 18.5 ml potassium thiocyanate with 1.48 mg of silver nitrate. (*b*) 18.5 ml of barium thiocyanate and (*c*) 18.5 ml of 0.086N potassium thiocyanate and (*d*) 0.1850 gm of sodium chloride?

[Ans: 6.7×10^{-3} N ; 0.13 N ; 0.066 N ; 0.132 N

REFERENCES

7.1 S.M. Khopkar, *Basic Concepts of Analytical Chemistry,* 3rd edn., New Age International Publishers (P) Ltd. (2007).

7.2 D.J. Pietrzyke and C.W. Frank, *Analytical Chemistry: An Introduction*, Academic Press, pp. 202, 216-17, 240, 245-6 (1979).

7.3 D.A. Skoog and D.M. West, *Analytical Chemistry,* 3rd edn., Holt Reinhard and Winston, pp. 172, 183 (1979).

7.4 V. Alexeyev, *Quantitative Analysis,* 2nd edn., Mir Publishing House, Moscow, pp. 393 (1979).

7.5 J.S. Fritz and G.H. Schenk, *Quantitative Analytical Chemistry,* Allyn and Bacon Inc., pp. 207-9, 213 (1979).

7.6 J. G. Dick, *Analytical Chemistry,* McGraw Hill and Co., pp. 324-5 (1973).

7.7 H.A. Latinen and W.E. Harris, *Chemical Analysis: An Advanced Text,* 2nd edn., McGraw-Hill and Co. p. 186 (1973).

7.8 G.H. Ayers, *Quantitative Chemical Analysis,* Harper and Row Brothers, New York (1980).

7.9 H.F. Walton, *Principles and Methods of Chemical Analysis*, 2nd edn. Prentice Hall Publishers (1960).

7.10 G.D. Christen, *Analytical Chemistry – Introduction to Instrumental Analysis,* 5th edn. John Wiley (1997).

Chapter 8

Complexometric Titrations

INTRODUCTION

The titration involving EDTA as the titrant are thoroughly studied. EDTA is hexadentate ligand and is written as:

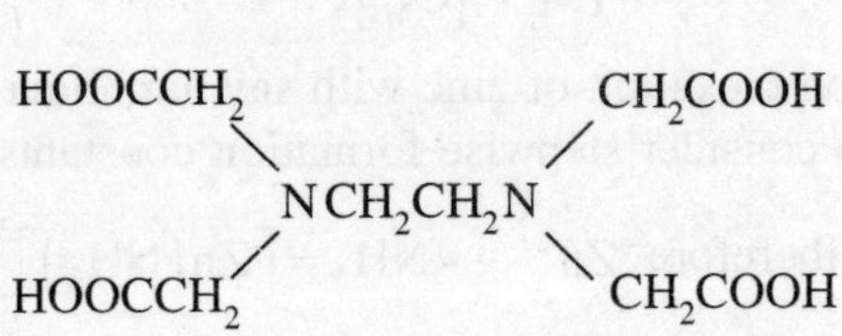

EDTA

However during titration we use disodium salt of EDTA which is written as Na_2H_2Y. It forms complexes with quadridentate coordination. In EDTA we have oxygen as well as nitrogen, as the donor atoms to form complexes with metals. When EDTA reacts with metal 'M' it forms following kinds of complexes in successive stages as:

(*i*) $M^+ + H_2R^{2-} \Leftrightarrow MHR^{2-} + H^+$; (Complex with monovalent metal, *e.g.*, Ag^+)
(*ii*) $M^{2+} + H_2R^{2-} \Leftrightarrow MR^{2-} + 2H^+$; (Complex with divalent metal, *e.g.*, Mg^{2+})
(*iii*) $M^{3+} + H_2R^{2-} \Leftrightarrow MR^- + 2H^+$; (Complex with trivalent metal, *e.g.*, Al^{3+})
(*iv*) $M^{4+} + H_2R^{2-} \Leftrightarrow MR + 2H^+$; (Complex with tetravalent metal, *e.g.*, Th^{4+})
(*v*) $M^{n+} + HR^{3-} \Leftrightarrow MR^{(4-n)-} + H^+$; (Complex with '*n*' valent metal, *e.g.*, M^{n+})

Now the pH at which EDTA forms complex with metal vary widely from pH = 1.5 [for Fe (*III*)] to pH = 10.0 (for Magnesium). This in turn provides a selectively scale for the determination of metals by EDTA titrations at different pH. The absolute stability constant is evaluated from expression as:

(*i*) $M^{n+} + R^{4-} \Leftrightarrow MR^{4-n}$; $K_{ab} = \dfrac{\left[MR^{(4-n)}\right]}{\left[MR^{n+}\right][R]}$. This is overall value of absolute or stability constant

while four dissociation constants for comple:ing ligand H_4R are given as follows:

(*ii*) $H_4R + H_2O \Leftrightarrow H_3O^+ + H_3R^-$; $K_1 = \dfrac{\left[H_3R^-\right]\left[H_3O^+\right]}{\left[H_4R\right]\left[H_2O\right]}$; $K_1 = 1.02 \times 10^{-2}$

(*iii*) $H_3R^- + H_2O \rightarrow H_3O^+ + H_2R^{2-}$; $K_2 = \dfrac{[H_2R^{2-}][H_3O^+]}{[H_3R^-][H_2O]}$; $K_2 = 2.1\times10^{-3}$

(*iv*) $H_2R^{2-} + H_2O \rightarrow H_3O^+ + HR^{3-}$; $K_3 = \dfrac{[HR^{3-}][H_3O^+]}{[H_2R^{2-}][H_2O]}$; $K_3 = 6.9\times10^{-7}$

(*v*) $HR^{3-} + H_2O \rightarrow H_3O^+ + R^{4-}$; $K_4 = \dfrac{[R^{4-}][H_3O^+]}{[HR^{3-}][H_2O]}$; $K_4 = 5.5\times10^{-11}$

$$\alpha_4 = \left[\frac{R^{4-}}{C_M}\right] = \frac{K_1K_2K_3K_4}{[H_3O^+]^4 + [H_3O^+]K_1 + [H_3O^+]K_1K_2 + [H_3O^+]K_1K_2K_3 + K_1K_2K_3K_4}$$

$$K_{ab} = \frac{MR^{(4-n)}}{[M^{n+}][\alpha_4 C_M]} \text{ and } K_{eff} = \frac{[MR^{4-n}]}{[M^{n+}][C_M]} = (K_{ab}\times\alpha_4) \text{ are stability constants.}$$

Now let us consider the complexation of zinc with say ammonia as auxiliary complexing ligand. It is interesting at this stage to consider stepwise formation constants for both zinc with ammonia as well as with EDTA; we have therefore $Zn^{2+} + 4NH_3 = [Zn(NH_3)]_4^{2+}$. This is overall reaction it is accomplished in four stages as:

$$Zn^{2+} + 4NH_3 = [Zn(NH_3)]^{2+};\ K_1 = 1.9\times10^2 = \frac{[Zn(NH_3)^{2+}]}{[Zn^{2+}][NH_3]^4}$$

$$[Zn(NH_3)]^{2+} + NH_3 = [Zn(NH_3)_2]^{2+};\ K_2 = 2.1\times10^2 = \frac{[Zn(NH_3)_2]^{2+}}{[Zn(NH_3)^{2+}][NH_3]}$$

$$[Zn(NH_3)]_2^{2+} + NH_3 = [Zn(NH_3)_3]^{2+};\ K_3 = 2.5\times10^2 = \frac{[Zn(NH_3)_3]^{2+}}{[Zn(NH_3)_2^{2+}]_2[NH_3]}$$

$$[Zn(NH_3)]_3^{2+} + NH_3 = [Zn(NH_3)_4]^{2+};\ K_3 = 1.10\times10^2 = \frac{[Zn(NH_3)]_4^{+2}}{[Zn(NH_3)]_3^{2+}[NH_3]}$$

So $K_f = K_1K_2K_3K_4 = 4.25\times10^8$ is overall formation constants.

Similarly the acid dissociation constant of EDTA is evaluated as follows:

$$H_4R + H_2O = H_3O^+ + H_3R^-;\ K_1 = 1\times10^{-2} = \frac{[H_3O^+][H_3R^-]}{[H_4R]}$$

$$H_3R^- + H_2O = H_3O^+ + H_2R^{2-};\ K_2 = 2.16\times10^{-3} = \frac{[H_2R^{2-}][H_3O^+]}{[H_3R^-]}$$

$$H_2R^{2-} + H_2O = H_3O^+ + HR^{3-};\ K_3 = 6.92\times10^{-7} = \frac{[H_3O^+][HR^{3-}]}{[H_2R^{2-}]}$$

$$HR^{3-} + H_2O = H_3O^+ + R^{4-};\ K_4 = 5.0\times10^{-11} = \frac{[H_3O^+][R^{4-}]}{[HR^{3-}]}$$

$$\alpha_4 = \frac{K_1K_2K_3K_4}{[H^+]^4 + K_1[H^+]^3 + K_1K_2[H^+]^2 + K_1K_2K_3[H^+] + K_1K_2K_3K_4} = \left[\frac{R^{4-}}{C_M}\right]$$

Overall dissociation constant (Kc) = $K_1K_2\,K_3\,K_4$. Another term we use is (α_4) which means $\alpha_4 = [R^{4-}]/C_T$. If C_T = Total concentration of uncomplexed EDTA which is given as

$$C_T = [R^{4-}][HR^{3-}] + [H_2R^{2-}] + [H_3R^-] + [H_4R]$$ When α_4 = uncomplexed reagent.

Table 8.1: Values of α_4 for EDTA in solutions at various pH

pH	α_4	*pH*	α_4
2.0	3.7×10^{-14}	7.0	4.8×10^{-4}
3.0	2.5×10^{-11}	8.0	5.4×10^{-3}
4.0	3.6×10^{-9}	9.0	5.2×10^{-2}
5.0	3.5×10^{-7}	10.0	3.5×10^{-1}
6.0	3.7×10^{-5}	11.0	8.5×10^{-1}
		12.0	9.8×10^{-1}

In above expressions K_{ab} = absolute stability constant while $K_1\,K_2\,K_3\,K_4$ ...... K_n are all called stepwise constants. Now (α_4) represents the fraction of all forms of uncomplexed EDTA present as R^{4-} K_{eff} is called effective or conditional stability constant. During titrations $[H_3O^+]$ competes with metal ion for EDTA. All complexation reaction are pH dependent as invariably during complexation more protons are released. The effect of other complexing ligands on EDTA is significant due to hydrolysis on precipitation as hydroxides. To prevent such hydrolysis the sequestering agents are used to keep metal in the solution. The influence of the auxiliary complexation agent is studied in terms of (β_4) defined as the fraction of metal ion uncomplexed form.

$\beta_4 = \frac{[M^{n+}]}{C_M}$ *i.e.*, $[M^{n+}] = \beta_4 \cdot C_M$. If C_M = meta' ion concentration, in such case K_{eff} is effective stability constant is given by $K_{eff} = K_{ab}\,a_4\,\beta_4$.

The complexometric titrations are carried out at fixed pH (*e.g.*, Ca at pH 7.3, Mg at pH 10.0, Co at pH 4.1, Ni at pH 3.2, Cd, Zn, at pH 4.1, Pb at pH 3.3 and Hg at pH 2.2). The buffer solutions of optimum pH are used. The end point is detected by making use of metallochromic indicators such as mureoxide, xylenol orange, eriochrom black, calmagite. In such indicators (K_{ln}) is having less value than (K_M) so they undergo dissociation. *i.e.*, $K_{ln} < K_M$.

The complexometric titration can be carried out for mixture of two or more metals by utilisation of the masking, demasking reactions, for example, Mg, Cu, Zn, can be directly titrated with EDTA then by complexation of Zn with KCN solutions is titrated and if formaldehyde HCHO is added when Zn complex is broken. In all successive reaction titration with EDTA can be carried out for metals which have been put into three groups depending upon the magnitude of K_M stability constant. If $pK = \log K_M$.

Table 8.2: pH for EDTA titrations

Group	*log K_M*	*pH*	*Name of metal*
Group I	log *K* 7.6 - 10.70	8-11	Mg (*II*), Ca (*II*), Sr (*II*). Ba (*II*)
Group II	log *K* 12.7 - 18.7	4-7	Mn (*II*), Fe (*II*), Co (*II*), Ni (*II*), Cu (*II*), Zn (*II*), Cd (*II*), Al (*III*), Pb (*II*), V(*II*)
Group III	log *K* 17.7 - 36	1-4	Hg (*II*), Bi (III), Co (*III*), Cr (III), Fe (III), Ga (III), In (*III*). Sc (*III*), Th (*IV*),V (*V*).

Disodium Salt of EDTA is represented as Na_2H_2 R when 1:1 complex is formed with metal like sodium.

Now in numerical problems we have to calculate the *pM, i.e.,* log *kM* at particular point in course of *pM* titration. The *pM* after addition of titrant, *viz.*, EDTA is calculated by using equation as stated in earlier sections,

$\dfrac{(V_R M_R - V_T M_T)}{(V_R + V_T)} = pM$. While after equivalence point the metal ion concentration is evaluated by using following expression.

$\dfrac{(V_T M_T - V_R M_R)}{(V_R + V_T)} = pM$. In such calculations we should know the value of the K_{ab} *i.e.,* absolute stability constant and α_4 fraction of uncomplexed EDTA in all forms. K_{eff}, *i.e.,* effective value of the stability constant. Sometimes magnitude of ionisation constant is also furnished.

SOLVED PROBLEMS

8.1 Ascertain the minimum pH needed for the titration of magnesium zinc and iron assuming one part in 10^4 remains uncomplexed by EDTA when 0.05 ml excess 0.02 M EDTA is added at end point?

Ans: For all three kinds of complexes $k_f = \dfrac{[MR]}{[M][R]^{-4}} = \dfrac{10^{-2}}{10^{-6}[R]^{-4}}$

$$[R^{-4}] = \frac{10^{-4}}{K_f} = \alpha_0\ [EDTA],\ \alpha_0 \frac{(0.05)(0.02)}{100} = 10^{-5} \text{ as } \alpha_0 = \frac{10^9}{K_f} \text{ if } \alpha_0 = \alpha_4$$

$$\text{for } Mg^{2+},\ \alpha_0 = \frac{10^5}{10^{8.7}} = 10^{0.3} \times 2 \text{ if } \alpha_0 = 1 \text{ at pH} \sim 12.0$$

For Zn^{2+}, $\alpha_0 = 10^9/10^{-16.5} = 10^{-7.5}$ at pH 4.5; For Fe^{3+}, $\alpha_0 = 10^9/10^{25} = 10^{-16}$ at pH ~ 2-0.

8.2 If 50 ml of 0.01 M calcium chloride buffered at pH 10.0 was titrated with 0.01M EDTA solution, evaluate value of pM, *i.e.*, pCa at different stages of titration. (Given K_{ab} = 5.0 × 10^{-10}, α_4 at pH = 10 was 0.35, so also $K_{eff} = 1.8 \times 10^{10} = \alpha_4 \times K_{ab}$).

Ans: The various points for titration were calculated as:

(*a*) Initial point: $[Ca^{2+}]$ = 0.01, pCa = 2.0

(*b*) After addition of 10ml of EDTA

$$[M]_R = \frac{(V_R M_R - V_T M_T)}{(V_R + V_T)} = \frac{(50 \times 0.001) - (40 \times 0.001)}{(50 + 40)} = 0.0067 \text{ M, pCa} = 2.17$$

(*c*) At the end point we have

$$[M]_R = \frac{V_R M_R - V_T M_T}{(V_R + V_T)} = [Ca^{2+}] = [EDTA^{4-}]$$

$$[CaR^{2-}] = 0.005 - [Ca^{2+}] = 0.005$$

$$[CaR^{2-}] = \frac{0.05}{100} = 5 \times 10^{-3} \text{ M as } K_{eff} = 1.8 \times 10^{10} = \frac{K_{ab}}{[Ca^{2+}]}$$

$$\text{or} \quad [CaR^{2-}] \times Ca = 1.8 \times 10^{10};\ [Ca^{2+}] = 5.2 \times 10^{-7} \text{ pCa} = 6.28$$

(*c*) Finally after addition of 60 ml excess of EDTA we have

$$Ca^{2+} = [R]_T = \frac{V_T M_T - V_R M_R}{(V_T + V_R)} = \frac{(60 \times 0.01) - (50 \times 0.01)}{(50 + 60)} = 4.55 \times 10^{-3} \text{M}$$

$$\text{Since } Ca_{EDTA} = \frac{10 \times 0.01}{110} = 9.1 \times 10^{-4}.\ \text{Hence } [CaR^{-2}] = \frac{0.5}{110} = 4.55 \times 10^{-3} \text{M}$$

$$K_{eff} = 1.8 \times 10^{10} = \frac{4.55 \times 10^{-3}}{[Ca^{2+}] 9.1 \times 10^{-4}} = \frac{[CaR^{2-}]}{[Ca^{2+}] Ca_{EDTA}};\ [Ca^{2+}]$$

$$= 2.8 \times 10^{-10}\ pCa = 9.55.$$

8.3 Calculate pZn of solution prepared by mixing 40, 50, 60 ml of 0.001 EDTA with 50 ml of 0.001M Zn^{2+} solution and hence draw titration curve. (Zn = 0.1 M, EDTA = 0.1 M in ammonia and 0.17 M NH_4Cl is added to get pH = 9.0). The formation constant was given as: K_1 = 190, K_2 = 210, K_3 = 250, K_4 = 110.

Ans: (*a*) Let us first consider calculation of conditional or effective stability constant

$$\beta_4 = \frac{1}{1 + K_1(NH_3) + K_1K_2(NH_3)^2 + K_1K_2K_3(NH_3)^3 + K_1K_2K_3K_4(NH_3)^4}$$

If we take $\beta_4 = \dfrac{1}{1+(19+420)+(1.04\times10^4)+(1.14\times10^3)} = 8\times10^{-6}$

If $\quad K_{ab} = 3.2\times10^{16}$ and α_4 at pH ~ 9.0 as 5.2×10^{-2}

$$K_{eff} = 5.2\times10^{-2}\times8\times10^{-16}\times3.2\times10^{16} = K_{ab};\alpha_4\,\beta_4 4 = 1.33\times10^{10}$$

(*b*) Now we calculate pZn after addition of 40 ml of EDTA as

$$C_{Zn} = \frac{(V_R M_R - V_T M_T)}{(V_R + V_T)} = \frac{(50\times0.001)-(40\times0.001)}{(50+40)} = 1.11\times10^{-4}\,\text{M}$$

$$\beta = \frac{[Zn^{2+}]}{C_{Zn}} \text{ or } [Zn^{2+}] = \beta\times C_{Zn} = (8\times10^{6})(1.11\times10^{-6}) = 8.9\times10^{-10}$$

If $\quad Zn = 8.9\times10^{-10}$ $pZn = 9.05$.

(*c*) The calculation of pZn after addition of 50ml of EDTA at end point

$[ZnR^{2-}] = 5\times10^{-4}$ M. If $\beta = [Zn^{2+}]CZn$

$[Zn^{2+}] = \beta\times C_{Zn} = (8.0\times10^{-6})(1.94\times10^{-7}) = 1.55\times10^{-12}$ $pZn = 1.81$

(*d*) Finally calculation of pZn after addition of 60ml of EDTA

$$C_{EDTA} = \frac{V_T M_T - V_R M_R}{(V_T + V_R)} = \frac{(60\times0.001)-(50\times0.001)}{(60+50)} = 9.1\times10^{-5}\,\text{M}$$

$$[ZnR^{2-}] = \frac{50\times0.001}{110} = 4.55\times10^{-4}\,\text{M. Since } [Zn^{2+}] = b\times C_{Zn}$$

$$K_{eff} = \frac{[ZnR^{2-}]}{C_{Zn}C_T[Zn^{2+}]} = (3.76\times10^{-10})(8\times10^{-6}) \text{ as } C_{Zn} = 3.76\times10^{10-},$$

$Zn = 3.01\times10^{-15}$.

Hence $\quad pZn = 14.52$.

8.4 What is optimum concentration for formation constant of nickel EDTA complex in buffer containing 0.05 M ammonia and 0.10 M ammonium chloride?

[Given $\log K_1 = 2.75$, $\log K_2 = 2.20$, $\log K_3 = 1.69$, $\log K_4 = 1.15$, $K_5 = 0.714$ and – 0.01, $K_a = 5\times10^{-10}$]

Ans: Since $\quad \alpha_M = \dfrac{1}{1+\beta_1[NH_3]+\beta_2[NH_3]^2 \,\ldots\ldots + \beta_n[NH_3]^n}$

So $\quad \alpha_M = \dfrac{1}{1+28+223+546+385+99+5} = 7.8\times10^{-4}$

From $K_a = 5\times10^{-10}; [H^+] = 1\times10^{-9}$ M and since

$$\alpha_R = \frac{K_1K_2K_3K_4}{[H^+]^4 + K_1[H^+]^3 + K_1K_2[H^+]^2 + K_1K_2K_3[H^+] + K_1K_2K_3K_4} = 0.052$$

$$\log K_{NiR} = 18.6 \text{ and } K_{NiR} = 10^{18.6}\times7.8\times10^{-4}\times0.0052 = 1.6\times10^{14}.$$

8.5 For nickel solution (10^{-3}M), Calculate when p Ni at following volumes of EDTA was added, as 0, 50, 99, 99.9, 100, 100.1, 101 and 110 ml?

Ans: Before end point $[Ni^{2+}] = C_{Ni}(1-x)$, if C_{Ni} was initial concern of nickel and x = fraction of EDTA was added.

At end point $[Ni] = [R^-] = \sqrt{\dfrac{C_{Ni}}{K^1_{NiR}}}$ and $[R^-] = C_{Ni}(x-1)$ but as

$$K^1_{NiR} = \frac{[NiR]}{[Ni][R]};\ [Ni] = \frac{1}{(x-1)K^1_{NiR}} \text{ so } pNi = -\log[Ni^{2+}]$$

$= -\log Ni[Ni^+] = 3.11 - \log[Ni]$ so corresponding values for nickel are

pNi = 6.1 (0), 6.4 (50), 7.1 (90), 8.1 (99), 8.4 (99.9), 11.7 (100), 14.36 (100.1) 15.36 (101) and 16.3 (110) m/s.

8.6 What is the concentration of zirconium, if 20 ml of the zirconium solution required 10.15 ml of 0.1 N of Na_2 (EDTA) ?

Ans: $20\times x = 10.15\times0.1$; So $x = \dfrac{10.15\times0.1}{20.0} = 0.05075$ N

eq. wt = 1/2 mol wt.

8.7 What is the amount of aluminium in a solution when 15 ml of 0.111.N EDTA was added as excesses and unconsumed EDTA solution was back titrated with 3 ml 0.101 N zinc solution?

Ans: (15 × 0.11 – 3 × 0.101) = amount of EDTA consumed.

The amount of Al = (15 × 0.11 – 3 × 0.101) = 2 × Al/2 = 18.20 (Since Eq wt. = 1/2 Mol wt.)

8.8 What is the concentration of magnesium and calcium if 20 ml of it was titrated with 0.1120 N EDTA, 18.5 ml EDTA was consumed, again to 20 ml of solution 19 ml was added (0.112 N) of EDTA when 12 ml was consumed?

Ans: Amount of Mg = 18.5 × 0.1120 = 20 × x

Amount of Ca = 19 × 0.112 = 20 × x

So we calculate as

$$(19\times0.1120 - 12\times0.1021)\times\frac{1000}{25} = 45.1 \text{ of Ca meq/L}$$

$$\left(18.15\times0.1120\times\frac{1000}{25} - 45.1\right) = 56.5 \text{ of mg meq/L.}$$

8.9 Evaluate the (α_4) fraction of the EDTA in solution that exist as anion at pH 5.0 if ionisation constants are $pK_1 = 2.07$, $pK_2 = 2.75$, $pK_3 = 6.24$, $pK_4 = 10.34$.

Ans: Since $\alpha_4 = \dfrac{[R^{2-}]}{[H_2R]+[HR^-]+[R^2]}$ by substituting values of $pK_1 \ldots\ldots pK_4$

$$\frac{1}{\alpha_4} = \left\{\frac{10^{-20}}{10^{-21.4}} - \frac{10^{-15}}{10^{19.3}}\right\} + \left\{\frac{10^{-10}}{10^{-16.6}} + \frac{10^{-5}}{10^{-10.3}} + 1\right\} \text{ on simplification}$$

$$\frac{1}{\alpha_4} = [10^{1.4} + 10^{4.3} + 10^{6.6} + 10^{5.3} + 1]. \text{ So } \alpha_4 = 10^{-6.6}.$$

8.10 Evaluate (α_{cd}) for cadmium in a solution containing 0.1M ammonia and 0.10 M unbound ammonia. The log of formation constant (log K_f) was $\beta_1 = 2.00$, $\beta_2 = 4.65$, $\beta_3 = 6.04$ and $\beta_n = 6.60$.

Ans: Since $\dfrac{1}{\alpha_M} = 1 + \beta_1(R) + \beta_2(R)^2 + \beta_3(R)^2 \ldots\ldots\ldots \beta_n(R)^n$

By substituting the values

$$\frac{1}{\alpha_{cd}} = 1 + (10^{2.6})(10^{-1}) + (10^{4.65})(10^{-2}) + (10^{6.92})(10^{-4}) + (10^{6.60})(10^{-5})$$

$$\frac{1}{\alpha_{cd}} = 1 + 40 + 450 + 1100 + 830 + 40 = 2461$$

$$\alpha_{cd} = \frac{1}{2461} = 10^{-0.4}.$$

8.11 What is the value of pCd when 0.01M cadmium solution was buffered with 0.10 M NH_4^+ and 0.10M free NH_3 and 50% of the cadmium was titrated with EDTA solution?

Ans: Ignoring dilution from EDTA we have cadmium concentration as

$$[Cd] = \frac{1}{2} \times 0.0100 = 0.0050 \text{ M}.$$

Since CdR was stable Cd from dissociation was ignorable we have concentration if Cd has dissociation cation

$$\alpha_{cd} = 10^{-0.39} \quad \alpha_{cd} = [Cd]/[Cd]$$

$$[Cd] = \alpha_{cd}[Cd^1] = 10^{-0.39} \times 5 \times 10^{-3} = 5 \times 10^{-6.39}$$

pCd = (6.39 – log5) = 5.69 no complexation then Cd at 50% titration was 0.0050M and pCd = 2.30.

8.12 What is the ratio of calcium ligand to free ligand at the stoichiometric point in the titration of 0.001 M calcium with 0.01 M EDTA buffered at pH = 10 (K_f = 105.68 and pCa = 6.6)?

Ans: Since pCa = 6.6 [Ca] = $10^{-6.6}$

$$K_f = \frac{[Ca]}{[Ca^+][R]} \text{ On putting the values we have}$$

$$10^{5.68} = \frac{[\text{CaR}]}{(10^{-6.6})(R)} \text{ or } \frac{[\text{CaR}]}{[\text{R}^-]} = 10^{5.68} \times 10^{-6.6} = 10^{-0.9} \text{ i.e., } \frac{1}{8} \text{ ratio.}$$

8.13 Find out the transition range of indicator used for the titration of nickel at pH 5.0 if formation constant for indicator = 10^6 ? Do you think this indicator can be used for titration of nickel at pH = 5.0 if pNi = 7.15 at end point?

Ans: Formation constant of indicator $= K = \frac{[\text{M}^+\text{ In}^-]}{[\text{M}^+][\text{In}]} = 10^6$ as $\text{M}^+\text{ In}^- \rightleftharpoons \text{M}^+ + \text{In}^-$

As we can note 1 part in 10 parts colour when [NiIn]: [Zn] = 10

$$10^6 = \frac{10}{[\text{Ni}]} \text{ so pNi} = 7.0.$$

Last colour change would be for ratio (NiIn] : [In] = 0.1

$$10^6 = \frac{0.1}{[\text{Ni}]} = \text{pNi} = 9.0$$

The transition range is (7.0 – 9.0) at pH = 2.0.

8.14 Nickel chloride (1×10^{-2} M) when buffered at pH 5.0 and titrated with 0.01M EDTA, Calculate the pNi at 50%, 100% and 200% of titration?

Ans: (1) At 50% (Ni^{2+}) = 50% of 0.010M = 0.005M 10g

$[Ni^{2+}] = 0.005 \times 2/3 = 3.3 \times 10^{-3}$ So pNi = (3 – log 3.3) = 2.48

(2) For 100% titration, at pH 5.0 $\alpha_N = 10^{-6.6}$

$$K_{\text{Ni}}\alpha_R = K^1_{MR} = \frac{[\text{MR}]}{[\text{M}][\text{R}^1]} \text{ so } K_{\text{NiR}} = K^1_{\text{NiR}}\alpha_R = 10^{6.6} \times 101^{8.6} = 10^{12}$$

$$[\text{NiR}] = 0.010 \times \frac{1}{2} = 0.0050 \text{ M}$$

$$10^{12} = \frac{[\text{NiR}]}{[\text{Ni}][\text{R}]} = \frac{0.0050}{[\text{Ni}][\text{Ni}]} = \frac{0.0050}{[\text{Ni}]^2} \text{ so } [\text{Ni}] = \sqrt{50 \times 10^{-16}} = 7.1 \times 10^{-8}$$

Therefore $\quad \text{pNi} = -\log 7.1 \times 10^{-8} = 7.15$

(3) For 200% titration

The ratio of [NiR]: [R] = 100: 100 *i.e.*,

$$K_{\text{Ni}} = 10^{12} = \frac{[\text{NiR}]}{[\text{Ni}][\text{R}]} = \frac{100}{[\text{Ni}]100}$$

$$[\text{Ni}] = 10^{-12}; \text{ pNi} = 12.0$$

8.15 What is the value of a_{Zn} for zinc in solution having 0.10 M NH_4 and 0.10 M unbound ammonium? The log formation constant for Zinc-ammonia complex were given as:

$$\beta_1 = 2.60,\ \beta_2 = 4.65,\ \beta_3 = 6.04,\ \beta_4 = 6.92 \text{ and } \beta_5 = 6.92.$$

Ans: Now stepwise formation constant for *MRn* will be given by

$$\frac{1}{\alpha_M} = 1 + \beta_1 (R)^2 + \beta_2 (R)^2 + \beta_3 (R)^2 + \beta_4 (R)^2 + \beta_5 (R)^2 \ ......\beta_n (R)^n$$

We subtitle these values in above equation.

$$\frac{1}{\alpha_{Zn}} = 1 + (10^{2.6})(10^{-1}) + (10^{4.65})(10^{-2}) + (10^{6.04})(10^{-3}) + (10^{6.92})(10^{-4}) + (10^{6.60})(10^{-5})$$

$$\frac{1}{\alpha_{Zn}} = 1 + 40 + 450 + 1110 + 830 + 40 = 2461.$$

Hence $$\frac{1}{\alpha_{Zn}} = \frac{1}{2461} = 10^{-3.4}.$$

8.16 What is the concentration of calcium if 50 ml of 0.01M calcium chloride was titrated with 0.01 M EDTA buffered at pH ~10.0?

What was pCa at pre equivalence and post-equivalence point?

Ans: The condition constant was obtained from K_f and α_4

$$K_{Ca} = \alpha_4 K_f = (3.5\times10^{-1})(5\times10^{10}) \text{ if } \alpha_4 = 3.5\times10^{-1} \text{ and } 5\times10^{10}$$

$$K_{Ca} = 1.75\times10^{10}$$

Now concentration at equivalence point was

$$[Ca^{2+}] = \frac{50\times0.01 - 25\times0.01}{(50+25)} = \frac{V_R M_R - V_T M_T}{(V_R + V_T)} = 3.33\times10^{-3} \text{ so pCa} = 2.48$$

Now concentration after equivalence point was

$$[Ca^{2+}] \text{ so } F_{CaR} = \frac{50\times0.01}{110} = 4.55\times10^{-3},\ F_{EDTA} = \frac{10\times0.01}{110} = 9.1\times10^{-4}$$

$$[Ca^{2+}] = 4.55\times10^{-3} - [Ca^{2+}] = 9.1\times10^{-4}$$

$$C_T = 9.1\times10^{-4} + [Ca^{2+}] = 9.1\times10^{-4}$$

$$\frac{4.55\times10^{-3}}{[Ca^{2+}]\times9.1\times10^{-2}} = 1.75\times10^{10} = K^1 \text{ Ca};\ [Ca^{2+}] = 2.86\times10^{-10}$$

$$\text{pCa} = 9.54.$$

PROBLEMS FOR PRACTICE

8.17 If nickel (1×10^{-2} M) solution was adjusted to pH = 5.0 and then titrated with 0.01 M EDTA (disodium salt), what is concentration *i.e.*, pNi at 50%, 100% and 200% titration?

[Ans: 2.48; 7.15 and 12.0]

8.18 A calcium chloride (1×10^{-2} M) solution was adjusted to pH 5.0 solution and was titrated with 0.01 M EDTA. Calculate the concentration at initial and 200% titration? [**Ans:** 2.0 and 4.1]

8.19 What is the pH if 0.01M cadmium solution when titrated with 0.01 M EDTA solution K_{sp} for Cd $(OH)_2$ is 6×10^{15} if conditional constant for cadmium was 1×10^8?

[**Ans:** pH 4 and pH = 8]

8.20 The hard water (50 ml) was titrated with 15 ml of 0.0100 M EDTA. While another sample of 50 ml water was made basic to give precipitate. This solution was titrated with 10 ml of 0.012M EDTA. What is concentration of calcium and hardness of water?

[**Ans:** 0.00240; 300 pm hardness]

8.21 Calculate the weight of EDTA so as to have 1.5 liter of 0.025 M of the titrant?

[**Ans:** 13.96g]

8.22 A 0.1148gm of magnesium carbonate was dissolved and pH was adjusted to 10.0 and then it was titrated with EDTA when end point was 17.61 ml. Calculate the concentration of EDTA solution? [**Ans:** 007974M]

8.23 What is the conditional constant of calcium complex of EDTA at pH = 10 if 50 ml of 0.01 M salt was titrated with 0.01M EDTA K_f for Ca = 5×10^{10} while $a_4 = 3.5 \times 10^{-1}$?

[**Ans:** 1.75×10^{10}]

8.24 A complexon solution with titer of 5 mg/ml of calcium carbonate was required. Indicate how it was prepared from 1.5 liter disodium salt? [**Ans:** 21.9 g]

8.25 Ascertain the conditional constant for Mn $(EDTA)_2$ complexes at pH 7.0, 9.0 and 11.0.

[**Ans:** 3×10^{10}; 3.2×10^{12} and 5.3×10^{13}]

8.26 What is the magnitude of a_4 for EDTA at pH 12.0 containing $[Na^+] = 0.1$ if activity coefficient was unity? [**Ans:** 0.17]

8.27 If a ferric ion (1×10^{-3}) M was masked with 0.05 M EDTA. Calculate the masking index at pH 2.0 if no hydroxy complex was formed? [**Ans:** 10.4]

8.28 Draw various graphical point on titration curve of 25 ml of 0.02 M zinc chloride by 0.02 M EDTA complexed at pH 10 and with 0.1 M free ammonia?

8.29 A lake water (50 ml) was titrated against 0.0102M EDTA when end point was as 29.85 ml. What is the hardness of water? [**Ans:** 300ppm]

8.30 A 500 mg of sample of German silver containing zinc was dissolved in 100 ml of mineral acid. When 25 ml was titrated with 0.01M EDTA gave end point at 10.25 ml. What is the percentage of Zinc in the German silver? [**Ans:** 53.6%]

8.31 A sample containing 500 mg calcium chloride, 300 mg of magnesium chloride was dissolved in acid and was made up to 2 liter. It was titrated with EDTA, what is the hardness of water?

[**Ans:** 407.6ppm]

8.32 A standard solution of EDTA was prepared as usual and 1 ml of this solution was titrated with magnesium in 10 ml solution containing 300 mg/L. Now if 100 ml required 8.60 ml of standard EDTA solution what was the hardness of water? [**Ans:** 271 ppm]

8.33 A sample of stainless steel containing iron and nickel was analysed and if 0.02M EDTA was used for titration when 30% nickel was encountered, what size of sample should be taken to accomplish accuracy?

8.34 A 500 mg of gun metal contaning zinc was dissolved in 10 ml of acid and if 25 ml of an aliquot required 10.25 ml of 0.01 M EDTA solution, what is the percentage of zinc in gun metal?

8.35 Depict titration curve for 50 ml of 0.015 M barium chloride with 0.03M EDTA solution buffered at pH 10.0. What is the pBa valves after addition of 10, 25, 30 ml of the titrant?

[**Ans:** 2.12, 5.09, 7.48]

REFERENCES

8.1 S.M. Khopkar, *Basic Concepts of Analytical Chemistry,* 3rd edn. New Age International Publishers (P) Ltd. (2007).

8.2 D.J. Pietrzyke and C.W. Fank. *Analytical Chemistry: An Introduction,* Academic Press p. 469 (1974).

8.3 H.A. Latinen and W.E. Harris, *Chemical Analysis: An Advanced Text,* McGraw-Hill and Co. p. 196-7; 215 (1975).

8.4 A. Muskin, *Problems in Quantitative Analysis,* Mir Publishers, Moscow, p. 140. (1984).

8.5 D.A. Skoog and D.M. West, *Analytical Chemistry,* 3rd edn. Holt, Reinhart and Winston, p. 255, 264, 267 (1970).

8.6 J.S. Fritz and G.H. Schenk, *Qualitative Analytical Chemistry,* Allyn and Bacon Inc. p. 223-5; 227-9, 233, 237-9 (1979).

8.7 C. Schwarzenbach, H. Flaschka, *Complexometric Titrations,* Methun and Co. (1969).

8.8 F.J. Welcher, *Analytical Uses of Ethylenediamino Tetracetic Acid,* D.Van Nostrand and Co. (1958).

8.9 A. Ringbom, *Complexation in Analytical Chemistry.* Interscience N.Y. (1963).

8.10 A.L. Vogel, *The Textbook of Quantitative Inorganic Analysis,* Longman and Green (1980).

Chapter 9

Solvent Extraction Separations

INTRODUCTION

Solvent extraction is one of the versatile method of separation, what one needs is good separatory funnel and pH meter for the adjustment of pH of the aqueous phase. The separations can be carried out both at macro and microgram concentration. The total time required for the extraction and determination is very little, one could accomplish clean cut separations by this method with greater accuracy.

The most essential component is the organic phase which must be immiscible with aqueous phase. The organic solvent must possess donor group like oxygen as in diethylether or as phosphorus in tributyl phosphate or may be it is a chelating extractant like oxime $|\bar{M}|$ or 8-hydroxyquinoline, dithizone, cupferron. The efficiency of the extraction is evaluated by ascertaining the magnitude of distribution ratio. When extraction occurs, the species from the aqueous phase get transferred into the organic phase, we evaluate the overall distribution ratio (D) as follows:

$$D = \frac{\text{Total concent of species in the organic phase}}{\text{Total concent of species in the aqueous phase}} = \text{Distribution } (D) \text{ ratio}$$

When we consider individual species in both phases we consider it in terms of (Kd)

$$Kd = \frac{\text{Con. of species in the organic phase}}{\text{Con. of species in the aqueous phase}} = \text{distribution coefficient } (K_d)$$

Now under idealised conditions $D = K_d$ such ideal conditions include when there is no dissociation, association or polymerisation of the extracted species in both phases. For practical purposes one uses (%) percentage (E) extraction if volume of organic phase (V_0) or volume of the aqueous phase (V_w) is known in that case we assume

When $V_w = V_0$ we have $D = \dfrac{E}{(100 - E)}$ however if

If $V_w \neq V_0$ we have $D = \left(\dfrac{V_W / V_0}{100 - E}\right)$

Now the magnitude of distribution ratio is infinity if there is a quantitative extraction. If one knows

V_w, V_0 and E it is quite easy to calculate distribution ratio. From the knowledge of V_w, V_0 and D it is quit easy to calculate % E.

The distribution ratio (D) depends upon several parameters for instance in chelate extraction system. If [HR] is extracting agent then

$$D = \frac{K_f K_a^n K_{D_x}}{K_{D_R}^n} \frac{[\mathrm{HR}]_0^n}{[\mathrm{H}^+]^n}$$

Where K_f = formation constant of complex, K_a = acid dissociation constant of the ligand, K_{D_x} = dist coefficient of complex, K_{D_R} = distribution coefficent of the ligand in the organic phase. We use pH1/2

$$\mathrm{pH}1/2 = 1/n \ \log K^* + \log(\mathrm{HR}) \text{ if } \left[\frac{K_f K_a^n K_{D_x}}{K_{D_R}^n}\right] = K^*$$

Now one can evaluate pH 1/2 *i.e.*, pH for the 50% extraction from pH 1/2 = 1/n logK^* + log [HR]

when, $\overset{*}{K} = \left[\frac{K_f K_a^n K_{D_x}}{K_{D_R}^n}\right]$.

The larger is the difference in pH 1/2 for two metals with same extractant, better is the separation. This can be also evaluated in terms of (α), *i.e.*, the separation factor (α) = D_1/D_2 if D_i dist coeff of first metal D_2 = dist coeff of second metal. For ion association or extraction by solvation we have $D = \overset{*}{K}[x]^2 \ r \pm 3$. If x = anion, r = activity, K^* same significance as given above.

Further it has been observed that instead of having one single extraction with larger volume of the organic phase by equilibration with fixed volume of aqueous phase, one should use small portions of organic solvent, number of times to get complete extraction.

Thus we have $W_n = \left[\frac{V}{DS+V}\right]^n$ where V = vol. of aqueous; S = vol. of organic phase W = amount of solute, W_n = Wt. of solute unextracted. This shows 'S' should be as small as possible for 'n' number of extractions. From this expression one can evaluate half extraction volume (K) as $K = \frac{0.693W}{V}$. The half extraction volume is V ml. And W is volume of the original solution. We define the half extraction volume as the volume of an extractant to decrease the amount of extracted constituent to one half of its original value.

Finally we can consider Craig extraction system. Here in the multiple extractions D V_0, V_N are interrelated as:

$$\left[\frac{D(V_0/V_w)}{1+D(V_0/V_w)} + \frac{1}{1+D(V_0/V_w)}\right]^n = 1.0. \text{ For } n > 20 \text{ the equation is}$$

$$\gamma_{\max} = np = \frac{nD(V_0/V_w)}{1+D(V_0/V_w)}. \text{ If } \gamma \text{ is number of tubes in continuous extractions.}$$

$$D = \frac{\gamma_{max} + 1}{(n - \gamma_{max})(V_0/V_w)} \text{ so}$$

$$D = \frac{(r+1)Tnr + 1}{(n-r)(V_0/V_w)Tnr} = \frac{rTnr}{(n-r)(V_0/V_w)Tn, (r-1)}$$

If Tn, r is fraction of solute in the 'r' th tube.

Finally let us consider extractions where HR is extractable as

$$[HR];\ D = \frac{(HR)_o}{(HR)_w} \quad [HR]_w = \frac{[H^+][R^-]}{K_a}.$$

Now the degree of extraction of substance from solution is given as Q = amt. of substance in respective phases, as $E = \frac{Q_{org} \times 100}{(Q_{org} + Q_{aq})} = \frac{D \times 100}{(D + V_W/V_O)}$.

If Q_{org} and Q_w are amount of substance in organic or aq. phase resp V = volume of corresponding phase. After 'n' extractions concen of substance left in solution is $C_n = C_{org}\left(\frac{100-E}{100}\right)^n$ and $\alpha = D_A/D_B$

and in multiple extractions $T_{nr} = \frac{n^1}{r^1(n-r)l}\left(\frac{100-E}{100}\right)^n\left(\frac{E}{100-E}\right)$. If R = serial number of the stage, n = number of transfers.

In this chapter several problems based on evaluation of $\%E$ or distribution ratio are presented. Further the evaluation of pH 1/2, α is also included in few problems.

SOLVED PROBLEMS

9.1 When 200 mg of iodide was extracted from 100 ml of an aqueous solution with 50 ml portions of carbon tetrachloride the first lot contained 4.6 mg while second extraction with 50 ml of carbon tetrachloride had unextracted species as 0.1057 mg. Calculate the relative errors for both extraction. What conclusion is drawn from this experiment? (Given D = 85)

Ans: Since D = 85 we have relative error as

$$\frac{0.0046 \times 100}{0.2} = 2.3\% \text{ also } \frac{0.0001057 \times 100}{0.2} = 0.05\%.$$

The conclusion is that double extraction is quite adequate for the extraction of iodide from the solution.

9.2 If magnesium was extracted (33%) with 8-hydroxyquinoline, in 100 ml of 10^{-5} M solution at pH = 6. With 20 ml of 10^{-3} M solution of 8-hydroxyquinoline in chloroform, what is the percentage of extraction under identical condition with 50 ml of 0.0005 M of the 8-hydroxyquinoline?

Ans: Since

$$K_{ex} = D\frac{[H^+]^2}{[HR]_{org}^2} = \left(\frac{E}{100-E}\right)\frac{V[H]^{2-}}{[HR]_{org}^2} = \frac{33}{67} \times \frac{100}{20}\left(\frac{10^{-5}}{10^{-3}}\right)^2 = 2.5 \times 10^{-4}$$

as $\left(\dfrac{E}{100-E}\right) = V_{ex}\dfrac{[HR]^2_{org}\,V_{org}}{[\mathrm{H}^+]^2\,V} = 2.5\times10^{-4}\dfrac{(5\times10^{-4})^2}{(10^{-6})^2} : \dfrac{50}{100} = 31.2$ & $E = 96.9$.

9.3 When copper was extracted with dithizone in the organic solvent following values were obtained $K_a = 3\times10^{-5}$, $K_{D_R} = 1.1\times10^4$, $K_{D_x} = 7\times10^4$, $\beta n = 5\times10^{22}$, $n = 2$. When 100 ml of a solution containing 10^{-7} M copper in 1M and 0.1 M hydrochloride was extracted with 10 ml of 0.01% dithizone in the organic phase (HR = 0.0004M) What was pH 1/2 and the amount copper unextracted?

Ans: At pH = 0, pH 1/2 = $\log\left(\dfrac{K_{DR}}{K_a}\right) - \dfrac{\log K_D \times \beta n}{n} - \log[HR]$ org substitute values

$$\mathrm{pH}\frac{1}{2} = -5.21 - \log[HR]_0$$

$$n\left(\mathrm{pH}\frac{1}{2} - \mathrm{pH}\right) = -\log D = \log\left(\frac{100-E}{E}, \frac{V_0}{V_w}\right) \text{ substituting the values,}$$

$$2\left(\mathrm{pH}\frac{1}{2} - \mathrm{pH}\right)\log[(100-E)]V_0/EV_w$$

$$\mathrm{pH}\frac{1}{2} = -5.21 - \log(4\times10^{-4}) = -1.8 \text{ as } \frac{V_0}{V_w} = 0.1, \log(100-E)/E = -2.62.$$

So $\mathrm{pH}\frac{1}{2} = 0(100-E)/E = 2.4\times10^{-3}$ and $(100-E) = 0.24\%$ unextracted.

9.4 A certain extraction system has a distribution ratio (D) = 10. If 300 mg of solute was dissolved in 100 ml of solvent A, find out the amount of solute extracted by one extraction with 100 ml of solvent B and by two extractions with 50 ml of solvent B which was immiscible with the solvent A if molecular weight of solute = 71.0. Comment which one of the extraction is better?

Ans: Moles of solute present $= \dfrac{0.30}{71} = 4.22\times10^{-3}$. Now x – moles is extracted then (4.22×10^{-3}) is unextracted solute. So $D = 10 = \dfrac{x/100}{\left(\dfrac{4.22\times10^{-3} - x}{100}\right)} = 3.84\times10^3$

$$\%\text{ Solute extracted} = \frac{3.84\times10^{-3}\times100}{4.22\times10^{-3}} = 91.0\%$$

Now in Second part a = moles be extracted in first 50 ml of solvent
b = moles be extracted in second 50 ml of solvent

$$D = 10 = \frac{x/50}{\left(\dfrac{4.22\times10^{-3} - a}{100}\right)} \text{ so } x = 3.52\times10^{-3} \text{ moles of solute unextracted}$$

$$D = 10 = \frac{y/50}{\left(\frac{0.70 \times 10^{-3}}{100}\right)} \text{ since } y = 0.58 \times 10^{-3} \text{ moles total } 4.10 \times 10^{-3} \text{ moles of solute}$$

So percentage extraction in the second extraction is high and is preferable

$$E = \frac{4.10 \times 10^{-3} \times 100}{4.22 \times 10^{-3}} = 97.2\%.$$

9.5 In extraction of Cerium (IV) with oxine in chloroform, the volume of aqueous and organic phase each were 25 ml with percentage of extraction 99.8%. Calculate the distribution ratio?

Ans: Since $V_W = V_0 = 25$, $E = 99.8\%$ we have

$$D = \left(\frac{E}{100 - E}\right) = \frac{\left(\frac{V_w}{V_0}\right)E}{(100 - E)} = \frac{\left(\frac{25}{25}\right)98.8}{(100 - 99.8)} = \frac{1 \times 99.8}{0.2} = \frac{99.8}{0.2} = 499.$$

9.6 Uranium (VI) when extracted with 2 - thenoyltrifluoroacetone in benzene had D = 999, $V_0 = 10$, $V_w = 25$ ml. Calculate the % extraction of uranium?

Ans: $D = 999$, $V_0 = 10$, $V_w = 25$ ml.

$$D = \frac{(V_w/V_0)E}{(100 - E)} = \frac{(25/10)E}{(100 - E)} = 999. \text{ So } E = 99.75\%.$$

9.7 What is the amount of iodine remaining unextracted in the aqueous phase if it was extracted with 100 ml of organic solvent with $D = 85$?

Ans: Since $x = \left(\frac{V_w}{V_0 D + V_w}\right) \times x_0$ if x = amount of unextracted iodine.

$$x = \left(\frac{100 \times 0.10}{(100 \times 85 + 100)}\right) \text{ as } V_w = 100, \quad V_0 = 100, D = 85$$

$$= 1.16 \times 10^{-3} \text{ moles.}$$

9.8 What is the amount of iodine unextracted if V_w = 100 with concentration 1 × 10^{-3} M after extraction with 50 × 2 ml of the solvent when distribution ratio was 85?

Ans: Since x = 100 × 1.0 × 10^{-3} = 0.100 M mole

$$x = \left(\frac{100}{50 \times 85 + 100}\right)^2 \times 0.100 = 5.28 \times 10^{-3} \text{ M moles after two extractions.}$$

9.9 Prove that single extraction with 10 ml of aqueous phase and 20 ml of organic phase was not efficient as compared to twice extraction with 10 ml from of aqueous phase with 10 ml organic solvent.

Ans: In first case let us assume x = % extraction

$$D = \frac{(V_w/V_0)E}{(100 - E)} = 20 = \frac{(10/20)x}{(100 - x)}$$

i.e., $20 = \dfrac{0.5x}{(100-x)}$ or $20(100-x) = 0.5x$ or $(20{,}000 - 20x) = 0.5x$

So $$20{,}000 = 20.5x. \text{ So } x = \frac{20{,}000}{20.5} = 97.6\%.$$

In Second case however 10 ml of aqueous phase with 10 ml of organic phase gives

$$D = \frac{(V_w/V_0)E}{(100-E)}. \text{ Since } V_w = V_0 = 10,\ D = 20,\ x = E$$

$$20 = \frac{x}{(100-x)} \text{ or } 20(100-x) = x, \textit{ i.e.}, (2000-20x) = x$$

$$20{,}000 = 21x \text{ so } x = \left(\frac{20{,}000}{21}\right) = 95.23\%.$$

Repeating another step, we get x = 99.8%.
This shows second extraction is better as it is large.

9.10 Evaluate the magnitude of the distribution ratio of iodine between organic phase and water if total extraction is 98.5% of iodine in 50 ml of water extracted with 50 ml of the organic solvent.

Ans: Since E = 98.5, $V_w = V_0 = 50$

$$D = \frac{(V_w/V_0)E}{(100-E)} = \frac{(50/50)98.5}{(100-98.5)} = \frac{98.5}{1.5} = 65.66.$$

9.11 Prove that a single extraction with equal volume of solvent in preference to five extractions with same total volume of organic solvent concludes that fraction of unextracted solute is negligible in latter extraction. (Given D = 2).

Ans: $D = \dfrac{C_{org}}{C_{aq}} = \dfrac{(x_A - y_A) \times V_{org}}{y_A/V}$ if x_A, y_A moles of solutes in organic phase and latter one unextracted resp.

So α_W = fraction remaining unextracted = $\dfrac{y_A}{x_A} = \dfrac{1}{(1+D(V_w/V))} = \dfrac{1}{(V+DV_{org})}$

If V_w = V the $DV_0/V = D$ for n-successive extraction

$$\alpha_n = \left(\frac{1+DV_{org}}{V}\right)^{n-}. \text{ Now since } D = 2,\ V_w = V \text{ in first case, unextracted.}$$

$$\alpha_w = \left(\frac{V}{V+D_{V_{org}}}\right) = \frac{1}{(1+2)} = \frac{1}{3} \textit{ i.e., } 0.333. \text{ unextracted species. In latter case } V \textit{ i.e., } V_{org} = \frac{V}{5}$$

so unextracted species shall be

$$a_w = \left(\frac{V}{V+2\times\frac{V}{5}}\right) = \frac{1}{(V+2V/5)} = \frac{1}{\left(1+\frac{2}{5}\right)} = \frac{1.0}{1.2} \text{ so } \alpha = 0.186.$$

9.12 What is the percentage of extraction of iron (III) from 100 ml of a 6M hydrochloric acid with 20 ml of diethylether assuming the distribution ratio is 100?

Ans: $E = \dfrac{100 \times 100}{\left(100 + \dfrac{100}{20}\right)} = \dfrac{DV_w}{(D + V_w/V_0)} = 95\%$ if $V_w = 100$, $V_0 = 20$ ml, $D = 100$

9.13 Calculate the percentage extraction of copper with sodium diethyl dithiocarbamate (cupral) in chloroform at pH = 8.9 if the ratio of the volume of organic to aqueous phase is 2:1. (Given $K_a = 10^{-8.7}$ and $D = [H^+]/K_a$)

Ans: Since $HR = H^+ + R^-$, $K_a = \dfrac{[H^+][R^-]}{[HR]}$ if HR represents sodium diethyl dithiocarbamate

The extent of extraction $E = \dfrac{[HR]V}{[HR]V^1 + [R]V} \times 100$ if $V^1/V = 2$ and $V = 1, V^1 = 2$

After extraction $E = \dfrac{V^1/V \times 100}{\left[\dfrac{V^1}{V} + \dfrac{K_a}{[H^+]}\right]}$ so at pH at 8 $E = 91\%$, pH at 9 $E = 50\%$, pH 10 $E = 9.1\%$.

9.14 What is the magnitude of separation factor (α) for extraction of gallium and indium with Aliquot 336S if $\left(E_{Ga} = 97\%, E_{\text{In}} = 23\%, V_w = V_0 \text{ and } \dfrac{[Ga]}{[\text{In}]}\right) = 5$?

Ans: $D_{Ga} = \dfrac{97}{3} = 32.3$, $D_{\text{In}} = \dfrac{23}{77} = 0.298$ if $D = \dfrac{\text{Conc. in organic phase}}{\text{Conc. in aqueous phase}}$

Now separation factor $(\alpha) = \dfrac{32.3}{0.298} = 109.$

9.15 How many times should extraction and rextraction process be repeated in above problem in order to separate gallium from indium in solution until the concentration ratio in the organic phase is 100:0.1?

Ans: In above problem concentration ratio is 5:1 Now
$Ga = 0.97$ and $ln = 0.23$ we have concentration ratio

$\dfrac{5 \times 0.97x}{1 \times 0.23x} = \dfrac{100}{0.1}$ 21. If x = nos. of times extraction is carried out

So $\dfrac{5 \times 0.97 \times x}{1 \times 0.23x} = \dfrac{100}{0.1}$. This gives value of $x = 3.7$ *i.e.* 4.0.

9.16 What is the magnitude of distribution ratio (D) if % $E = 90$ and volume of organic and aqueous phases is same?

Ans: $D = \dfrac{E}{(100 - E)}$. If $V_w = V_0$ same. If $E = 90\%$.

$$D = \frac{90}{(100-90)} = 9.0.$$

PROBLEMS FOR PRACTICE

9.17 A sample of water is polluted with 3, 4 benzopyrene from autoexhaust emissions. When 100 ml of water was shaken with one ml solvent, 53% of 3-4 benzopyrene was extracted, what was the percentage of 3, 4 benzopyrene when 10 ml of water was shaken with one ml of the organic solvent? **[Ans:** 91.9%]

9.18 Calculate the distribution ratio when 99% of solute was extracted from 50 ml of the aqueous phase with *(a)* (20 × 2 ml) of the organic phase and *(b)* (5 × 10) ml of the organic phase? **[Ans:** 22, 5.8]

9.19 Calculate the concentration of solute remaining unextracted after 75 ml of 0.05M solute was treated by extraction with 60 ml, (10 × 6 ml), (2 × 12 ml) and (30 × 2 ml) of the organic solvent assuming distribution ratio of 23.5?

[Ans: 2.53×10^{-3}, 4.62×10^{-4}, 1×10^{-5} M and 6.12×10^{-7} M]

9.20 Calculate the distribution ratio of strontium between water and carbon tetrachloride if reagent 8-Hydroxyquinoline (1M) was used at pH = 11.0? **[Ans:** 11^{-6}]

9.21 Calculate the distribution ratio if aqueous phase and organic phase volume is same and is extracted with 0.1 M reagent at pH = 10? **[Ans:** 7.1×10^{-5}]

9.22 Calculate the distribution ratio of solute extracted with extractant volume of 10 ml from 25 ml of the aqueous phase, with the magnitude of the percentage extraction was 99.8%? **[Ans:** 1250]

9.23 If the total extraction is 99.9% of iodine from 50 ml of water by (50 × 5ml) of the immiscible solvent, what would be the magnitude of the distribution ratio? **[Ans:** 2.0]

9.24 A 100 ml of aqueous phase containing 0.5M of iron and 5M hydrochloric acid was shaken with 10 ml ethylacetate gave distribution ratio as 17.6, calculate the decrease in the concentration of the aqueous phase after one extraction? **[Ans:** 65]

9.25 What will be the percentage extraction of vanadium (V) when extracted with mesityl oxide if the volume of organic phase was 10 ml and the volume of aqueous phase was also 10 ml with the distribution ratio 1247? **[Ans:** 92%]

9.26 What is the concentration in each phase after extraction of 100 ml of 10^{-2} M of organic nitrogenous compound in water with 100 ml of ether assuming distribution ratio = 3.0? Will 1000 ml of ether would improve the extraction?

[Ans: 7.5×10^{-4} and 7.5×10^{-3}m. The efficiency of extraction is improved to 75-97% with 1000 ml ether.]

9.27 Gallium and Indium were extracted from 8M hydrochloric acid with tributyl phosphate for ten times. Assuming total extraction of gallium and indium was resp 97% and 23% and if phase ratio was unity, Calculate fraction for optimum extraction of gallium and indium?

[Ans: 10th fraction Ga, 2nd fraction Indium]

9.28 Ferric (III) iron was extracted with tributyl phosphate (TBP) from 10 M nitric acid with the phase ratio of unity and distribution ratio 1.0 with % *E* = 50, What is the concentration of iron in the organic phase if original amount of iron was also one?

[Ans: Maximum in fraction 2 and 3]

9.29 What is the distribution ratio for the extraction for the copper diethyldithiocarbamate with 1×10^{-4} M of sodiumdiethyl dithiocarbamate at pH 4.0 in the presence of 0.01 EDTA as sequestering agent?
(Given log $K = 18.8$, $-\log \alpha_4$ for EDTA = 8.44) **[Ans:** $D = 130$]

9.30 What is the efficiency of the separation for a volume ratio (V_0/V_w) of 10, 1, and 0.1 for the extraction assuming distribution ratio of 10? **[Ans:** 99%, 90.9% and 50%]

9.31 What is the amount of solute extracted in 100 ml of solution with three 10 ml portions of the organic solvent if the distribution ratio is 50? **[Ans:** 0.167, 0.0278, 0.0046 gm]

9.32 If iron (III) was extracted from mineral acid with diethylether when volume of aqueous phase was 25ml while volume of the organic phase was 10 ml with 99.8% extraction. What was the distribution ratio? **[Ans:** 1247.5]

9.33 What is the percentage extraction of cerium (IV) when extracted from hydrochloric acid with methyl isobutyl ketone by single, double or triple extraction with 100 ml of ceric solution with 30ml of methylisobutyl ketone? (Given $D = 200$) **[Ans:** 98.37]

9.34 When nitric acid solution of uranyl nitrate was equilibrated with twice its volume of ethylacetate saturated with nitric acid, 99% of uranium was extracted, what was the distribution ratio of uranium (VI) between nitric acid and saturated ethylacetate? **[Ans:** 49.5]

9.35 If 90% of zinc was extracted with trioctylamine in xylene phase and if 5% was left behind in the organic phase, calculate the separation factor when the phase volume ratio was 1 and 27.

9.36 What are the total number of extractions required if 90% of substance with distribution ratio as unity was to be extracted if phase ratio was 1/9? **[Ans:** $n = 3.3$ *i.e.*, 4-extraction.]

REFERENCES

9.1 S.M. Khopkar, *Basic Concepts of Analytical Chemistry,* 3rd edn., New Age International Publishers (P) Ltd. (2007).

9.2 D.A. Skoog and D.M. West, *Analytical Chemistry,* 3rd edn. Holt Reinhart and Winstone pp. 301, 500, 525, 529 (1979).

9.3 J.S. Freitz and G.H. Schenk, *Quantitative Analytical Chemistry,* Allyn and Bacon, pp. 357, 8 (1979).

9.4 D.J. Pietrzyke and G.W Frank, *Analytical Chemistry: An Introduction,* Academic Press, p. 503 (1974).

9.5 D.G. Peter J.M. Heycs, G.M. Hjeftje, *Chemical Separations and Measurements,* Saunder Golden Series p. 961 (1980).

9.6 A.K. De, S.M. Khopkar, R.A. Chalmers, *Solvent Extraction of Metals,* Van Nostrand Reinhold and Co. (1970).

9.7 G.H. Morrison and H. Freiser, *Solvent Extraction in Analytical Chemistry,* Wiley Interscience (1958).

9.8 T. Sekinc Y. Hasegawa, *Solvent Extraction Chemistry,* Marcell Dekker and Co. (1980).

9.9 J. Starry, *Solvent Extraction of Metal Chelates,* Pergamon Press (1961).

9.10 Y. Marcus, and A.S. Kertes, *Ion Exchange and Solvent Extraction of the Metal Complexes,* Wiley, Interscience (1969).

9.11 T. Braun and G. Ghersini, *Extraction Chromatography,* Elesvier Pub. Co. (1975).

9.12 A. Zolotov, *Extraction by Compound Formation,* Michigan Press (1972).

9.13 S.M. Khopkar, *Solvent Extraction Separation of Elements with Liquid Ion Exchange,* New Age International Publishers (P) Ltd. (2007).

9.14 S.M. Khopkar, *Analytical Chemistry of Macrocyclic and Supramolecular Compounds,* 2nd edn., Springer-Verlag (2006).

Chapter 10

Supramolecules in Solvent Extraction

INTRODUCTION

The solvent extraction chemistry of *s*-block elements, *viz*., alkali and alkaline earth's showed a great promise with the discovery of crown ethers, while cryptands proved to be versatile extractants for quantitative separation of metal pollutants from the aquatic environment while supramolecular compounds like calix (*n*) arene or Calix (*n*) resorcinarene and Rotaxane were useful for the extractive separation of chemically similar elements.

On account of their complex structure IUPAC system of the nomenclature was not used. Pederson and his co-workers devised a novel system of the nomenclature. As regards classification they were termed as macrocyclic polyether, *e.g*., DB18C6, or Macrobicyclic polyethers such as cryptand –222 and supramolecular compounds like calix (*n*) arene or calix (*n*) resorcinarene. The methods of synthesis of these compounds was simple and generally yield was > 70%.

There were various factors which influenced the process of extraction, *e.g*., size and charge of the ion in reaction that determined which metal could be extracted by a particular crown ether. What was needed was that the cavity of crown ether should match with the ionic size of the metal. Then metal was embedded in such cavity carrying on it a positive charge which in turn was neutralised by equivalent charge of large size counter anions like picric acid, rose bengal, erythrosine which were needed during extractions. The new important parameter influencing the extraction was the solubility of such complexes in the organic solvents. Those of diluents which had high dielectric constant favoured better extraction. Nitrobenzene, dichloroethane, dichlormethane were best diluents as they promoted metal ligand interaction resulting into formation of the more complex which in turn was extracted into organic solvent. Finally the geometry and hence stereochemistry of the complex also led to the stability of the complexes.

Several newer extractants were developed for the solvent extraction of transition metals, however with the discovery of crown ethers a new era had dawned for the extraction of alkali and alkaline earths while cryptands were best for separation metal pollutants but calixarenes would prove in future to be promising extractants. This is because it was possible to synthesis these extracting ligands with variable cavity or inner diameter of the cone-shaped compounds with effective substitution in lower or upper limb of the extractant. This in turn facilitated the phenomena of the complexation and extraction of metals.

A new class of reagents called 'Rotaxane' has come into forefront in last decade. These are interesting compounds which have two wheels and central axis with stoppers to prevent escaping of wheels. There are many compounds. So far selenium (VI) was extracted quantitatively by Rotaxane.

Therefore in this chapter large number of multiple choice questions are provided which cover important aspect such as nomenclature, classification, synthesis, complexation, role of the various counter anions, kind of diluents used during extraction.

PROBLEMS

Multiple Choice Questions

Tick mark the correct answer from the following multiple choices:

10.1 One of the following ligand has a planner structure:

(*a*) 18 Crown6 (*c*) Calix (6) arene
(*b*) cryptand - 2.2,2 (*d*) Calix (6) resorcinarene.

10.2 One of the following extractant is designated as macrobicyclic polyether.

(*a*) cryptand - 2,2,2B (*c*) 30C10
(*b*) 24C8 (*d*) calix (4) arene.

10.3 Crown ethers had readily formed complexes with metals belonging to group of

(*a*) alkalimetals and alkaline earths (*s*-block metals)
(*b*) transition elements, (*d*-block metals)
(*c*) main group elements, (*p*-block metals)
(*d*) inner transition elements, (*f*-block metals).

10.4 The extractant which is most effective for the separation of lead from waste water is

(*a*) 18C6 (*c*) cryptand - 2.2 2B
(*b*) B 18C6 (*d*) cryptand - 2.2, 2.

10.5 Calix (*n*) arene was discovered by one of the following chemist in 1985:

(*a*) Pederson (*c*) Lehn
(*b*) Guitche (*d*) Cram.

10.6 The extractant obtained by base catalysed reaction between formaldehyde and tertiary butyl phenol is

(*a*) calix (6) arene (*c*) Lariat ether
(*b*) calix (6) resoranrene (*d*) Proton ionsable crown ether.

10.7 One of the compounds which can be considered as supramolecular compound is

(*a*) Calix (6) arene (*c*) Cryptand - 2, 2, 1
(*b*) 30C10 (*d*) Aza-18C6.

10.8 The parent compound which was used for synthesis of DB18C6, with dichloroether, butanol and base by Pendersen was:

(*a*) catachol (*c*) resorcinol
(*b*) formaldehyde (*d*) t-butyl phenol.

10.9 Calix (*n*) resorcinescarene is obtained by acid catalysed reaction between

(*a*) resorcinol and acetaldehyde (*b*) catachol and formaldehye

(*c*) tertiary butyl phenol and paraformaldehyde
(*d*) tertiary butyl phenol and acetaldehyde.

10.10 If the cavity in the crown ether, complex matches with one of the following item it readily forms complex

(*a*) ionic radii of metal ion (*c*) charge of cation
(*b*) ionic size of metal ion (*d*) number of protons in cation.

10.11 The counter anion used in extraction of potassium with large molar volume with 18C6 as extractant is

(*a*) picric acid (*c*) perchlorate
(*b*) tetraphenyl boron (*d*) zincon.

10.12 The ionic diameter of potassium readily matches with the cavity of

(*a*) DB24C8 (*c*) Rotaxane
(*b*) calix (6) arene (*d*) DB18C6.

10.13 The crown ether with cavity of 4.00 A unit forms readily complexes with

(*a*) lithium (*c*) potassium
(*b*) sodium (*d*) rubidium.

10.14 The cavity size in 15C5 is of the order of

(*a*) 1.7°A (*c*) 3.4°A
(*b*) 2.6°A (*d*) 4.0°A.

10.15 The cation dipole interaction is increased if the crown ether contains a large size donor atom such as

(*a*) sulphur (*c*) nitrogen
(*b*) oxygen (*d*) phosphorus.

10.16 In extraction of copper with 1-10 phenanthroline the counter anion used is

(*a*) perchlorate (*c*) Rose Bengal
(*b*) picric acid (*d*) Tropelion -000.

10.17 The role of counter anion during extraction of metals with crown ether is to

(*a*) neutralise the charge of the complex
(*b*) increase dielectric constant of aqueous phase
(*c*) increase molecular weight of the complex
(*d*) enhance the charge of the cationic complex.

10.18 The counter anion most popularly used for the solvent extraction of alkali metals with crown ethers is :

(*a*) picric acid (*c*) zincon
(*b*) HDEHP (*d*) EDTA.

10.19 The counter anions with following characteristics promotes extraction

(*a*) high molar volume and plannar structure
(*b*) low molecular weight
(*c*) low molar volume and three-dimensional structure
(*d*) very high concentration of counter anion.

10.20 Diluents with high dielectric constant give better extraction because of the

(*a*) strong interaction between oxygen and hydrogen
(*b*) less solubility of extractant
(*c*) solvation of cations during extraction
(*d*) dehydration of environment around metal and ligand.

10.21 The extraction with crown ether or cryptand increases with

(*a*) increase in concentration of anion
(*b*) decrease in concentration of extractant
(*c*) decrease in the concentration of cation
(*d*) long period of equilibration.

10.22 Crown ether embedded with cationic potassium in its cavity readily extracts:

(*a*) anionic complex like $[Sb\ I_4]^-$,
(*b*) cationic complex like $[Cu\ (phen]^{2+}$,
(*c*) A complexes like $[Ca\ (EDTA_2)]^{2-}$
(*d*) oxyanion like $[MoO_4]^{2-}$

10.23 In calix (6) arene the substitution of lower limb is effected with

(*a*) acetate
(*b*) nitrate
(*c*) sulphate
(*d*) chloride.

10.24 The diluents with high dielectric constant were used for extraction of caesium with 21 crown 7 such as:

(*a*) benzene
(*b*) xylene
(*c*) dichloromethane
(*d*) chloroform.

10.25 The concentration of the counter anion as well as extractants should be

(*a*) identical for best extraction.
(*b*) different exchange
(*c*) high counter anion concentration
(*d*) high reagent concentration.

Ans: 10.1 (*a*), 10.2 (*a*), 10.3 (*a*), 10.4 (*c*), 10.5 (*a*), 10.6 (*a*), 10.7 (*a*), 10.8 (*a*), 10.9 (*a*), 10.10 (*b*), 10.11 (*a*), 10.12 (*d*), 10.13 (*d*), 10.14 (*a*), 10.15 (*a*), 10.16 (*a*), 10.17 (*a*), 10.18 (*a*), 10.19 (*a*), 10.20 (*a*), 10.21 (*a*), 10.22 (*a*), 10.23 (*a*), 10.24 (*c*), 10.25 (*a*).

REFERENCES

10.1 S.M. Khopkar, *Basic Concepts of Analytical Chemistry,* 3rd Edn. New Age International Publishers (P) Ltd. (2007).

10.2 S.M. Khopkar, *Analytical Chemistry of Macrocyclic and Supramolecular Compounds.* Second Edition. Springer Verlag Publication (2006).

10.3 M. Hiraoka, *Crown Ether and Analogus Compounds,* Elsevier Publishing Co. N.Y. (1992).

10.4 M. Hiraoka, *Crown Compounds: Their Characteristics and Applications,* Elsevier Amsterdam (1982).

10.5 C.J. Gutsche, *Calixarene,* Royal Society of Chemistry, London, Second Edition (1995).

10.6 G.A. Melson, *Coordination Chemistry of Macrocyclic Compound,* Plenum Press Ltd. (1979).

10.7 S.M. Khopkar, *Solvent Extraction Separations of Elements with Liquid Ion Exchanger,* New Age International Publishers (P) Ltd. (2006).

10.8 G.W. Gokel, S.H. Korzeniowski, *Macrocyclic Polyethers Synthesis*, Springer Verlag (1982).

10.9 R.M. Izatt, G.D Christen, *Synthetic Multidentale Macrocyclic Compounds*, Academic Press, London (1978).

10.10 A.K. De, S.M. Khopkar, R.A. Chalmers, *Solvent Extraction of Metals,* VonNostrand Reinhold Co. (1970).

Chapter 11

Principles of Chromatography

INTRODUCTION

Although there is spectacular advancement in the instrumental methods of analysis, the success or failure of instrumental method largely depends upon how pure is the sample for such analysis. This is because a slight impurity in sample will lead to erroneous results. So large number of separation methods were discovered to isolate analytical species before any instrumental method is resorted. Such separation methods include not only chromatographic methods but also other non-chromatographic techniques like solvent extraction, ring oven; zone refining, froath floation, dialysis, reversed osmosis and precipitation methods.

However, chromatographic methods have become most popular because of the simplicity and cost of analysis. In any chromatographic technique, we have stationary phase and mobile phase with stationary support packed into the column. We can accomplish good separation by exploiting difference in migration rate of two species on column. Such difference in the migration front originates due to difference in some physical property with which solute is attached to the stationary support. Such physical properties include, adsorption, partition, ion exchange, size difference or the difference in potential. We are able to separate large number of species by just exploiting, difference in their physical exchange properties responsible for binding solute to the stationary support.

In the course of separation, one uses a reagent which removes the species from column. It is called as 'eluant'. The volume of an eluant for total removal of a component is called '*Vt*' while volume when 50% of the component has been removed is called 'V_{max}' or peak elution volume. V_0 stands for the volume when elution starts if C_M, C_S are concentration of solute in the mobile and stationery phases, resp. and V_M and V_S are volumes of mobile and stationary phase then $\left(\frac{R}{1-R}\right) = \frac{C_M V_M}{C_S V_S}$ if R = equilibrium form of solute in mobile phase while $(1 - R)$ is solute fraction in the stationery phase we then relate it to K_d, *i.e.*, partition coefficient as

$R = \left[\frac{V_M}{V_M + K_d V_S}\right]$. This equation is similar to one used in solvent extraction method viz.

$$D = \frac{(Vw/V_0 - E)}{(100 + E)} \text{ or } E = \frac{100D}{(D + Vw/V_0)} \text{ if } V_0 = V_w.$$

Instead of volume one can also use time spent on column as t_m, and t_s with same notations. The column capacity is expressed in terms of K as:

$K = \left(\frac{C_S V_S}{C_M V_M}\right)$ if $\beta = (V_M / V_S)$ phase ratio and if $Kd = Cs/Cm$, then we have (K) column capacity as

$$K = \frac{K_d}{\beta} = K_d \beta^{-1} = \left[\frac{V_R - V_M}{V_M}\right]$$

In majority of chromatographic work, we use Van Deemeter equation it states mathematically that $A + \frac{B}{u} + Cu = H$. If H = Height of theoretical plate in chromatography A, B, C are constants, u = rate of flow of liquid.

If a column consists of (N) number of plates and if (H) is height of one plate, if (L) is length of column then

$N = L/H$ or $H = L/N$ is height equivalent of theorised plate (HETP)

It is interesting at this stage to relate (N) with peak evaluation volume (V_{max})

In typical elution curve. V_{max} = peak elution volume, Vt = total volume of an eluant, V_0 = commencement volume for elution W = represent base of the elution curve plot while We - width of elution curve, N = nos. of theoretical plates then N is related to W and V_{max} as, (Fig. 11.1).

$$N = 16\left(\frac{V_{max}}{W}\right)^2 \text{ or } N = 8\left(\frac{V_{max}}{We}\right)^2.$$

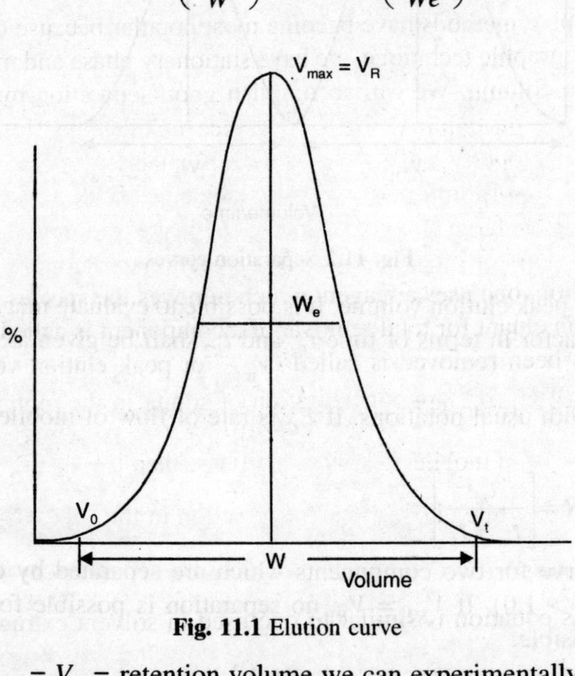

Fig. 11.1 Elution curve

In these equations $V_{max} = V_R$ = retention volume we can experimentally calculate the numbers of plates in a column. Larger is the numbers of plates better is the efficiency of the separation. From (N) we can calculate height of plate (H) = L/N. It has been observed that smaller is the height and larger are number of plates we expect better efficiency of the separation.

In paper or thin layer chromatography we use term R_f (retardation factor)

$R_f = \dfrac{\text{distance travelled by solute front}}{\text{distance travelled by solvent front}}$ from the knowledge of R_f one can find out efficiency of separation (R).

This efficiency is generally measured in terms of resolution factor (R) it is $R = \dfrac{Z}{4\sigma}$ if Z = separation of zone centers, σ = standard deviation of the zones so $R = \dfrac{(V_{R_2} - V_{R_1})}{0.5(W_2 + W_1)}$. Here V_{R1}, V_{R2} are peak elution volumes for two components. Then the resolution factor can be evaluated from the knowledge of these terms. The number of plates *(N)* can be also calculated from t_M and t_S *i.e.*, time solute spends (Fig. 11.2) in two phase $N = 16\left(\dfrac{t_R}{W}\right)^2 = 16\left(\dfrac{V_{\max}}{W}\right)^2 = 8\left(\dfrac{V_{\max}}{W_e}\right)^2$, Also $R = \dfrac{t_{R_1} - t_{R_2}}{0.5(W_2 + W_1)} = \left(\dfrac{t_m}{t_m + t_s}\right)$.

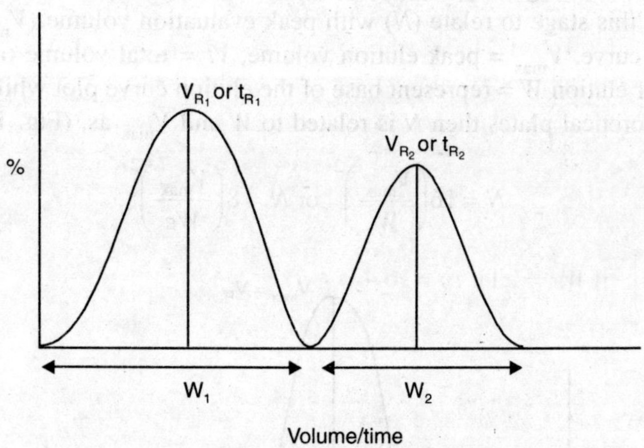

Fig. 11.2 Separation curves

Here t_R = time of the peak elution volume. It is possible to evaluate magnitude of (V_R) peak elution volume. The resolution factor in terms of time t_{R1} and t_{R2} shall be given as:

$R = \dfrac{(t_{R_1} - t_{R_2})}{0.5(W_2 + W_1)}$ with usual notations. If F_C is rate of flow of mobile phase on the column,

$V_R = (T_R \times F_C)$ and $R = \left[\dfrac{t_m}{t_m + t_s}\right]$.

In typical elution curve for two components which are separated by chromatography with very good resolution ($V_{R2}/V_{R1} > 1.0$). If $V_{R2} = V_{R1}$ no separation is possible for best separation the ratio should be as large as possible.

SOLVED PROBLEMS

11.1 Calculate minimum height *(H)* of plate if $A = 0.01$, $B = 0.30$ cm^2/ml, $C = 0.155$ and $u = 0.287$ ml/mt with usual notations in Van Deemeter's equation.

Ans: As per Van Deemeter's equation

$$H = A + \frac{B}{u} + C \quad \text{substituting the given values}$$

$$H = 0.01 + \frac{0.30}{0.287} + (0.155 \times 0.287)$$

$$H = 0.01 + 1.045 + 4.305 = 5.36 \text{ is plate height.}$$

11.2 If a substance x and y had retention time of 16.40 and 17.63 mt on column with height of 30 cm, the unretained species had retention time of 1.30 mt if width of the elution curve was 1.11 for W_x and 1.21 for W_y. Calculate the column resolution the average number of plates, height *(H)* and overall time required to elute component *(y)* on the length of column?

Ans: The resolution $(R) = \frac{2\Delta Z}{Wx + Wy}$ where in Wx and Wy are width of elution peak of two species and ΔZ is difference in time of their arrival at detector $\left(t_R^x - t_R^y\right) = 1.23$. If (w) is the width of the elution curve and ΔZ is difference in time of their arrival at detector.

Thus if $Wx = 1.11,\ Wy = 1.21,\ \Delta Z = (17.63 - 16.4) = 1.23$

$$\text{So } Rs = \frac{2\Delta Z}{Wx + Wy} = \frac{2 \times 1.23}{(1.11 + 1.21)} = 1.06 \text{ resolution.}$$

The number of theoretical plates are evaluated as:

$$Nx = 16\left(\frac{t_R^x}{W_x}\right)^2 \quad \text{if } Wx = 1.11,\ t_R^x = 16.4$$

$$Nx = 16\left(\frac{16.4}{1.11}\right)^2 = 3493 \text{ if } Wy = 1.21,\ t_y^R = 17.63$$

$$Ny = 16\left(\frac{17.63}{1.21}\right)^2 = 3397$$

The average number of plates is calculated

$$Nav = \frac{(Nx + Ny)}{2} = \frac{(3493 + 3397)}{2} = 3445 \text{ is average number of plates}$$

Now $H = L/N$ if $L = 30$, $N = 3445$ then H = height of plate

$$H = \frac{30}{3445} = 8.708 \times 10^{-3} \text{ cm is height equivalent of theoretical plate}$$

The overall time required for elution of component (y) is

$$(t_R)_y = \frac{16R^2H}{u}\left(\frac{\alpha}{\alpha - 1}\right)\frac{(1 + K^1 y)}{(K^1 y)^2} \quad \text{if } (t_R^x) \text{ is time needed for elution of 'Y'}$$

$$\text{So } (t_R^1)(t_R^2) = (Rs_1)^2/(Rs_2)^2 = 17.63/t_R^y = (1.06)^2/(1.5)^2 = 35.3 \text{ mts.}$$

11.3 Calculate the retention volumes of two solutes X and Y when volume of stationary phase (Vs) is 1.5 ml, volume of mobile phase (V_M) = 2.5 ml, distribution coefficient of ions resp., Dx = 5.0 and Dy = 15?

Ans: Since $V_R = (D\ Vs + V_M)$ if V_R = retention time, we have $V_R^x = (Dx\ Vs + Vm)$. Substituting these values $V_R^x = (5 \times 1.5) + 2.5 = 10$ ml for x.

$V_R^y = (15 \times 1.5) = 2.5 = 25$ ml is eluted latter for y.

11.4 Calculate total number of plates in the gas chromatograph if flow rate is 25 ml/mt, $t_M = 0.20$ mt, $t_R = 3.64$ mt and peak width of curve W is = 0.36 mts.

Ans: $N = 16\left(\frac{V_R}{W}\right)^2$ or $16\left(\frac{t_R}{W}\right)^2$ since $t_R = 3.64, W = 0.36.$

$$N = 16\left(\frac{3.64}{0.36}\right)^2 = 1636 \text{ total plates.}$$

11.5 Assuming if $N = 1636$, evaluate peak width at t_R^{max} = 6 mts. in above problem.

Ans: $N = 16\left(\frac{t_R}{W}\right)^2$ if $N = 1636$, $t_R = 6$ mt.

$$1636 = 16\left(\frac{6}{W}\right)^2 \quad \therefore\ W = \sqrt{\frac{16 \times 6^2}{1636}} = 0.59 \text{ mt.}$$

11.6 If peak elution time of compound x is t_R^x = 4.4 mt and for compound y is t_R^y = 5 mt on the chromatographic column with length 90 cm (L) and height of plate (H) = 0.03. State whether separation is possible?

Ans: Since $N = L/H = 90/0.03 = 300$

$$N = 16\left(\frac{t_R}{W}\right)^2 \quad N = 300, \text{ as } t_R^x = 4.4,\ t_R^y = 5.0$$

$$N^x = 300 = 16\left(\frac{4.4}{W}\right)^2 \text{ and } N^y = 300 = 16\left(\frac{5}{W}\right)^2$$

$Wx = 1.02$ $Wy = 1.15$

First peak and second peak overlap. Since there is overlap no separation is possible.

11.7 Calculate the resolution of compounds if N = 900, $t_R^x = 4.4$ mt., $t_R^y = 5$ mt., Wx = 0.59, Wy = 0.67?

Ans: $Rs = \frac{2\left(t_R^y - t_R^x\right)}{\left(W_x + W_y\right)}$

$$Rs = \frac{2(5 - 4.4)}{(0.59 + 0.67)} = 0.95 \text{ is resolution.}$$

11.8 If N = 1350 instead of 900 by increasing column length by 50% *i.e.*, 90 to 135 cm and if $t_R^x = 6.6$ mt. and $t_R^y = 7.5$ mt calculate the resolution?

Ans: Since $N = 16\left(\frac{V_R}{W}\right)^2 = 16\left(\frac{t_R}{W}\right)^2$ and as $Nx = 1350$, $t_R^x = 6.6$

$$Nx = 1350,\ N = 16\left(\frac{6.6}{W_x}\right)^2 \text{ so } W_x = 0.72 \text{ mt.}$$

$$Ny = 1350\ t_R^y = 7.5 \text{ mt.}$$

$$1350 = 16\left(\frac{7.5}{Wy}\right)^2. \text{ So } Wy = 0.82 \text{ mt. now resolution}$$

$$Rs = \frac{2(ty - tx)}{(Wx + Wy)} = \text{On substituting values as } ty = 7.5,\ tx = 6.6,\ Wx = 0.72,\ Wy = 0.82.$$

$$\text{we have } Rs = \frac{2(7.5 - 6.6)}{(0.72 + 0.82)} = 1.17 \text{ resolution factor.}$$

11.9 Is it possible to compare resolution of component '*a*' and of '*b*' if N = 900? Given (*i*) Da = 1.0, Db = 1.2 also (*ii*) Da = 10, (Db = 12), α = 1.2.

Ans: $R = \frac{1}{4}\left(\frac{\alpha - 1}{\alpha}\right)\sqrt{N}\left(\frac{Db}{1 + Db}\right)$ substitute the values

$$\text{For } (i),\ R_a = \frac{1}{4}\left(\frac{0.2}{1.2}\right)\sqrt{900}\left(\frac{1.2}{1 + 1.2}\right) = 0.68 \text{ (poor resolution for component '}a\text{') as } (a - 1) = 0.2$$

$$\text{For } (ii),\ R_b = \frac{1}{4}\left(\frac{0.2}{1.2}\right) \times 30\left(\frac{12}{12 + 1.2}\right) = 1.15 \text{ (resolution reasonable for the component '}b\text{')}$$

11.10 In HPTLC separation, length of the plate used for migration was 3 cm, having 4000 theoretical plates in 10 mt time, on the contrary in TLC for migration of 12 cm distance took 25 minutes having total of 2000 theoretical plates. What is the height equivalent of theoretical plate and state whether the separation is good with TLC or HPTLC technique?

Ans: Since $N = L/H$ or $H = L/N$ we substitute the values in HPTLC L = 3cm, N = 4000,

$$H = \frac{30 \text{ mn}}{4000} = 0.0075 \text{ mm plate for TLC, we have } L = 12,\ N = 2000,$$

$$H = \frac{120}{2000} = 0.6 \text{ mm. This shows HPTLC technique is good if height of plates as small as}$$

0.0075 mm.

11.11 In certain chromatographic separation on column with W and V_R were resp. being 3.7 and 42 ml; Calculate the number of theoretical plates as well as the height equivalent of each plate?

Ans: Since $N = 16\left(\frac{V_R}{W}\right)^2$ if W is width of elution curve,

$N = 16\left(\frac{42}{3.7}\right)^2 = 16\left(\frac{V_R}{W}\right)^2$ as $V_R = 42$, $W = 3.7$, we substitute to get

Hence N = 2030 number of plates.

PROBLEMS FOR PRACTICE

11.12 Is peak width obtained from a column with $N = 400$ and $t_R = 3.17$ mt. comparable with one obtained when $t_R = 5.86$ mt? **[Ans:** 0.63]

11.13 Because *N, R, D,* are correlated with width of two peaks (W_1) equal to (W_2) state equation for resolution in terms of $N\, t_R$ or V_R if the peaks are to be resolved assuming *W* peak being unequal. **[Ans:** $R = N/2\,(t_{R1} - t_{R2})/(t_{R1} + t_{R2})$]

11.14 Calculate the average number of plates and average plate height of the column if the retention times (t_R) of compounds *A, B, C, D* are 2.5, 10.7, 11.6 and 14 minutes resp. with width of peak base as (*w*) being resp, 0, 1.3, 1.4 and 1.8. What is the number of plates from each peak as well as plate height?

11.15 In gas chromatographic separation if column length is 22.6 cm, volume of mobile phase (V_m) = 1.26 ml, volume of stationary phase (V_s) = 0.148 ml and corresponding, retention times (t_R) are 0, 0.45, 1.07 and 1.16 resp, what is the number of plates from each peak as well as plate height for the column?

11.16 With Van deemter's equation in above problem calculate values for *A, B, C, D* capacity factor and partition coefficient similarly for *B* and *C* give resolution factor length of column to give *Rs* = 1.5 finally calculate for compounds *C* and *D* the resolution length of the column if *Rs* = 1.5 and t_R for separation of last two *C* and *D* compounds is 1.5 mt?

11.17 The mobile phase needed 25 ml to pass through the column. Calculate the resolution of solute if $t_R = 261$ mt.? **[Ans:** *R* = 0.0958]

11.18 A column (10 cm) was used with speed of the mobile phase as 0.1 cm/sec, when component (*x*) requires 40 mt for elution, calculate the fraction of time needed for its elution if *x* spends its time in mobile phase. What is the resolution for system? **[Ans:** 0.4167, 0.4167]

11.19 The retention time of (t_R) for two components is 15 and 20 mt. On other column if the first component has retention time of 22 mt. Calculate the retention time for other component on the other column? **[Ans:** 29.33 mt]

11.20 Calculate total number of plates if the width of elution curve is as (*w*) 40 sec and retention time of 25 mt.? **[Ans:** *N* = 22,500]

11.21 Calculate the height equivalent of theoretical plate, if column is 2000 mm long with width of curve as 40 and retention time of 25 minute? **[Ans:** 0.089mm]

11.22 Assuming the retention time to be 10^2, 10^3 or 10^4 seconds, with theoretical plates of 10^5 in number, what is the peak width of the each curve? **[Ans:** 1.265, 12.65 and 126.5 second]

11.23 Calculate the magnitude of t_s as well as resolution (*R*) if time of mobile phase is 25 mt and $t_R = 261$ mt. **[Ans:** *R* – 0.0958, t_s = 236 mt]

11.24 Calculate the time spent in stationary phase if $R = 0.1$, $Vm = 2$ ml with flow rate as 10 ml/mt?
[Ans: 1.08 second]

11.25 What are the number of theoretical plates for resolution if $N = 4200$ with retention time for compound (*x*) is 15.05 mt and compound (*y*) is 14.82 mt.? **[Ans:** $N = 67{,}500$]

11.26 Calculate the peak width for column if $N = 400$ $t_R = 3.17$ mt and $t_R = 5.86$ mt is expected time?
[Ans: $W = 0.63$, $N = 1.7$ mt]

11.27 Evaluate the magnitude for *N* and *H* if the elution volume is 10.20 ml with base of elution curve as 2.6 and column height of 36 cm.?

11.28 What is an expression for calculating resolution in terms of (*N*) and retention time (t_R) or retention volume (V_R), if the peak widths were not identical? **[Ans:** $R = N \div \dfrac{2(t_B - t_A)}{(t_B + t_A)}$]

11.29 Calculate *N* and *H* for the 30 cm length if in paper chromatography the distance on the paper tape from the initial spot to peak maximum was 5 cm and the peak width was 5 cm.
[Ans: $N = 1600$, $H = 0.002$]

11.30 Evaluate magnitude of R_f for a TLC work if the distance from the centre of spot to the initial point is 3 cm and distance from the beginning to the solvent is 15 cm. **[Ans:** $R_f = 0.2$]

REFERENCES

11.1 S.M. Khopkar, *Basic Concepts of Analytical Chemistry,* 3rd edn., New Age International Publishers (P) Ltd. (2007).

11.2 J.S. Fritz & G.H. Schenk, *Quantitative Analytical Chemistry,* Allyn & Bacon Inc, pp. 368-71, 409 (1979).

11.3 D. J. Pietrzyke & C.W. Frank, *Analytical Chemisty: An Introduction,* Academic Press (1979).

11.4 D.G. Peters, J. M. Hayes & G.M. Hieftje, *Chemical Separations and Measurements,* Saunder Golden Series, pp. 344-5 (1974).

11.5 D.G. Peters, J.M. Hayes, G.M. Hieftje, *A Brief Introduction to Modern Chemical Analysis,* Saunders & Co. (1980).

11.6 B.L. Krager, L.R. Snyder, C. Horwath, *An Introduction to Separations Science,* Marcel Dekker & Co. (1980).

11.7 J.A. Dean, *Chemical Separation Methods,* Van Nostrand Reinhold Co. (1970).

11.8 E.W. Berg, *Physical & Chemical Methods of Separation,* McGraw Hill & Co. (1963).

11.9 E. Lederer, G.M Lederer, *Chromatography—Theory & Practice,* 2nd edn., Van Nostrand & Co. (1957).

11.10 E. Heftman, *Chromatography,* Reinhold (1969).

Chapter 12

Ion Exchange Separations

INTRODUCTION

Amongst the various methods used for the separations specially for elements, the best known method is ion exchange chromatography. A remarkable progress has been made in the separation of lanthanides by cation exchange methods. The chemically similar elements like zirconium, hafnium; or niobium tantalum were best separated by cation exchange chromatography.

For this purpose synthetic ion exchange resins were easily made available. The resin with varying cross linkage ranging from 4-16 % DVB in the matrix of copolymers of styrene and divinly benzene was available. Further resins with varying particle size from 50-200 mesh were also marketed. With these resins one could accomplish any kind of difficult separations in inorganic chemical analysis.

The capacity of the ion exchange resin was always expressed in terms of weight distribution coefficient (designated as K_d or D_w) it is defined as:

$$D_w = \frac{\text{Concentration of ion in the resin phase}}{\text{Concentration of ion in the solution phase}}$$

However in columnar work, another term called volume distribution coefficient (D_v) was used. It was defined as (D_v)

$\dfrac{1}{D_v} = E = \dfrac{d,\ A}{V_{\max}}$ if d = column height. A = area of cross-section of the column, $V_{\max}$ = peak elution volume, E = elution constant and D_v = volume distribution coefficient.

This expression doesn't account for volume of resin which is occupied as resin has trapped solution in interstial void called fractional (γ) void we therefore deduct it from peak elution volume V_R. We write

$E = \dfrac{d.A}{(V_{\max} - \gamma)}$ here γ is called as the free column volume.

If we consider ions with their charge, *viz.* valancy then we have to consider

$$bA + aB_R = aB + bA_R;\ K_B^A = \frac{[A]_R^b\,[B]^a}{[A]^b\,[B]_R^a};\ K_B^A = \frac{[A]_R^b\,[B]^a}{[A]^b\,[B]_R^a} = K_d^{b/a}\,\frac{[B]^{a/b}}{[A]}$$

This expression is not used in practice similarly for anionic complexes of metals with anion '*X*', we have $D_M = K\left[\frac{[X]^c}{[X]}\right]^a \cdot F$ if F = fraction of sorbable anionic metal.

As described earlier we could calculate (N) number of plates and (H) plate height, we have

$$N = 8\left(\frac{V_{\max}}{W_e}\right)^2$$

$$= 16\left(\frac{V_{\max}}{W}\right)^2 \text{ if } W = \text{base and } W_e = \text{width of elution curve.}$$

Now if 'L' is height of column then height equivalent of theoretical (HETP) (HETP) = $[L/N] = H$.

In terms of retention time (t_1) and (t_2) for two ions, one can easily evaluate resolutions (R) by following expression.

$$R = \left(\frac{t_1 - t_2}{4\sigma}\right) = \left(\frac{t_1 - t_2}{\frac{w_1}{2} + \frac{w_2}{2}}\right) = \frac{2(t_1 - t_2)}{(w_1 + w_2)}$$ we can easily value of σ which represents standard deviation while the value of W- which represents base of the elution curve for ions 1 and 2. It is interesting at this stage to find out interrelationship between D_w and D_v. If ρ is density of resin then $D_v = \rho D_w$. The elution constant is related to D_v as $E = D_v^{-1} = \rho D_w^{-1} = \frac{d \cdot A}{V_{\max}}$

The separation factor (α) for two ions is evaluated as

$$\alpha_2^1 = D_1/D_2 = D_{w2}/D_{w1} = DV_1/DV_2 = E_2/E_1 = \frac{V_{\max(2)}}{V_{\max(1)}} \text{ if } d, A \text{ are common for both ions.}$$

If α = 1.0 no separation is possible.

We can correlate volume of mobile phase (V_m) with the volume of stationary phase (V_s) and distribution coefficient (D) as $V = D.V_s + V_M$ or in terms of weight distribution (D_w) coefficient

$$V_R = (D_w + 1)V_m \text{ and } D_w = \frac{V_R - V_M}{V_M}.$$

In chromatographic work, F the force of attraction between anion exchange site of resin and oppositely charged an ion separated by distance (v) is given as

$$F = \frac{\rho(\text{ion}) \times \rho(\text{resin})}{v2} = \frac{\rho(\text{ion}) \times \rho(\text{resin})}{v(\text{hydrated ion})2} \text{ if } \rho = \text{charge of ion.}$$

This is reflected in inverse square law in electroanalytical chemistry. Now suppose ion exchange of ions 'A' occurs on resin in form 'B'

$$A^+ + B_R \rightleftharpoons A_R + P \quad \text{so } K_B^A = \frac{[A]_R[B]}{[A][B]_R}$$

This K_B^A is called selectivity coefficient.

In order to get reasonably good separation K_B^A should be as large as possible. Further under non-idealised conditions we consider

$$K_B^A = \frac{[A]_R[B]}{[B]_R[A]} = \frac{X_{A_R}}{X_{B_R}} \frac{\gamma_{A_r}}{\gamma_{B_R}} \frac{C_B}{C_A} \frac{\gamma_A}{\gamma_B}$$ if γ represent activity coefficient of ion in resin phase and in solution.

We can easily derive selectivity scale for mono or multivalent metal ions based upon the knowledge of K_B^A for two ions or more.

Finally one more term generally used for, calculation of the concentration of an ion is given as per following equation in batchwise studies.

$$\frac{m\,\text{std}}{Q\,\text{std}} \frac{m - \text{resin}}{V\text{soln}} = K_d,$$ if m = amount of ion in solution.

V = volume of the solution.

Q = amount of resin,

K_d = distribution coefficient

This expression also facilitates evaluation of optimum volume of eluant.

It would be thus seen ion exchange chromatography gives separation in quantitative terms, we can verify experimentally with measurement of various factors to calculate value of others in the same expt. It is possible to theoretically prepare a model to predict possible expression for separation of two or more ions.

SOLVED PROBLEMS

12.1 Calculate the density of the resin (Dowex-1X8) when in typical separation peak elution volume was 75 ml, with 18 cm of column length with diameter of 0.80 cm. (Assume the weight distribution coefficient of resins is $D_w = 2.60$).

Ans: Since $V_{\max} = 75$ ml, $d = 18$ cm, $V = 0.08$ cm $A = \pi r^2 = 3.14 \times (0.40)^2$

$$E = \frac{dA}{V_{\max}} = \frac{18 \times 3.14 \times (0.40)^2}{75} = 0.48, \text{ now } \frac{1}{E} = D_v = \frac{1}{0.48} = 2.60$$

$$D_v = \rho D_w,\ \rho = \frac{D_v}{D_w} = \frac{2.60}{2.08} = 1.25 \text{ gm/ml.}$$

12.2 In anion exchange separation of lead as its anionic malonate complex on Dowex 21K resin in column with height and diameters of column as (1.4 × 24) cm, the elution constant was 0.60 calculate the peak elution volume in given system if $r = 0.70$ is radius.

Ans: Since $d = 24$, $A = (\pi r^2) = 3.14 \times (0.70)^2$, $A = 1.54$

Since $$E = \frac{d \cdot A}{V_{\max}} \text{ or } V_{\max} = \frac{d \cdot A}{E}$$ we substitute

$$V_{\max} = \frac{24 \times 1.54}{0.60} = 61.6 \text{ ml is peak elution volume.}$$

12.3 In cation exchange separation of copper on column with height 20 cm with diameter = 1.4 cm, packed with Dowex 50W X8 resin, the peak elution volume was 25 ml. If the free column volume (γ) was 10.3 ml. With density of resin as 0.80 gm/ml, calculate the elution constant as well as weight and volume distribution coefficients?

Ans: As $E = \dfrac{d.A}{V_{max}}$ or $E = \dfrac{d \cdot A}{(V_{max} - \gamma)}$ if γ = free column volume, d = 20 cm. As free column volume is 10.3 ml (25 – 10.3) = 14.7 ml is V_{max} we substitute;

$$A = (0.70)^2 \times 3.14 = 1.538 = \pi^2$$

$$E = \frac{20 \times 1.538}{14.7} = 20.089,\ D_v = \frac{1}{E} \text{ so } D_v = 0.478,\ D_w = \frac{D_v}{\rho} = \frac{0.478}{0.80} = 0.597$$

since V_{max} = 14.7.

12.4 Calculate the magnitude of N-nos. of theoretical plates and H-height of plate if column exit time was 9.34 mt after sample sorption with the width of elution curve W_e = 0.43 mt and height 18.3 cm?

Ans: $N = \dfrac{L}{N}$, $N = \left(\dfrac{d}{w}\right)^2$ now d = 9.34, w_e = 0.43, $N = 16\ (9.34/0.43)^2 = 7.55 \times 10^3$, $H = L/N$, if

L = 18.3 so $H = L/N$, = 18.3/7.55 × 10^3 plates = 2.42 × 10^{-3}.

12.5 Calculate the standard deviation of single band if standard deviation of band broadening process were 6.7 × 10^{-3}, 2.4 × 10^{-3}, 4.7 × 10^{-2}, 1.3 × 10^{-2}. Also ascertain portion in percentage each contribute to total value?

Ans: Since $\sigma^2 = \sigma_1^2 + \sigma_2^2 + \sigma_3^2 + \sigma_4^2 + ...\sigma_n^2$ and if portion contribution $\dfrac{100 \times \sigma_1^2}{\sigma^2}$.

we calculate both terms $\sigma^2 = \sigma_1^2 + \sigma_2^2 + \sigma_3^2 + \sigma_4^2 + ...$ ...(1)

$$\sigma^2 = (6.7 \times 10^{-3})^2 + (2.4 \times 10^{-3})^2 + (4.7 \times 10^{-2})^2 + (1.3 \times 10^{-2})^2 = 2.43 \times 10^{-3} \text{ cm}^2$$

$$\sigma = (2.43 \times 10^{-3})^{1/2} = 4.9 \times 10^{-2} \text{ cm.}$$

$$\text{Portion contributed} = \frac{100 \times \sigma_1^2}{(4.9 \times 10^{-2})^2} = 1.8\% \text{ if } \sigma_1 = 6.7 \times 10^{-3},\ \text{if } \sigma_2 = 2.4 \times 10^{-3}$$

$$\text{then } \frac{100 \times \sigma_2^2}{(4.9 \times 10^{-2})^2} = 0.2\%.$$

12.6 Calculate the value of weight and volume distribution coefficients if one gm dry Dowex 50W X8 resin packed with 20 ml of 0.1 M zinc chloride in column and then eluted when 10 ml of aliquat on titration with 0.01M EDTA showed end point at 2.40 ml (given ρ = 0.68 gm/ml)?

Ans: $$D_w = \frac{\text{moles of Zn in resin}}{\text{moles of Zn in soln}} = \frac{(20 \times 0.1) - (2 \times 2.4 \times 0.01)}{(2 \times 2.4 \times 0.01)} = 40.67$$

Since 10 ml aliquot was one half of total solution of 20 ml so in total solution

so $D_w = 40.67 \times 20 = 813$ gm/ml.

$D_v = \rho D_w$ if ρ = density of resin so $\rho = \dfrac{\text{wt. of resin}}{\text{vol. of resin}} = 0.68$

$D_v = 813 \times 0.68 = 5.6.$

12.7 What is the elution volume for two ions (*A*) and (*B*) if D_{c_A} and D_{c_B} are resp. 2.0 and 8.00 with bed volume. of band as 1.3 and 7.0 for ions *A* & *B*?

Ans: $V_R = (D_c V_S + V_m)$, we substitute the values

$V_{R_A} = (2 \times B) + 7 = 33$ ml, since $V_m = 7, D_{C_A} = 2, D_{C_B} = 8.0, V_S = B$

$V_{R_B} = (8 \times B) + 7 = 111$ ml, $V_A = 2$ and $V_B = 8$ ml.

12.8 The separation of alkaline earth on cation exchange column gave peak elution volumes as 1.5 (Mg), 3.68 (Ca), 6.10 (Sr), and 11.45 (Ba) ml. If $V_m = 1.57$ and with a column of (0.2×50) cm, what was weight distribution ratio and separation factor alkaline earths?

Ans: $D_w = \left(\dfrac{V_R - V_M}{V_M}\right)$ so weight distribution is 0 (Mg), 1.34 (Ca), 2.88 (Sr) and 6.29 (Ba) resp.

Now separation factors are

(α) $D_{sr}/D_{ca} = 2.15$, $D_{Ba}/D_{Ca} = 2.18$, $D_{Sr}/D_{Ba} = 2.88$, D_{Ba}/D_{mg} is best = 3.0

A good separation is expected for barium and magnesium.

12.9 What is the number of theoretical plates with column dead space of 100 and 1 ml/gm for two solutes having specific retention volume of 1.0 and 0.9 of stationary phase resp.?

Ans: As $\left(\dfrac{V_2^1}{V_1^1} - 1\right)\dfrac{V_2^1}{V_1^1} = 0.11/1.1.$ The partition ratio of second is 1.0

$N = \left(4 \times \dfrac{1.11}{0.111} \times \dfrac{2}{1}\right)^2 = 6400$ for second partition ratio is 0.01

$N = \left(4 \times \dfrac{1.11}{0.111} \times \dfrac{10.1}{1}\right)^2 = 1.6 \times 107.$ This high value is not favourable in practice.

12.10 Calculate retention volumes of two components if $V_s = 1.5$ and $V_m = 2.5$ ml with $D_1 = 5$ and $D_2 = 15$.

Ans: Since $V_{R_1} = (D_1 V_s) + V_m$ so $V_{R_1} = (5 \times 1.5) + 2.5 = 10$ ml.

$V_{R_2} = (D_2 V_S) + V_m$ so $V_{R_2} = (15 \times 1.5) + 2.5 = 25$ ml.

component (1) is first eluted and separated from component (2).

12.11 If numbers of plates is 1636 with column length of 20 cm. What is the magnitude of height of theoretical plate (HETP)?

Ans: $H = L/N$ if $L = 20$cm, $N = 1636$

$H = 20/1636 = 0.00122$ cm.

12.12 Calculate the volume capacity of the strongly basic anion exchange resin with 8% DVB having weight capacity of 3.8 mcq/g dry resin with its density as 1.11 g/ml and the water content is 42.48% with the void fraction is 0.4?

Ans: Since $Q_r = (1 - \beta)\, \rho Q_w\, (100 - S/100)$ = volume capacity

If $\beta = 0/V_b\ V_b$ – bed volume V_0 = void fraction, ρ density of resin = 1.11

S = water contents = 42.48 we substitute all values if Q_w = 3.8 wt. capacity

$$Q_v = (1-0.40)(1.11)\left(\frac{100-45}{100}\right)(3.8) = 1.4 \text{ m meq/ml.}$$

12.13 How much amount of metal will be sorbed by one gram of resin if cation exchange resin has capacity of 3 Meq/g when shaken with 50 ml of 0.001 M metal salt when it has selective coefficient of 2.567?

Ans: $K_a = \dfrac{[M^+]_R}{[M^+]} = 2.56\left(\dfrac{[H^+]_R}{[H^+]}\right)$

excess in resin is 50 × 0.001 = 0.05 Meq if $[H^+]_R$ = 2.95.

maximum value for $[H^+]$ = 0.001

the minimum value is K_d = (2.56) (2.95)/0.001 = 7550

$$\text{now} \quad \frac{\text{amt. of } M^+ \text{ in resin}}{\text{amt. of } M^+ \text{ in soln}} = 7550\left(\frac{1}{50}\right) = 151.$$

12.14 Ascertain the number of theoretical plates and height equivalent of a theoretical plate (HETP) for 50 cm column with peak elution time of 60 seconds, with 50% response and the width of elution curve as 8.3 seconds, with retention time 59 seconds.

Ans: We use following equation

$$N = \left(\frac{4t_R}{W_e}\right)^2 \quad \text{if } W_e = \text{plate width and } t_R = \text{retention time so}$$

$$N = \left(\frac{4\times 59}{8.3}\right)^2 = 8.0\times 10^2 \text{ theoretical plates. Now HETP is}$$

$$H = \left(\frac{L}{N}\right) = \frac{50}{8\times 10^2}\ (\text{since } L = 30,\ N = 8\times 10^2)\ H = 6.3\times 10^2 \text{ cm/plate.}$$

12.15 What is the asymmetric factor at 10% peak height? What are total nos. of theoretical plates and HETP of column with height of 180 cm, retention time is 6.40 mt with height = 3 units?

Ans: The symmetry factor is evaluated from equation

$af = \dfrac{B}{A}$ if af = symmetry factor, if B– portion of peak with width to the right if retention time to peak width which is left as A of retention time.

We have B = 0.64 A = 0.42

Symmetry factor $= \dfrac{0.64}{0.42} = 1.5$ as at 10% peak height peak with width on left is 0.42 and width on right it is 0.64 and W_1 is 1.06 mt.

The number of theoretical plates (HETP) if W_0 its peak height at 10% rate

$$N = \frac{41.7(t/W_1)}{af\,0.1+1.25} = \frac{41.7(6.40/1.6)}{1.5+1.25} = 5.5\times10^2 \text{ plate}$$

Now height of plate $= (H) = \dfrac{L}{N} = \dfrac{180}{5.5\times10^2} = 0.33$ cm/plate.

12.16 Evaluate the resolution factor between acetylsalicylic acid and salicylic acid peak if retention times were resp. 7.42 and 8.92 mt when widths of plate resp. were 0.87 and 0.91?

Ans: The resolution calculated from the following equation

$$R = \frac{t_2-t_1}{4\sigma} = \frac{t_2-t_1}{(W_1/2+W_2/2)} = \frac{2(t_2-t_1)}{(W_1+W_2)}$$ since we know all terms we have

$$R = \frac{2(8.92-7.42)}{(0.87+0.91)} = 1.69$$ there is very good separation as $R > 1.0$.

12.17 Ascertain the proportion of scandium and lanthanum in a solution in an ion exchanger if one ml of resin Dowex 50WX8 in equilibrium with one ml of solution (0.1 N) (w.r.t. metals) and 2N w.r.t. acetic acid. The value of $\log K_a$ = 3.5 for Sc and 2.7 for La (III)?

Ans: Since *M*stand × *M*soln is 3160a for Sc and 500a for La resp.
If a = *Q*std/*V*soln = 1; also (*M*std + *M*soln) = 0.1
From *M*std, *M*soln = 0.1 ÷ 3160 for Sc
*M*soln = 0.1 ÷ (500 + 1) for La.
In the solution ratio of [Sc]/[La] = 500/3160 = 1/63
so in Dowex 50WX8,

$$\frac{[\text{Sc}]}{[\text{La}]} = \frac{\left(1+\dfrac{1}{500}\right)}{\left(1+\dfrac{1}{3160}\right)} = 1.0$$

The ratio is 1 : 6.3 and other is 1 : 1.

12.18 What is concentration of ferrous ion in solution in batch experiment with equilibrium after shaking 100 ml of 0.1 N solution of ferric solution (2N w.r.t. CH_3COOH) with one gm cation exchanger Dowex 50WX8? (*D* evaluated from $\log k_d$ equal to 3).

Ans: We have for *m*—

$$\frac{M\text{std}}{1}\times\frac{M\text{soln}}{100} = 100m = \text{amt. of ion in resin or solution and also if}$$

$$\frac{M\text{std}}{Q\text{std}}\times\frac{M\text{soln}}{V\text{soln}} = K_d$$ if Q = amt. of resin, V = volume of solution K_d = dist. coefficient

so we have $\frac{M\text{std}}{1} \times \frac{M\text{soln}}{100} = 1000$ so we have = 0.1 *i.e.*, 10% it will be in solution Msoln

$M\text{std} = 100 \times 0.1 = 10$ mg/eq, $M\text{std} = 9.1$

$M\text{soln} = 0.9,\ C = \frac{0.9}{100} = 0.009$ N.

12.19 What is the amount of resin needed to separate calcium from 1000 ml of 0.1 N of chloride solution (Sp. volume of resin = 2.8 – 3.0 and exchanger capacity w.r.t. 0.1 N chloride was 4.3 – 4.99?

Ans: We calculate amount of resin from formulas

$(C_B/C_A)\text{std}(C_B/C_A)\ \text{sin} = K$ and $\frac{M\text{std}\ M\text{soln}}{Q\text{std}\ V\text{soln}} = K_d$ so we have

$0.1 \times 1000 \times 4.5 = 22.2$ gm; $22.2 \times 2.9 = 65$ ml.

12.20 What is the volume of a 0.07N solution of barium chloride which can be passed through 100 ml of resin Dowex 50W X 8 to breakthrough barium given exchange capacity = 1000-1300?

Ans: Volume of 0.07 ml barium chloride is calculated as

$$\frac{1200 \times 100}{0.07 \times 1000} = 1715 \text{ ml. } (120 \text{ mg eq. Ba}).$$

12.21 How many times should one gm of Dowex 50W X 8 containing 0.010 gm of ferric ion be washed by 20 ml of 2N hydrochloride so that not more than 0.05 mg of ferric ion was left in column (K_d = 3.16)?

Ans: Since $\frac{M\text{std}}{Q\text{std}};\ \frac{M\text{soln}}{V} = K_d = 3.16$

The first elution $\frac{100 - x_1}{1}\left(\frac{x_2}{20}\right) = 3.16$ so $x_1 = 8.6$ mg

$x_2 = (10 - 8.6) = 1.4$

Second elution $x_2 = 1.21$, $(1.4 - 1.21) = 0.19$

Third elution $x_2 = 0.16$, $(0.19 - 0.16) = 0.03$

Therefore three washings are necessary for the complete elution.

PROBLEMS FOR PRACTICE

12.22 What is the bed volume of resin and void column volume for a column (1 × 10 cm) containing 5gm of resin with density of 1.5? **[Ans:** $V_b = 10$, $V_r = 3.3$ ml, $V_0 = 6.67$]

12.23 What is volume capacity if resin of cation exchanger which has density of 1.25 gm/cm with the water content of 47%? **[Ans:** 2.2 mcq/ml of bed]

12.24 What is volume capacity for exchanger Dowex50W 2X8 with density of 1.09 gm/ml and water content (s) of 77? **[Ans:** 0.83 meq/ml of bed]

12.25 For separation of alkali metals like rubidium and cesium on resin (8% DVB) with diameter of 0.05 m, Calculate numbers of plates necessary for 0.1% DVB resin and state also kind of column required. **[Ans:** $\alpha = 1.03$, $N = 40,000$, $L = 10$ m]

12.26 State how many plates are needed for separation of sodium and potassium in equal amount and find out also column height if $H = 0.025$ mm? **[Ans:** $N = 23{,}000$, $L = 57.5$ cm**]**

12.27 With 2-4 hr as retention time on passing through column for 23 minutes it had 0.08 M of sodium. When column was washed with 0.080 M solution then traces of radioactive sodium was detected indicate time when this solution would come out of column? **[Ans:** 2.4 hr after Sorption.**]**

12.28 The specific retention volumes for two solutes is 19 and 20 ml/g of stationary phase, if with dead space of 1 ml/g, and 200 ml/g, what would be the plate number required to procure resolution of at least 2 and 0.75? **[Ans:** (*a*) 2800, 3×10^6 (*b*) 3970, 4.4×10^5**]**

12.29 If distribution ratio is 8.4 what is the peak elution volume (V_R) on 60 × 100 mm ion exchange column volume of stationary phase (Vs) 1.5 ml with volume of the mobile phase (V_m) as 1.3 and the flow rate being 2 ml/mt? **[Ans:** 13.9 ml**]**

12.30 Calculate the amount of potassium nitrate if 2gm of it with soap solution was dissolved in water and if 10 ml of this solution when titrated with 0.111 M potassium hydroxide required 15 ml for end point? **[Ans:** 70.8% KNO_3**]**

12.31 On passing 40 ml of 0.15 M sodium chloride through Dowex 50WX8 and collecting 250 ml of an effluent find out concentration of hydrochloric acid after elution in the flask? **[Ans:** 9.3 M hydrochloric acid**]**

12.32 Calculate the distribution coefficient of ions A and B gm of resin if one gram was taken with 10 ml of 0.25 and if two solutes A and B were added to separate bottle concentration of 10^{-4} M after shaking 58-84% of *x* and 32.26% B remains in solution? **[Ans:** $K_{dx} = 7.0$. $K_{dy} = 21$**]**

12.33 What is the pk_a of ionogenic group if end point in titration was at 9.2 mg/g, if at half conversion point pH of external solution was 6.4 with 0.1 M sodium chloride with water contents of 60%? **[Ans:** = 4.9**]**

12.34 A silicate (0.3gm) solution on passing on Dowex resin (25 × 2.4 cm), of mesh 0.005 cm exchange capacity 1.7 meg/cm and subsequently eluted with 0.7 M acid at flow of 0.55 cm/mt. What is the V_{max} for alkali metals on column? **[Ans:** V_{max} 140 (Li), 283 (Na^+) 522 (Mg^{2+})**]**

12.35 Show that elution curve for Mg-K would overlap with 1M acid as eluant instead of 0.7M acid in above problem? **[Ans:** $D_{Mg} = 3.8$, $D_K = 3.9$ with 1MHCl**]**

12.36 If an ion exchange resin had 1.65 meg/cm column bed. Calculate bed total number of exchange group per methyl of bed? **[Ans:** 10^{21} exchange group**]**

12.37 A column 108 cm long with resin radius of 2.4×10^{-3} cm D_v, for an organic acid was 1.37 at pH 2.9, $D_v = 3.3 \times 10^{-7}$, $D_f = 1 \times 10^{-5}$, $v = 0.4$.
What is plate height and plate number if velocity of flow is 0.00 783 cm/sec? **[Ans:** $H = 0.00545$, $N = 19.800$**]**

12.38 (*a*) Arrange the following ions according to their probable order of elution from cation exchange column containing Dowex 50WX8.
Na^+ K^+ Li^+ Cs^+ Rb^+ **[Ans:** Li^+ Na^+ K^+ Rb^+ Cs^+**]**

(*b*) Predict the probable order of elution of the following ions from cation exchanger filled column containing Amberlitee IR-120:
Ca^{2+}, Ba^{2+}, Mg^{2+}, Be^{2+}, Sr^{2-} **[Ans:** Be^{2+} Mg^{2+} Ca^{2+} Sr^{2+} Ba^{2+}**]**

12.39 (*a*) Predict the order of attraction of the following anions to an anion exchange column Dowex-21K for Cl^- Br^- I^- F^- **[Ans:** F^- Cl^- Br^- I^-**]**

(*b*) Predict the order of sorption of following anions on the anion exchange column filled with Dowex 1X8:

NO_3^-, Cl^-, SO_4^{2-}, PO_4^{3-}, I^- ? **[Ans:** PO_4^{3-}, SO_4^{2-}, NO_3^-, Cl^- and I^-]

12.40 (*a*) The amount of free acid in aluminium salt solution is hard to analyse as it hydrolyses and precipitate if titrated with sodium hydroxide. How will you analyse free HCl in $AlCl_3$ of known aluminium contents?

[Ans: Determine Aluminum by EDTA titration, pass solution on cation exchange resin and titrate liberated acid.]

(*b*) Titanium (IV) forms a complex anion with fluoride as TiF_6^{2-} copper is cation and does not form complex with fluoride. Show how titanium and copper can be separated from one another using anion exchange column? In case of all copper with traces Titanium (IV) how you would separate them?

[Ans: The reaction on anion exchanger column shall be

$$\underset{\text{(sorption)}}{RNH_3\,X + TiF_6^{2-} \rightarrow RNH_3 - TiF_6^{2-}} + HX^1,\ \underset{\text{Resi soln}}{RNH_3 - X + CuCl_2}$$

$$\rightarrow \underset{\text{resin}}{RNH_3 - X} + \underset{\text{soln}}{CuCl_2}\ (\text{no reaction})$$

In case of traces of titanium it is better to use cation exchange separation method as

$$H_2R + CuCl_2 \rightarrow \underset{\text{(sorption)}}{Cu - R} + 2HCl,\quad H_2R + TiF^{6-} \rightarrow H_2R + TiF_6^{2-}\ (\text{no reaction})$$

REFERENCES

12.1 S.M. Khopkar, *Basic Concepts of Analytical Chemistry,* 3rd edn., New Age International Publishers (P) Ltd. (2007).

12.2 D.A. Skoog & D.M. West, *Analytical Chemistry,* 3rd edn., Holt Reinhart & Winston, pp. 368-70, 507 (1979).

12.3 H.A. Lattinen & W.E. Harris, *Chemical Analysis: An Advanced Text,* 2nd edn., McGraw-Hill & Co., pp. 468, 489, 496 (1975).

12.4 J.S Fritz & G.H Schemt, *Quantitative Analytical Chemistry,* Allyn & Bacon In, pp. 417-19, 427-8 (1979).

12.5 J.A. Dean, *Chemical Separation Methods,* Van Nostrand Reinhold & Co., pp. 96-8, 120-5 (1970).

12.6 D.G. Peter, J.M. Hayes & G.H. Hjeftje, *Chemical Separations & Measurements,* Saunders Golden Series, p. 598 (1974).

12.7 D. J Pietrzyke & C.W. Frank, *Analytical Chemistry: An Introduction.* Academic Press, p. 583 (1979).

12.8 O. Samuelson, *Ion Exchange Separations in Analytical Chemistry,* John Wiley, 2nd edn. (1963).

12.9 H.F. Walton & W. Reiman, *Ion Exchange in Analytical Chemistry,* Pergamon Press & Co. (1970).

12.10 R. Kunin, *Ion Exchange Resin,* 3rd edn., John Wiley and Sons (1958).

12.11 Y. Marcus & A.S. Kerles, *Ion Exchange and Solvent Extraction of Metals,* John Wiley (1969).

12.12 A. Inczedy, *Analytical Applications of Ion Exchange Resin,* Pergamon Press Ltd. (1970).

12.13 S.M. Khopkar, *Solvent Extraction Separation of Elements with Liquid Ion-Exchangers*, New Age International Publishers (P) Ltd. (2006).

Chapter 13

Ion Chromatography

INTRODUCTION

Analytical chemistry witnessed rapid growth of instrumental chromatographic methods of separation such as gas chromatography; high performance liquid chromatography; capillary electro-phoresis: and supercritical fluid chromatography. However the separation of the cations on large scale was not possible by conventional ion exchange chromatography in spite of the fact that such methods were selective. There was paucity of rapid methods for the analysis of cations after elution. In gas chromatography of the volatile compounds were analysed by, ECD FID and TCD detectors. While in HPLC technique most commonly used methods are consisted photometry, amperimetry or refractive index. However with the discovery of suitable detectors for analysis of cations, in ion chromatography excellent separation of multielement mixtures was possible in few minutes.

Ion chromatography represents the separation of cations as well as of anions in short time. The eluants used are very dilute. The detectors for measurement of electrical conductivity meter, colour of complex with photometric detector or measurement of diffusion current with an amperometric detector was possible. Since the signals provided by the weak eluants are considerably small eluants subsequently passed on suppressor column which is capable of releasing eluant having measurable physical property. These in turn are automatically recorded on plot to know actual concentrations. The analysis of signals metals as well as anions was neglected. In fact, methods for quantitative analysis were unknown. Thus only recently electroanalytical methods such as ion selective electrodes facilitated analysis of ions. With discovery of ion chromatography a new era in field of environment speciation was opened with brisk activity in study of toxicity and health. The separation of as many as 10 elements could be accomplished in just 6-8 minutes. The cations and anions were separated at nanogram concentrations in record time say in fraction of seconds.

In separations it is absolutely essential to use designed pelliculer resins which would withstand pressure in the column during separation. The resins were chemically similar to copolymer of styrene and 8–16% divinly benzene. However best resins were obtained with one containing 1-5 percent DVB linkage. Such resins were hydrophobic. The eluants in separation of anions were very weak while in cation exchange separation these eluants consisted of organic carboxylic acids with varying pK values. The most important component of system was a suppressor column which produced eluants containing cation with high electroconductivity. For organic anions fluorimetric detector was used but for the analysis of cations conductivity or photometric detectors were used. The chromogenic ligand, *e.g.*, PAR *i.e.*, 4-(2 pyridyl azo) resorcinol was most favoured reagent for complexation due to its selectivity to complexation of metals at different pH. Very useful separations were carried out like separation of five components of anions or six to eight components of cations. Most interesting feature was the facility of environmental specification enabling one to understand toxicity of metals.

In this chapter several multiple choice questions are provided to test the knowledge of basic principles, classification, structure of resin, kind of eluant used, role of suppressor column, different kinds of the detectors available for the separation and many illustrations for the separation of cations as well as anions from complex mixture. The application mainly cover on environmental speciation.

PROBLEMS

Multiple Choice Questions

Tick (√) mark the correct answer from the following multiple choices given below:

13.1 Ion exchange technique was less popular in comparison with GLC or HPLC because of the
(*a*) lack of detectors for analysis of ions after elution.
(*b*) non-selective methods of separation.
(*c*) unstable ion exchange resin.
(*d*) separation process was slow.

13.2 Ion chromatography facilitates rapid separation of
(*a*) cations
(*b*) anions
(*c*) cations and anions both simultaneously
(*d*) neutral compounds.

13.3 Ion chromatography separation involves one of the followings process
(*a*) exchange of ions (*b*) exclusions of ions
(*c*) pairing of ions (*d*) adsorption of ions.

13.4 The technique involving sorption of ions on stationary or neutral supports is called as
(*a*) ion exchange chromatography (*c*) ion pair chromatography
(*b*) ion exclusion technique (*d*) ion retardation method.

13.5 Ion chromatography is most popular because
(*a*) technique is very cheap.
(*b*) it can handle constituents at milligram level.
(*c*) it needs large volume of eluant for separation.
(*d*) it permits simultaneous analysis of multielements from sample.

13.6 The ion exchange resin used in ion chromatography
(*a*) is same as one used in columnar work
(*b*) is specially treated resins
(*c*) having very high cross-linking in the structure
(*d*) are pellicular with coating on glass balls with conventional resins.

13.7 Most of the resin used in ion chromatography contains
(*a*) 4% cross-linking of DVB (*c*) 8% cross-linking of DVB
(*b*) 16% cross-linking of DVB (*d*) less than 5% of DVB.

13.8 The eluant used in separation of anions in ion chromatography are:
(*a*) very dilute (*c*) highly viscous
(*b*) very concentrated (*d*) very mobile.

13.9 The eluants used in separation of cations by ion chromatography are:
(*a*) complexing agents needing no pH control
(*b*) eluants workable only at elevated temperature
(*c*) diluted complexing ligands like organic carboxylic acids
(*d*) mineral acids.

13.10 The most important criterion for selection of eluant for purpose of separations of cations by complexation is
(*a*) solubility (*c*) molecular weight
(*b*) pk value (*d*) number of carboxylic groups present on ring.

13.11 The common eluant with highest pK_1 value of 3.04 is
(*a*) oxalic acid (*c*) tartaric acid
(*b*) benzoic acid (*d*) succinct acid.

13.12 An eluant after emerging out from exchange column is then passed through suppressor column to produce an eluant is
(*a*) pure solution (*c*) volatile eluant
(*b*) containing complexing ions (*d*) two eluants.

13.13 The mechanism involved in passage of eluant through suppressor column in cation separation leads to release of:
(*a*) strong anion with high conductivity
(*b*) strong cation with low conductivity
(*c*) weak anion with no conductivity
(*d*) a cation with high conductivity.

13.14 The kind of detector used during ion chromatography separation of amino acid is
(*a*) conductometry (*c*) fluorimetry
(*b*) photometry (*d*) amperiomity.

13.15 The chromogenic ligand used commonly in photometric detector consists of
(*a*) PAN, *i.e.* 1(2-pyridylazo napthol) (*c*) TTA (2 Thenoyltrifluoroacetone)
(*b*) PAR 4-(2-pyridylazo resorcinol) (*d*) HDEHP bis (2 ethylhexyl phosphoric acid).

13.16 The solutions emerging out of the suppressor column consists of:
(*a*) NaCl (*c*) $NaHCO_3$
(*b*) NaBr (*d*) HBr.

13.17 The column in ion chromatography is made of
(*a*) brass (*c*) glass
(*b*) teflon (*d*) steel.

13.18 The greatest contribution of ion chromatography was to facilitate separation in environment of:
(*a*) anions (*c*) isotopes
(*b*) cations (*d*) charged radicals.

13.19 The work on environment spaciation was greatly facilitated with the discovery of:

(*a*) ion chromatography (*b*) high performance liquid chromatography
(*c*) supercritical fluid chromatography (*d*) capillary electrophoresis.

13.20 Ion chromatography facilitated separation of anions in

(*a*) less time with small volume of eluant
(*b*) more time with large volume of eluant
(*c*) with any eluant with coloured ion
(*d*) eluant with high viscosity.

13.21 The quantitative analysis of anions with ion chromatography is comparable to their analysis by

(*a*) polarography (*c*) ion selective electrode
(*b*) spectrophotometry (*d*) conductometry.

13.22 The kind of detector used in separation of iron, copper, and nickel by ion chromatography is

(*a*) photometric detector (*c*) amperiometric detector
(*b*) conductometric detector (*d*) refractive index detector.

13.23 Environmental speciation facilitates the close study of the

(*a*) metabolism and toxicity of pollutants on health
(*b*) characterisation of anions
(*c*) detection of cations
(*d*) comparison between cations and anions.

13.24 Ion chromatography permits separation of

(*a*) two ions (*c*) many ions
(*b*) four ions (*d*) one ion.

13.25 Ion chromatography separation of twelve ions is possible in just

(*a*) one hour (*c*) one minute
(*b*) four hours (*d*) ten minutes.

13.26 The most useful reaction necessary for speedy detection by conductance measurement is one of the following:

(*a*) $HR + NaHCO_3 \rightarrow NaR + CO_2 + H_2O$

(*b*) $HR + NaCl \rightarrow NaR + HCl$

(*c*) $HR + Na_2CO_3 \rightarrow NaR + H_2O + CO_2$

(*d*) $HR + Na_2C_2O_4 \rightarrow NaR + H_2C_2O_4$

ANSWER

13.1 (*a*)	13.2 (*c*)	13.3 (*a*)	13.4 (*a*)	13.5 (*d*)	13.6 (*d*)
13.7 (*d*)	13.8 (*a*)	13.9 (*c*)	13.10 (*d*)	13.11 (*c*)	13.12 (*b*)
13.13 (*d*)	13.14 (*c*)	13.15 (*b*)	13.16 (*d*)	13.17 (*b*)	13.18 (*a*)
13.19 (*a*)	13.20 (*a*)	13.21 (*c*)	13.22 (*a*)	13.23 (*a*)	13.24 (*c*)
13.25 (*d*)	13.26 (*b*).				

REFERENCES

13.1 S.M. Khopkar, *Environmental Pollution Analysis,* Second Edition, New Age International Publishers (P) Ltd. (2008).

13.2 J. Wesis, *Handbook of Ion Chromatography*, Dionex Company Manual (1983).

13.3 J. Tarter, *Advanced Ion Chromatography*, Wiley Interscience (1989).

13.4 E. Sawicki, J. D. Mulik, E. Wittgen Stein, *Ion Chromatography Analysis Environmental Pollutants*, Vol. I, An Arbor Science Pub. (1978).

13.5 W. F. Rich, F. Smith, L. McVeill, G.T. Sidebottom, *Ion Chromatography Analysis of Environmental Pollutants*, Vol. II An Arbor Science Michigan (1979).

13.6 S.M. Khopkar and M.N. Gandhi, *Indian Journal of Environmental Protection*, **14**, 256 (1994).

13.7 S. Cherian, M.N. Gandhi, S.M. Khopkar, *Indian Journal of Environmental Protection*, **12**, 324 (1992).

13.8 S.M. Khopkar, *Basic Concepts of Analytical Chemistry*, Third edn., New Age International Publishers (P) Ltd. (2007).

13.9 J. Janta, *Principles of Chemical Sensor*, Plenum Press Ltd., New York (1990).

13.10 J.E. Saloman and D. K. Hale, *Ion Exchange*: *Laboratory Manual*, Butterworths, London (1959).

13.11 B.L. Krager, L.R. Snyder, C. Horwith, *An Introduction to Separation Science*, Interscience (1980).

13.12 D.G. Peters, J.M. Hayes, J.G.M. Hieftje, *Chemical Separation and Measurements*, Saunders Golden Series (1974).

13.13 S.M. Khopkar, *Environmental Pollution, Monitoring and Control,* New Age International Publishers (P) Ltd. (2006).

Chapter 14

Adsorption Chromatography

INTRODUCTION

Amongst various chromatographic methods used, the adsorption chromatography is less frequently practised in inorganic analysis, although it finds extensive applications in organic analysis for the separations of organic compounds.

There are three kinds of forces which are acting between solute and the stationary support. These forces originate from weak Van der Waals interaction, or are due to dipole induced dipole interaction or what is called as 'London forces'. By taking the advantage of difference in forces for different species it is possible to develop columnar separations.

This is mainly due to differential migration rates of various solutes on the stationary support in the column due to difference in adsorbility of ions on column.

The adsorption isotherm gives idea about the amount of solutes adsorbed per unit weight of adsorbent (x/w). From a solution of concentration 'c' at equilibrium so we have distribution ratio as:

$K_d = \frac{(x/w)}{c}$ here x = amount of solute adsorbed on the column, c = equilibrium concentration

w = weight of adsorbent. The equilibrium is attained as adsorbed layer consists of mono layer covering the entire adsorbent surface volume (V_a) as:

$V_a = (3.5 \times 10^{-8}$ cm × surface area in cm^2/g − 0.001% water vapour if present)

In above expression 'V_a' - represents monolayer covering the adsorbent surface volume.

Further according to Snyder, the surface energy function or adsorbent activity (ϕ) and solute eluant parameter (f) are correlated by an expression as:

$\log K_d = \log V_a + \phi f$ where f = solute eluant parameter and ϕ = adsorbent activity, K_d = dist. coefficient, V_a = Volume covering adsorbent.

The value $V_a = (3.5 \times 10^{-8}$ cm × surface area − 0.001% of water vapour) is used.

These expressions facilitate us to evaluate either the distribution ratio (K_d) or adsorbent surface volume (V_a) and adsorbent activity or surface energy (ϕ) 'n the system.

The classical example of adsorption chromatography is the gas-solid chromatography which exclusively operates on the principle of adsorption phenomena. Adsorption chromatography is experimentally easier to interpret as it provides more uniform, reproducible pattern of separation. Adsorption is sensitive to steric difference for similar molecules. One can easily alter the polarity of mobile phase to have separation. The technique can be used for separation of large amount of samples. One can exploit the difference in polarity or structure to develop chromatographic differential migration fronts. The kind of stationary support used in order of increasing adsorption power is: Cellulose > Starch > CaO > Silica > Charcoal > MgO > Al_2O_3.

The general adsorption sequence of functional groups in order of increasing polarity and consequently increasing adsorbility is as:

Acids < bases < hydroxyl compounds < aldehydes < ketones < esters < hydrocarbons. The eluotropic series of common solvent used as the mobile phase is as follows: cyclohexane > carbon tetrachloride > toluene > benzene > chloroform > acetone > acetaldehyde > acetic acid > water.

The large number of separations like the mixture of aniline with 80 per cent acetonitrile was separated by adsorption chromatography.

As regard the determination of (N) number of the theoretical plates or (*H*) plate height one can use same formula as discussed in the earlier chapters. One must know height of the column for this purpose. It is also possible to evaluate the resolution or separation factor (α) for separation of two complex mixture having more or less similar chemical characteristics.

SOLVED PROBLEMS

14.1 Calculate the adsorption activity (ϕ) of the fully activated alumina whose surface area was measured by nitrogen adsorption and was 155 m^2/g when $K_d = 68$ with pentane as the standard eluant in system (Given - deactivated alumina has 3% water and $K_d = 3.3$).

Ans: By Snyder the surface energy function or a adsorbent activity (ϕ) and eluant parameter (*f*) are related as .... $\log K_d = \log V_a + \phi f$.

If $V_a = 3.5 \times 10^{-8}$ cm = $(155 \times 10^{-4}$ $cm^2/g = 0.0542$ cm^3/g because V_a the adsorbent surface volume is given by $V_a = [3.5 \times 10^{-8}]$ (surface area) in cm^2/g for adsorbent when $V_a = [3.5 \times 10^{-8} \times$ surface area) – 0.001% water when so the quantity of water is added now with ϕ = 1.0 for an adsorbent eluate surface parameter is evaluated as $\log K_a = \log + \phi f$, *i.e.*, log 68 = log 0.0542 + ϕ × 3.10 (if *f* = 3.10). Now for activated alumina $V_a = (3.5 \times 10^{-8})(155 \times 10^4) - 0.001$ (3) = 0.0242 cm^2/g.

The experimental value of $K_d = 3.3$ for deactivated alumina so log 3.3 = log 0.0242 + 3.10 ϕ which gives ϕ = 0.69.

14.2 What was the height equivalent of theoretical plate (HEPT) for a column with 200 cm in length with net number of plates as 1636?

Ans: Since $H = \dfrac{L}{N}$ if *L* = 200 cm, *N* = 1636 we have $H = \dfrac{200}{1636} = 0.122$ cm is height equivalent of theoretical plate.

14.3 Calculate the efficiency of the column in terms of total plates and height if (V_{max} *i.e.*, peak elution volume was 9.34 when width of the base of the elution curve (*W*) was 0.43.

Ans: Since $N = 16\left(\dfrac{V_R}{W}\right)^2$ we have $V_R = 9.4$, W = 0.43 so $N = 16 = \left(\dfrac{9.4}{0.43}\right)^2 7.55 \times 10^3$ is the absolute number.

Now $H = L/N$ if $L = 18.3$ cm $N = 7.55 \times 10^3$ then we have

$H = 18.3/7.55 \times 10^3 = 2.42 \times 10^{-3}$ cm/ plate is HETP.

PROBLEMS FOR PRACTICE

14.4 What was the partition coefficient (K_d) of a particular solute-eluant water deactivated silica (V_a = 0.342 ml/g) from the data given as: adsorbed phase (g/g) 1×10^8, 549, 153, 38, 15.3, 3.82 non-adsorbed phase (g/ml) 1×10^6, 61, 52.5, 13.4, 5.36 and 1.32? [**Ans:** K_d = 8.35]

14.5 Calculate the K_d value for adsorbent (*a*) X = 4.6% H_2O- SiO_2 (*b*) X = 15% H_2O - SiO_2. The experimental data value for K_d with pentane as eluant from X = 6.9% H_2O - SiO_2 is 45.6 if V_a = 0.23 and ϕ = 0.69. [**Ans:** 7.3, 16.2, 21.4 ml]

14.6 If activated silica gel with surface area of 1.55 m^2/g and K_d = 68 with naphthalene as the solute and pentane an eluant was used had f = 3.10 while K_d values were 30, 17, 7.1 3.3 with corresponding water as 0.5, 1.0, 2.0 3.0%.

Find out the adsorbent surface volume (*i.e.*, V_a) as well as the adsorbent activity (ϕ) of the alumina.

[**Ans:** V = 0.049, ϕ = 90 (0.5%) gel, V_a = 0.44, ϕ = 0.84 (1% gel), V_a = 0.034ϕ = 0.75 (2%) gel, V_a = 0.024ϕ = 0.69 (2%) gel, V_a = 0.014ϕ = 0.63 (2%) gel.]

14.7 What was the number of layers equivalent to each step of Brackmann activity for a alumina sample with specific surface area of 123M^2/g?

[**Ans:** (*a*) 69.93 (*b*) 0.0946 (*c*) 6.61 cm/sec.]

14.8 Explain the reasons for the stronger adsorption of 2-amino anthriquiones on silica in comparison with 1-isomer. [**Ans:** Intermolecular H-bonding is feasible for 1-isomer]

14.9 Explain why there was preferential adsorption of o-dihydroxybenzene and o-hydroxyl and o-aminobenzene on alumina? [**Ans:** Chelate formation with aluminium atoms]

14.10 How would you explain that the pyrrole derivatives on alumina during a desorption indicate that the bulky groups adjacent to *N*-actually increases in a desorption?

[**Ans:** The acidity of these compound is increased when *N-H* bond is twisted or strained.]

14.11 Indicate which one between 2-vinyl, 2-napthyl, napthalene is preferably adsorbed on silica and alumina. [**Ans:** The last compound]

14.12 What were the reasons that the very weakly adsorbing solutes showed high linear, capacity but pure compounds showed lower linear capacity.

[**Ans:** This was due to the difference in polarity]

REFERENCES

14.1 S.M. Khopkar, *Basic Concepts of Analytical Chemistry,* 3rd edn., New Age International (P) Ltd. (2007).

14.2 J.S. Fritz and G.H. Schenk, *Quantitative Analytical Chemistry,* Allyn and Bacon Inc., pp. 3, 70, 507 (1979).

14.3 J.A. Dean, *Chemical Separation Methods,* Van Nostrand Reinhold Co. pp. 157-8 (1970).

14.4 D.G. Peters, J.M. Hayes and G.M. Hieftje, *Chemical Separations and Measurements,* Saunders Series, p. 389 (1974).

14.5 B.L. Karger L.R. Snyder, C. Horwith, *An Introduction to the Separation Science*, Wiley-Interscience (1974).

14.6 L.R. Snyder, J.J. Kirkland, *Introduction to Modern Liquid Chromatography,* Wiley-Interscience (1974).

14.7 L. Turkora, *Affinity Chromatography*, Elesevier Pub. Co. (1980).

14.8 A.V. Kiseler Y-1 Vashi, *Gas Adsorption Chromatography*, Plenum Press (1969).

Chapter 15

Partition Chromatography

INTRODUCTION

The entire credit for popularising chromatography technique for the separation goes to the technique partition chromatography. The simple technique like paper chromatography, thin layer chromatography (TLC), liquid-liquid partition chromatography (LLPC) reversed phase partition chromatography (RPPC), gas liquid chromatography (GLC), etc., were very useful. The basic principles underlying these techniques is to exploit the difference in partition coefficient of solute between stationary and the mobile phase to set up differential migration front so as to facilitate the separation. Those of substance with high K_d would remain on the column in stationary phase (which is generally hydrophobic) while those with least value for K_d would flow down (hydrophilic) mobile phase. However in case of reversed phase partition chromatography the phases are reversed, *viz.*, stationary phase is hydrophobic while mobile phase is hydrophilic. Therefore it is called as extraction chromatography.

Now the efficiency of a paper chromatography technique is generally evaluated in terms of R_f Value (*i.e.*, retardation factor) defined as

$$R_f = \frac{\text{distance travelled by solute front}}{\text{distance travelled by solvent front}}$$

For columnar methods if V_M = volume of mobile phase, V_s = volume of the stationary phase V_R = peak elution volume so $V_R = (V_M + K_d V_S)$ if K_d = distribution coefficient, we also have $(V_R - V_M)$ is adjusted retention volume and $[V_R] \times \frac{V_M}{W_L}$ is specific retention volume if W_L = wt. of solvent on stationary phase

$(K_d V_S = V_M)$ as $V_M/V_S = 0.5$ taken in column or work. For carbohydrate separations we use terms

$R_M = \log\left(\frac{1}{R_f} - 1\right)$ and $R_M = \left[\frac{-\Delta H}{2.3RT} + C\right]$ if

H = heat of partition, R = gas constant, T = absolute temp, C = constant.

When we measure area in terms of A_M, A_S then

$R_f = \left[\dfrac{A_M}{A_M + K_d A_S}\right]$ with usual notation A_M = area of mobile phase, A_S = area of stationary phase.

There are several advantages of using partition chromatography. The partition provides larger resolving power. It is more suitable for separation of low concentration of a mixture. The partition depends on solubility so small difference in molecular weight will promote good separation (*e.g.* sep. of homologus series of compounds). The partition coefficient is independent of the concentration there by providing linear isotherm. The chemical structure and partition coefficient are interrelated. The kind of stationary support used consists of kiselghur silicagel, cellulose, starch, diatomaceous earth, etc. The solvent combination used for normal partition work are in pair for stationary to mobile phase as water/alcohol; water/hydrocarbon; water/chloroform; water/ethyl acetate formamide/MEK; glycol/pyridine. In reversed phase extraction chromatography however we use extractants as the stationary phase. These are TBP, TBPO, TOPO, liquid ion exchangers, HDEHP, in combination with water or water containing mineral acids or salts as mobile phase. The organic solvents are used sometimes with suitable diluents. The support is made hydrophobic by exposure of stationary support to vapour of diethyl dichlorosilane and is subsequently coated with one of these extractants in rotary vacuum evaporator.

Apart from separation of element by partition chromatography or that of pesticide by TLC, large number of metals were separated by using extraction chromatography, as an analogy can be drawn with liquid-liquid extraction. Many of the separation which were not quantitative by solvent extraction were successful with extraction chromatography.

Finally let us consider inter relationship between V_M, V_S, V_R, K_d and R_f. It is $\dfrac{R_f}{V_m} = \dfrac{C_m V_m}{C_S V_S}$,

$$R_f = \frac{V_m}{(V_m + K_d V_s)} = \frac{t_m}{(t_m + K_d t_s)}.$$

If t_m, t_S is time spent in mobile and stationary phase resp, V_m, V_s are volume of mobile and stationary phases, we can carry out numerous separation of metals by PC, TLC, LLPC and extraction chromatography *i.e.*, reversed phase partition chromatography.

SOLVED PROBLEMS

15.1 In paper chromatography separation of silver, lead and mercury the solvent front was 18 cm while front due to respective metals were 16 (Ag), 12 (Pb) and 6 (Hg) cm. What is the R_f value of these metals?

Ans: In planner chromatography retardation factor is:

$R_f = \dfrac{\text{distance travelled by solute front}}{\text{distance travelled by solvent front}}$ as solvent front is 18 cm while solute front varies for each metal.

So $R_f = \dfrac{16}{18} = 0.88$ for Ag, $R_f = \dfrac{12}{18} = 0.66$ for lead, $R_f = \dfrac{6}{18} = 0.33$ for mercury.

15.2 In thin layer chromatography separation if R_f value of the unknown compound is 0.819. While front due to analogus compounds for A, B, and C are resp. 24, 28, 30 with solvent front of 34 cm can you identify the unknown compound whether it is A, B or C?

Ans: For $A = \frac{24}{34} = 0.70\ R_f$ for

For $B = \frac{28}{34} = 0.82\ R_f$ for

For $C = \frac{30}{34} = 0.88\ R_f$ for

Since unknown compound has $R_f = 0.82$ the compound is same as B.

15.3 In separation of gallium, indium, thallium by paper chromatography with tributyl phosphate has respective front 20, 30, 40 cm with solvent from as 50 cm. An unknown compound has R_f value of 0.6, what is the unknown compound?

Ans: We first calculate R_f value for all metals, as solvent front = 50

Gallium $\frac{20}{50} = 0.40$ Indium $\frac{30}{50} = 0.60$

Thallium $\frac{40}{50} = 0.80$ Unknown $\frac{x}{50} = 0.60, (ZP = 0.6 \times 50 = 30)$

This shows unknown ion is Indium.

15.4 What was value for R_f for paper chromatographic separation if distance from centre of spot to starting point is 3 cm and distance from start to solvent is 15 cm?

Ans: We calculate

$$R_f = \frac{\text{distance travelled by solute}}{\text{distance travelled by solvent}} = \frac{3}{15} = \frac{1}{5} \text{ i.e., so } R_f = 0.20.$$

15.5 Evaluate the R_f for thin layer chromatographic analysis on plate when the ratio of volume of mobile to stationary phase was 0.33, if $D = 5.0$.

Ans:
$$R_f = \frac{1}{1 + 5 \times 0.10/0.33} = 0.40 \text{ as } R_f = \frac{1}{1 + D\left(\frac{V_M}{V_S}\right)}$$

When $\left(\frac{V_s}{V_m}\right)$ = ratio of vol. of stationary to mobile phase $\left(\frac{0.10}{0.33}\right)$, $D = 5.0$.

15.6 In paper chromatography, 10 ml fraction were separated as follows:

Fraction No.: I, II, III, IV, V, VI

C (mg/ml): 0.1, 0.2, 0.3, 0.2, 0.18, 0.2

Fraction No.: VII, VIII, IX, X, XI, XII

C (mg/ml): 0.3, 1.7, 5.1, 7.0, 0.3, 0.2

Calculate the content of each component in the substance?

Ans: $Q_1 = (0.1 + 0.2 + 0.3 + 0.2 + 0.2 + 0.18) \times 10 = 35.9$ mg.

$Q_2 = (0.3 + 1.7 + 5.1 + 7 + 0.3 + 0.2) \times 10 = 94.9$ mg.

$$P_1 = \frac{35.9 \times 100}{(35.9 + 94.9)} = 27.5\%$$

$$P_a = (100 - 27.5) = 72.5\%.$$

15.7 What was the number of theoretical plates and height of plate for 30 cm column, if distance from the starting point is 5 cm and peak width (m) = 0.5 cm?

Ans: If $N = 16\left(\frac{l}{m}\right)^2$ l = dist from start to maximum spot

m = peak width, N = Nos. of theoretical plates

$16\left(\frac{5}{0.5}\right)^2 = 1600$ number of theoretical plate (N)

The height of theoretical plate is $H = \frac{L}{N}$ if L = 30 cm, N = nos. of plate = 1600

$H = \frac{30}{1600} = 0.002$, is height = 0.002 HEPT.

15.8 In thin layer chromatographic separation of four pesticide the fronts were 50 mm, 100 mm, 200 mm, and 150 mm. The solvent front was 250 mm. The unknown pesticide had R_f = 0.600 Calculate R_f for four pesticide and identify unknown pesticide?

Ans: We first calculate R_f for all pesticides

1st pesticide 50/250 = 5/25 = 0.20

2nd pesticide 100/250 = 10/25 = 0.40

3rd pesticide 150/250 = 15/25 = 0.60

4th pesticide 200/250 = 20/25 = 0.80

unknown compound had R_f = 0.60. Hence it is 3rd pesticide.

PROBLEMS FOR PRACTICE

15.9 What was the percentage recovery of lead in two solvents if separation of lead with hexanone is 99% lead was stripped with 50ml water, of (20 × 2) ml of solvent A or (10 × 5) ml of solvent B as eluant? **[Ans:** (*a*) 22% (*b*) 5.8%]

15.10 Can you identify the unknown aminoacid from the front developed in paper chromatography if R_f values were asparetic acid (0.14), glutamic acid (0.24), alaninc (0.59) and proline (0.90). If the solvent front was 100 cm with unknown front 90 cm? **[Ans:** Proline R_f = 0.90]

15.11 Calculate the retention index if an unknown compound had an adjusted retention time of 11-23 mt while the adjusted retention times of octadecane and eicosane were resp. 10.5 and 10.65 mt.? **[Ans:** 1840.4]

15.12 Calculate R_f value of a compound (*A*) if it migrated 7.6 cm in paper chromatography with solvent front of 16.2 cm. In case the solvent front was 14.3 cm instead of 16.2 cm indicate spot for location of the peak. **[Ans:** 0.47. 0.71 from start]

15.13 In partition chromatographic separation of organic compounds *W*, *X*, *Y*, *Z*, the fronts were 5.7, 16.6, 14.3, 10.2, while solvent front was 18 cm. Calculate R_f and identify a compound which had R_f of 0.79. **[Ans:** Unknown compound is second *Y*]

15.14 Uranium, magnesium, aluminium, were separated by paper chromatography and analysed as coloured oxinate complexes by spectrophotometer, with standards of 100 ppm the absorbance for complex of uranium was 0.720, for complex of magnesium was 0.550 and complex of aluminium was 0.800 with unknown mixture the absorbance of the complex for uranium was 0.150, for magnesium was 0.840 and for aluminium it was 0.750. What is the percentage composition of the uranium in the mixture? [**Ans:** 24.92%]

15.15 In the analysis of malonic acid by thin layer chromatography following results were obtained

malonic acid con (mg/ ml)	5	10	15	20	25
wt. of paper (mg/ml)	78	123	154	172	194

When 100 mg of sample was dissolved in 10 ml of distilled water and three (0.02 ml) samples were chromatographed the corresponding spots had masses of 105, 98, 109 mg. Calculate percentage of malonic acid in sample? [**Ans:** 3.7 0.4%]

15.16 It was observed that with phenol as mobile phase and water as the stationary phase proline migrated faster than valine but with *n*-butanol valine moved faster than proline. Can you explain this anomaly? [**Ans:** adsorption increases in order listed]

15.17 What was pellicular support? Is it advantages over usual support?
[**Ans:** hard sphere with porous outside layer. It is more efficient (lower) than ordinary supports of the similar particle size.]

15.18 Indicate the number of excitation wave length, which can be used when anthracene and phenanthrene were separated by TLC on non-fluorescent alumna plate, will the colour of phenanthrene spot appear to be greenish yellow on excitation?
[**Ans:** 36.5 mm greenish yellow]

15.19 Is it possible to improve the resolution when a dye was chromatograhed on a 0.5 cm column with 75% methanol and 25% water and if, all constituents were eluted within ten minutes with no satisfactory resolution? [**Ans:** Use eluant with less methanol or alternatively use gradient elution technique for the purpose of separation.]

15.20 How will you calculate the efficiency of column in terms of H and N if it was 9.3 mt after sample feeding with base of elution curve as 0.43 mt with column height 18.3 cm?
[**Ans:** $N = 7.5 \times 10^3$ and $H = 2.42 \times 10^{-3}$ cm/pt]

REFERENCES

15.1 S.M. Khopkar, *Basic Concepts of Analytical Chemistry,* 3rd edn., New Age International Publishers (P) Ltd. (2007).

15.2 D.A. Skoog and D. M. West, *Analytical Chemistry,* 3rd edn., Holt, Reinhart and Winston, p. 525 (1979).

15.3 D.G. Peter, J. M. Hayes and G. M. Hjeftje, *Chemical Separation and Measurement,* Saunders Golden Series pp. 361, 388 (1974).

15.4 J.A. Dean, *Chemical Separation Methods,* Van Nostrand Reinhold Co., p. 179 (1970).

15.5 E. Heftman, *Chromatography,* Van Reinhold and Co. (1970).

15.6 E.W. Berg, *Physical and Chemical Methods of Separation,* McGraw-Hill and Co. (1963).

15.7 J.M. Bobbitt, *Thin Layer Chromatography,* Van Nostrand, Reinhold and Co. (1963).

15.8 E. Stahl, *Thin Layer Chromatography – A Laboratory Handbook,* Academic Press (1964).

15.9 J.J. Kirkland, *Modern Practice of Liquid Chromatography,* Wiley Interscience (1971).

15.10 M. Leaderer, *Paper Electrophoresis,* Elsevier Pub. Co. (1955).

Chapter 16

Gas Chromatography

INTRODUCTION

Amongst the various chromatographic methods, gas chromatography has become most popular for several reasons. The first important reason is it can be applied for the separation of volatile and thermally stable compounds. Secondly we have now efficient detectors like ECD, FID or TCD to determine the components quantitatively. Thirdly there is good provision to alter temperature of column or modify rate of flow of gas through the column in programmed temperature or flow programing in gas chromatography resp. One can handle very small concentration of the components. We can get columns containing several kinds of stationary support to accomplish clean-cut separations. Since the entire method is based upon exploitation of difference in partition coefficient of solute between the stationary and liquid phase, we have choice to change the stationary liquid phase to create large difference in the partition coefficient to accomplish the clean cut separation by differential migration of the components. We have scope to vary length of the column which is in the form of coiled tube to increase *(N) i.e.,* number of theoretical plates; and height of the theoretical (Δ) plate. In gas chromatography, we get several peaks, the area under each peak represents the total concentration of a particular component. The sum of all areas in each peak leads to the calculation of the percentage composition in the mixture.

In gas liquid chromatography, several terms are very frequently used to express various parameters, we see their interrelationship. If V_R = retention volume and t_R = retention time, R_C = flow of gas in column, V_R' = retention volume of second component, t_{air} = retention time of air, V_M = volume of mobile phase, g = compressibility factor, K_d = distribution coefficient, K = is partition ratio, V_S = volume of stationary phase, C_M = con. of mobile phase, C_S = con. of stationary phase, W_L = weight of stationary phase, ρ_L = density of liquid phase, V_R = specific retention volume, R = gas constant, q = sharpness of peak, W = volume, N = nos. of plates, and t_C = column temp. of operation V_N = net retention adjusted volume.

We have various equations relating to each another as:

$V_R = t_R R_C$ (from flow rate of gas, we can calculate retention volume)

$V_R' = \left(t_R R_C - t_{air} R_C\right) = \left(V_R - V_M\right)$ gives retention volume of second component

$V_N = gV_R^1$ or $g = \left(\dfrac{V_R^1}{V_N}\right)$ if g = represent compressibility factor,

Since $K_d = (C_S/C_M)$ with usual notation the partition ratio (k) is

$k = K_d\left(\frac{V_s}{V_m}\right)$ as $V_R C_M = \left(V_M C_M + V_S C_S\right)$ from general equation we have

$\therefore V_R = \left(V_M + K_d V_S\right)$ or $\left(V_R - V_M\right) = K_d V_S = K_d\left(\frac{W_L}{\rho_L}\right)$. Now specific retention volume is

$V_R = \left(\frac{237}{t_c}\right)\left(\frac{V_N}{W_L}\right)$ is volume of carrier gas required to remove 50% of the component. The retention volume in terms of K_d shall be

$$K_d = V_r P_L\left(\frac{T_c}{273}\right) \text{ or } V_R = \left(\frac{273}{P/L}\right)\left(\frac{K_d}{t_C}\right)$$

The specific retention volume is $V_R = \left(\frac{KV_s}{W}, \frac{273}{t_c}\right)$ if $P_L = \left(\frac{W}{V_S}\right)$ the low temperature favours better retention on the column. Now for column efficiency we have

$q = \left(\frac{V^1 R_C}{W}\right)$ if q = relative sharpness of peak if W = volume so number of theoretical plates (N) in column are given by equation as

$N = 16\left(\frac{V_R^1}{W}\right)^2 = 16q^2$ we can also get plate height (H) as

$H = \frac{Q^2}{L}$ if Q^2 = standard deviation L = length of the column

$H = \frac{Q^2}{L} = \frac{L_f^2}{t_R}$ if f is variance in time.

The criteria followed for selection of stationary liquid phase is, the solvent must have high boiling point, with different solvent power for various components. The stationary liquid phase used be non-volatile and thermally stable and must be chemically inert. The inert gas as nitrogen/helium are used as carrier. The gas must be perfectly inert, cheap and should give good detector response. In case we use internal standard during analysis then chromatographic peak of standard should be different from that of internal standard. The retention time (t_R) for standard and analyt sample must be close to each another. The concentration of standard should be similar with that of analytical sample.

If F_m is corrected flow rate, it can be given by an expression $F_c = \left(\frac{F_m t_c}{t_R}\right)$ is adjusted flow rate

If P_o = pressure at outlet *i.e.*, atmospheric pressure and
P_W = vapour pressure of water, then we have flow rate (F_c) as

$$F_c = \left(\frac{F_m t_c (P_o - P_w)}{t_R P_o} \right)$$ But retention volume V_R can be evaluted from F_c

$V_R = F_c t_R$ if F_c is adjusted flow rate while $P = \left(\frac{P_o}{J} \right)$ is average pressure in column when pressure drop factor (j) is given as:

$$j = \frac{3}{2} \left[\frac{(P_1/P_o)^2 - 1}{(P_1/P_o)^3 - 1} \right]$$ if P_1 = inlet pressure, P_o = outlet pressure *i.e.*, atmospheric pressure

$$V_R = V_N = \left(\frac{273}{W_L t_C} \right) = \frac{273 F_C (t_R - t_M) j}{W_L t_C}$$ if W_s = wt. of stationary phase.

Another term very commonly used in gas liquid chromatography is "Koval retention index (I)". It is the ratio of the retention of the sample component to the retention of chain hydrocarbons which elutes before and after a sample component.

$$I = 100 \left[\frac{\log(t_R - t_M) - \log(t_N - t_M)}{\log(t_{N+1} - t_M) - \log(t_N - t_M)} \right] + 100n.$$

In terms of volume it is;

$$I = 100 \left[\frac{\log(V_R - V_M) - \log(V_N - V_M)}{\log(V_{N+1} - V_M) - \log(V_N - V_M)} \right] + 100n$$ which gives equation on simplification.

$$I = 100 \left[\frac{\log \alpha_s^n}{\log \alpha_{n+1}} \right] + 100n$$ if t_n, V_n refers to saturated hydrocarbons.

α = relative retention of sample w.r. to hydrocarbons.

This terms (I) is useful with column operating at fixed temp. This "Kovals index" is used to find out hydrocarbon number (n) which would elute from the column. If I = 6000 – 7000 m/s this shows compound elutes between 6-carbon and 7 carbons atom.

Large number of problems have been given based upon the above theoretical considerations. A reader is however advised to refer to number of monographs listed at the end of this chapter.

SOLVED PROBLEMS

16.1 When equal volumes of monochloro, dichloro and trichlorobenzene were chromatographed on gas chromatograph, the areas under each peak were resp. 6.36, 7.55 and 7.45cm^2. If the detector response was uniform which show that % area tally with the weight % composition of the solvent if their densities were 1.106, 1.305 and 1.288 g/ml? (The molecular weights are known)

Ans. 6.37 + 7.55 + 7.45 = 21.36 cm^2

$$\% \text{ monochlorobenzene} = \frac{6.36}{21.36} \times 100 = 29.0\% \text{ M.W.} = 112.0$$

$$\% \text{ dichlorobenzene: } \frac{7.55}{21.36} \times 100 = 35.34\% \text{ M.W.} = 147.0$$

$$\% \text{ trichlorobenzene: } \frac{7.45}{21.36} \times 100 = 34.87\% \text{ M.W.} = 181.5$$

The ratio is same for all three compounds.

16.2 What was the percentage composition of two species (x) and (y) if in gas chromatographic separation of two species x and y with density of 1.32 and 1.60g/ml, area under each peak was 348 and 532 mm^2/ml. For standard sample of one micro liter the corresponding peaks for x and y were 675, 1088mm^2?

Ans: A gives peak 675 so 348 gave $\frac{645}{348} \times 1 = 1.94$ ml.

Now species B gives peak 1088 so 532 gave $\frac{1088}{532} \times 1 = 2.5\,\mu l$

Since densities were given so weight can be calculated as (weight ÷ volume = density)

x = density 1.32, volume 1.94µl has weight (1.32 × 1.94) = 2.56 µg

y = density 1.60, volume 2.05µl has weight (1.60 × 2.05) = 3.28 µg

$$\% \ x = \frac{2.56 \times 100}{5.84} = 43.8\%, \quad \% \ y = \frac{3.28 \times 100}{5.84} = 56.2\%$$

16.3 What was the height equivalent of theoretical plate and total number of plates, if the elution peak (V_{max}) was 9.34 mt after sample injection with width of curve 0.43 mt (W_e) and column length of 18.3m?

Ans: Since $N = 16$ (d/w) if $d = 9.34$, $w = 0.43$, or $N = 16\left(\frac{V_{max}}{W_e}\right)$

$$N = 16\left(\frac{9.34}{0.43}\right) = 7.55 \times 10^3 \text{ number of plate.}$$

$$H = \frac{L}{N} = \frac{18.3}{7.55 \times 10^3} = 2.42 \times 10^{-3} \text{ cm/plt height of each plate } i.e., \text{ HETP.}$$

16.4 In gas chromatographic separation of benzene, toluene, xylene the areas were 31, 14.5 and 53.2 cm^2. Calculate % composition of the mixture?

Ans: $\% \text{ benzene} = \frac{31}{98.7} \times 100 = 31.4\%$ if total area is 98.7 cm^2.

$$\% \text{ toluene} = \frac{14.5}{98.7} \times 100 = 14.69\%$$

$$\% \text{ xylene} = \frac{53.2}{98.7} \times 100 = 53.9\%$$

16.5 What is Koval's retention index (I) with following data: retention time: air (0.22), hexane (4.13), sample (5.20) heptane (6.35) $n = 6$?

Ans: Koval's retention index is given by equation

$$I = 100\left[\frac{\log(t - t_m) - \log(t_n - t_m)}{\log(t_{n+1} - t_m) - \log(t_n - t_m)}\right] + 100\,n;\ t^m_{\text{air}} = 0.22,\ t^n_{\text{hexan}} = 4.13,\ t^{n+1}_{\text{heptan}} = 6.35$$

as $t_m = 0.22$, $t_n = 4.13$, $t = 5.20$ for sample

$$I = 100\left[\frac{\log(5.20 - 0.22) - \log(4.13 - 0.22)}{\log(6.35 - 0.22) - \log(4.13 - 0.22)}\right] + 100(6) = (653 - 8) = 645$$

16.6 What is the percentage of ethane, butane, propane, and pentane in mixture with following data?

Gas	*ethane*	*butane*	*propane*	*pentane*
Peak area (V)	5	7	5	4
K	0.60	0.77	1.0	1.11
V/K	5/0.6 = 8.43	7/0.77 = 9.1	5/1 = 5.0	4/1.11 = 3.6 **Total = 26**

Ans: We calculate V/K and hence % as shown in the above table, Total % = 26.0.

$$\%\text{ ethane} = \frac{8.43 \times 100}{26.0} = 32,\ \%\text{ butane} = \frac{9.1 \times 100}{26.0} = 35,\ \%\text{ propane} \frac{5 \times 100}{26} = 19,$$

$$\%\text{ pentane} = \frac{3.6 \times 100}{26.0} = 14.$$

Hence, ethane = 32%, butane = 35%, propane = 19%, pentane = 14%.

16.7 What was retention volume, corrected retention volume, relative corrected volumes w.r.t. propane with following data:
Peak dist. area ethane (14) propane (21.3), butane (35.5)
The velocity of carrier gas was 3 liters at temp of 2.5°C?

Ans: We calculate the values as:

Retention	*ethane*	*propane*	*butane*
V^1R (ml)	10.7	17.7	29.5
V_{RX}/V_{Ry}	0.65	1	1.67

$$14 \times \frac{3000}{3600} = 10.7, \left(\text{as } \frac{3000}{3600} = 21.3\right)$$

$$14 \div 21.3 = 0.65,\ 35.5 \div 21.3 = 1.67.$$

16.8 If number of theoretical plates were 7.55×10^3 with HETP = 2.42×10^{-3} cm/plate, what is the elution peak for solute at column exit if width of base peak is 0.43 mt and state the length of the column?

Ans: Since $N = 16\left(\frac{d}{W}\right)^2$ we have $N = 7.55 \times 10^3$, $W = 0.43$, $d = x$

$$7.55 \times 10^3 = 16\left(\frac{x}{0.43}\right) \text{ so } x = 9.34 \text{ mt elution peak}$$

also $H = L/N$ if $H = 2.42 \times 10^{-3}$ and $N = 7.55 \times 10^3$ we have $(H \times N) = L$ *i.e.,*

$7.55 \times 10^3 \times (2.42 \times 10^{-3}) = 18.3$ cm is length.

16.9 What was percentage composition of the mixture of ethane, propane and butane if in gas chromatographic separation, the peak areas were 53.2, 14.5 and 31.0 cm resp?

Ans: Total area 53.2 + 14.5 + 31 = 98.7

$$\% = \frac{53.2 \times 100}{98.7} = 53.9,\ \% = \frac{14.5 \times 100}{98.7} = 14.69,\ \% = \frac{31 \times 100}{98.5} = 53.9$$

Hence percentage were
ethane = 53.9%, propane = 14.69%, butane = 53.9%.

16.10 Calculate the volume at 760 mm pressure and at temp. 275° if gas at 27°C and 740 mm of mercury had a volume of 172.0 ml?

Ans: $PV = nRT \quad \frac{P_1V_1}{T_1} = \frac{P_2V_2}{T_2}$

$$\frac{740 \times 172}{(273 + 27)} = \frac{760 \times V_2}{(275)} \text{ so } V_2 = 152.4 \text{ ml.}$$

16.11 Calculate the volume of a gas at 760mm and constant temperature if the volume of gas was 200 ml at 742 mm pressure?

Ans: $P_1V_1 = P_2V_2$ so

$742 \times 200 = 760 \times V_2$

$V_2 = 195.3$ ml.

16.12 What will be the volume of gas at 57°C at constant pressure if at 37°C volume of gas was 620 ml?

Ans: $\frac{P_1V_1}{T_1} = \frac{P_2V_2}{T_1}$ since $P_1 = P_2$. Constant $\frac{V_1}{T_1} = \frac{V_2}{T_2}$

$$\frac{620}{(37 + 273)} = \frac{V_2}{(57 + 273)}. \text{ Hence } V_2 = 660 \text{ ml.}$$

16.13 What was the percentage composition of mixture of carbon dioxide and nitrogen when corresponding partial pressures were 340 and 560 mm?

Ans: (340 + 560) = 1000 Total pressure of the system.

$$\frac{340 \times 100}{1000} = 34\% \text{ for } CO_2;\ \frac{560 \times 100}{1000} = 56\% \text{ for } N_2$$

16.14 Ascertain the partition of calcium carbonate to be taken to prepare 12.3 ml of carbon dioxide at 18°C and 740 mm pressure?

Ans: At NTP 12.3 ml CO_2 will have volume of

$$\frac{P_1V_1}{T_1} = \frac{P_2V_2}{T_2}$$

$$V_1 = \frac{P_2V_2T_1}{P_1T_2}$$

$$V_o = \frac{740 \times 12.3 \times 273}{760 \times 291} = 11.2 \text{ ml.}$$

$$\text{Amount of calcium carbonate } \frac{\text{CaCO}_3 \times 11.2}{22.4} = 0.050 \text{ gm.}$$

16.15 What volume of ammonia gas was liberated at N.T.P. by decomposition of 1 gm of calcium nitride in water?

Ans: $Ca_3N_2 + 3H_2O = 3CaO + 2NH_3$ so $Ca_3N_2 = 2NH_3$. If x = vol. of ammonia

$$x = \frac{1 \times 2 \times 22.4}{\text{Ca}_3\text{N}_2} = 302 \text{ ml.}$$

16.16 What was the percentage composition of mixture by weight if mixture of gas had volume composition as: 18% CO_2, 13% O_2, and 69% N_2?

Ans: The composition by weight was calculated as

$$\text{CO}_2 = \frac{18 \times 44}{22.4} = 35.3; \quad \text{N}_2 = \frac{69 \times 28}{22.4} = 86.25$$

$$\text{O}_2 = \frac{13 \times 32}{22.4} = 18.6 \text{ Total amount in weight } (35.3 + 80.3 + 18.6 = 140.2)$$

$$\% \text{ CO}_2 = \frac{35.3 \times 100}{140.2} = 25.2\%, \ \% \text{ O}_2 = \frac{18.6 \times 100}{140.2} = 13.3\%,$$

$$\% \text{ N}_2 = \frac{80.3 \times 100}{140.2} = 61.5\%.$$

16.17 Indicate contraction in volume if 40 ml of hydrogen is burnt in oxygen?

Ans: For 1 vol of H_2 we need 1/2 vol of oxygen

$H_2 + \frac{1}{2}O_2 \rightarrow H_2O$ so total amt. of compression of the volume of mixture is equal to sum of vols. of H_2 and O_2 or $\frac{3}{2}$ volume of H_2 so $40 \times \frac{3}{2} = 60$ ml.

16.18 What will be contraction in volume of 25 ml of methane if burned in oxygen?

Ans: $CH_4 + 2O_2 = CO_2 + 2\ H_2O$ so 2 vol of O_2 need 1 vol of CO_2 to give 1 vol of CH_4 *i.e.*, the doubled volume of CH_4 is taken for analysis so 80/2 = 40 ml of gas expanded.

16.19 Calculate the percentage composition of the mixture of gases if 16 ml volume of mixture containing methane, hydrogen and ethylene after explosion in air was 26 ml?

Ans: If $CH_4 + H_2 + C_2H_6 = 16$ ml volume is decreased by $\frac{3}{2}(CH_4) + 2(H_2) + \frac{3}{2}(C_2H_6)$

$$CH_4 + \frac{4}{3}H_2 + C_2H_6 = 17.33\ (17.3 - 16) = \frac{1}{3}H_2 = 1.33 \text{ ml.}$$

$H_2 = 4$ ml *i.e.*, 25% of whole volume.

16.20 Calculate the retention index (*I*) of the gas (*x*) if the adjusted retention time (t_R) for a compound *X* was 11.23 mt while for compound *Y* and compound *Z* it was resp. 10.50 and 14.05 mt.

Ans: Since $\frac{a}{b} = \frac{A}{B}$ (similar triangles law) *I* unknown = $(1800 + b)$

$$\text{So } \alpha = \log 11.23 - \log 10.50 = 0.0292 = (\log A_{tR} - \log B_{tR})$$

$$A = \log 14.65 - \log 10.50 - 0.1446 = (\log B_{tR} - \log B_{tR}) \text{ if } t_R \text{ is retention time}$$

$$B = (2000 - 1800) = 200$$

$$B = \frac{ab}{A} = \frac{0.0292 \times 200}{0.1446} = 40.4.$$

I for unknown = $1800 + b = 1840.4$.

PROBLEMS FOR PRACTICE

16.21 What was the adjusted retention volume, net retention volume and specific retention volume at 0°C for benzene which when passed at flow rate of 48 ml/mt with retention time of 52 mt, with $P_0 = 75.8$ and $P_1 = 122.2$ on a column 25% dinonylphthalate (80 mesh) chromesorb column with 91.54 m column having 2.5dnpdinonyl pH – ($d = 0.9712$) with 7.5gm, chromesorb?

[**Ans:** $V_R^1 = 952$, $V_N = 739$, $V_a = 247$]

16.22 Calculate the number of plates needed to get fraction impurity content of 1% if two peaks have retention areas of 0.7 and 0.3 and relative retention ratio of 1.04 when fraction impurity is 0.01? [**Ans:** *H* = 14,000 Plates]

16.23 Calculate the net sum of the plates required to effect separation with 0.1% impurity in adjacent band when α is unity by using either ordinary column, or tubular column (*K* = 0.5) or open tubular column (*K* = 3)? [**Ans:** 4360-6240, 39,200, 7740 nos. of plate]

16.24 Calculate the carrier gas velocity and the minimum plate height if helium as the carrier gas had: *A* = 0.0 cm, *B* = 0.30 cm/sec and *C* = 0.05 in Van Deemeter's equation?

[**Ans:** *H* = 0.345 cm, *u* = 2.45 cm]

16.25 Calculate the column pressure if gas chromatograph which was operated with inlet pressure of 45 gauze pressure and outlet presssure with one atmospheric? [**Ans:** 2.84 atmosphere]

16.26 What will be the rate of the retention time if the amount of stationary phase applied to a solid support was doubled? [**Ans:** The t_R will be also doubled]

16.27 A sample of fruits weighing 3.2045 gm was extracted in hexane and made up to one liter. The aliquat on injecting in gas chromatograph gave area of 32.6 m^2. A standard (0.541 mg/ml) gave peak of 41.7m^2, if 5 μ*l* sample was injected. Calculate the amount of hexane in sample of the fruits? **[Ans:** 13.2 ppb]

16.28 What was the retention time of higher hydrocarbon sample, the retention time for methane is 25 sec, propane is 1.2 mt and higher hydrocarbon is 6.35 mt on column operating at 150°C? **[Ans:** 45.28 mts.]

16.29 Calculate retention index at peak if toxic compound was stripped at 155.2°C with temperature programming from 100-200 at 2° C/mt. The temp of stripping solution of *n*-tetradeacane and *n*-hexadecane were 141 and 162° C resp.? **[Ans:** 1535]

16.30 What was the retention index of compound which was injected with *n*-teracosane and n-hexacosane showing retention time of 10.2, 9.81 and 11.56 mt? **[Ans:** 2448]

16.31 Calculate the highest temperature of column when peak width was 25 second if the adjusted retention time was 24.21 and 17.21 mt at 180, 200 resp. if N = 5000 plates? **[Ans:** 259°C]

16.32 Was it possible to strip out ester in 45 mts at 438°K with 22 acid has t_R = 72 and 52mt at 180 and 200°C? **[Ans:** No as t_R^1 = 52.8>]

16.33 What was the percentage composition of the mixture if it contains oxalic, malonic, adipic and palmitic acids on chromatograph gave areas resp. 4.3, 3.8, 12.5 and 12 m^2? **[Ans:** 19.7, 17.4, 57,4. and 5.5% resp.]

16.34 Calculate the percentage composition of the mixture when mixture of acetone and diacetone when analysed gave area under each peak as 5.6 and 1.2 cm^2. Assume the major peak is due to diacetone. **[Ans:** 82.4% and 76.7%]

16.35 What was the percentage composition the mixture if 2 mg of sample gave peaks for hexane 35 cm^2, for toluene 80 cm^2, for methybutylketone 15 cm^2 and for the heptanol 20 cm^2? **[Ans:** 22.8, 12.3, 9.9, and 54.9% resp.]

16.36 Find out value of α for acetone and xylene with pentane as standard if relative retention time for acetone is 150, for xylene is 9000? **[Ans:** α for acetone = 1.06]

16.37 Calculate the corrected retention volume with retention time of 18.72 mt., when gas chromatograph was isothermally operated at 165°C with helium flow of 24 hr/mt at 23°C and 740 pressure with inlet pressure of 25 psi? **[Ans:** 335.7 ml]

16.38 Calculate the corrected retention volume time if column was loaded with 1.27 gm of stationary support as 3% on chromesorb when density of latter was 0.9 gm/ml and net retention volume was 273 ml? **[Ans:** 6779]

16.39 Calculate the retention index of decane on another column instead of given column, if the retention indext of 2-decane on a column is 1150 with same data given in earlier problem? **[Ans:** 1258]

16.40 Calculate the respective molar sensitivities in flame (FID) ionisation detector for decanane for trimethyl pentane, decanane, diethyl adipate? **[Ans:** 10, 10.5, 9.0 and 8.0 resp.]

REFERENCES

16.1 S.M. Khopkar, *Basic Concept of Analytical Chemi· ·ry,* 3rd edn., New Age International Publishers (P) Ltd. (2007).

16.2 J.A. Dean, *Chemical Separation Methods,* Van Nostrand Reinhold and Co., pp. 265, 268-9 (1970).

16.3 A.J. Willeth, D.Keaty, *Gas Chromatography,* John Wiley and Co. (1978).

16.4 M.L. Lee, F.J. Young, K. Bartic, *Open Tubular Gas Chromatography,* John Wiley (1984).

16.5 W. Jennings, *Gas Chromatography with Glass Capillary Column,* 2nd edn., Academic Press (1980).

16.6 R.L. Grob, *Modern Practise of Gas Chromatography,* John Wiley (1985).

16.7 G.M. Message, *Practical Aspect of Gas Chromatography—Mass Spectrometry,* John Wiley (1984).

16.8 S. Dal Nogare, R.S. Juvet, *Gas Liquid Chromatography*, Wiley Interscience (1982).

16.9 J.H. Purnell, *Gas Chromatography,* John Wiley (1962).

16.10 D.W. Grant, *Gas Liquid Chromatography,* Van Nostrand Reinhold Co. (1971).

16.11 D.Y. Pietzyke and C.W. Frank, *Analytical Chemistry,* Introduction Academic Press, pp 539, 547 (1979).

16.12 D.C. Peters, J.M. Hayes and C.M. Hjefije, *Chemical Separation and Measurement* Saunders, pp. 361, 389–90, 596–6 (1974).

16.13 D.K. Skoog and D.W. West, *Analytical Chemistry*, 3rd Ed., Holb. Reinholt & Winston, p. 497 (1979).

16.14 A. Muskin, *Problems in Quantitative Analysis,* Mir Publishing House, Moscow, pp. 182, 206 (1984).

Chapter 17

High Performance—Liquid Chromatography

INTRODUCTION

Gas Chromatography facilitated the separations of volatile and the thermally stable compounds with variety of detectors; but the separations of nonvolatile and thermally unstable compounds could not be analysed rapidly. This was possible only with the technique of the high performance liquid chromatography. It facilitated separation of organic, inorganic and biological compounds; including polymer and chiral compounds. The impurities could also be analysed with HPLC. The determination of the neutral, ionic or zwitterionic molecules was also possible. Further very complex separations could be accomplished by HPLC. It was non destructive method of analysis involving qualitative as well as quantitative analysis of compounds.

In biochemical analysis aminoacids, proteins of physiological origin could be analysed at level of ultratrace to preparative scale. The analysis of the pesticides and insecticides was very easily carried out. In environmental monitoring it was possible to analyse several pollutants by HPLC.

It was necessary to use the sample in liquid form which was then injected into the system. Solid samples have been brought into the solvent as the solution before analysis. Such solvent should be compatible with both the mobile and stationary phase. About 5-10μ*l* (in small lot) is injected by hypodermic needle into the system. The nature of the detector and its corresponding sensitivity determined the quantum of the sample. A sample preparation may or may not always involve dilution, preconcentrations, filtration, extractions, ultrafiltration or derivatizations. This process was adopted depending upon the nature of the sample to be analysed.

The overall process of analysis was rapid as within 2-5 mts one could get final results on the composition of the samples. If the sample preparation is specialised one then it needed 10 min to 2 hrs. The kind of the popular detectors used in HPLC is refractive index detector. The most critical step in operation was the retention of high pressure in the system. A pressure of 100 psi was employed during the separation. This in turn facilitates rapid elution. The assembly includes a precolumn which facilitates initial removal of the impurities before the fractionation process. The HPLC technique can be used for any kind of chromatographic separations discussed earlier. Few techniques like ion exchange needed some resin modification as it was essential to use pellicular resin. These are the resins which are coated on glass beads to maintain same height of the column. Irrespective of absence or presence of compression due to high pressure, the height remains same balancing some number of the theoretical plates in column. One could also use silica or alumina as the stationary support in separations.

There are few limitations of this technique which must not be ignored. In the absence of mass spectra in the hyphenated methods, the quantitative analysis and identification of compounds is cumbersome because the resolution is poor. The best results are possible if only one sample at a time is analysed. As compared to capillary electrophoresis or super critical fluid chromatography, time of analysis is relatively longer, and in most of the separations of the samples good preparation is must to get reproducible results.

This hyphenated technique in combination with mass spectrometer gives excellent results. It not only facilitates identification of the compound but also permits its quantitative analysis. Apart from this, elucidation of structure as well as molecular weight determination is quite possible using HPLC-MS. In addition to the refractive index one can also use uv-visible detector. The fluoroscence and electrochemical detectors like conductivity detectors gives reliable results. Variety of solvents like benzene, chloroform, dioxane, acetone, ethylalcohol and methylalcohol are commonly used in the separations. In combination with the conventional chromatographic methods one could also use HPLC in ion pair chromatography, ion chromatography; affinity chromatography and chiral chromatography. It is used in biotechnology and pharmaceutical analysis. Good separation of as many as 20 components could be carried out in just 30 minutes. Interesting separations of lithium, sodium, ammonium, potassium, calcium and magnesium was carried out with EDTA solution (0.1mm)/3M nitric acid as the eluants.

PROBLEMS

Multiple Choice Questions

Tick mark the correct answer from the following multiple choices indicated:

17.1 HPLC technique is used for the separation of

(*a*) ionic substances (*b*) non-ionic substances
(*c*) neutral compounds (*d*) volatile compounds

17.2 HPLC is resorted for the separation of the

(*a*) non-volatile compounds (*b*) pure compounds
(*c*) chiral compounds (*d*) cationic compounds

17.3 To perform HPLC separation, sample must be in the

(*a*) liquid state (*b*) solid state
(*c*) gaseous state (*d*) pure condition

17.4 The volume of the sample injected in HPLC column is as small as:

(*a*) 10-100 μ*l* (*b*) 10-100 ml
(*c*) 1-2 liters (*d*) 5-10 μ*l*

17.5 An average time of analysis for sample by HPLC is approximately

(*a*) 4-6 hours (*b*) 1-10 seconds
(*c*) 5-50 minutes (*d*) 10-20 hours

17.6 HPLC permits analysis of only

(*a*) single sample (*b*) multiple samples
(*c*) fixed number of samples (*d*) no restriction on number of samples.

17.7 HPLC can be used in combination with following chromatographic techniques:

(*a*) ion exchange and ion exclusion
(*b*) adsorption and partition chromatography
(*c*) adsorption partition and ion exchange chromatography
(*d*) gas chromatography

17.8 HPLC permits one of the separations of following compounds:
(*a*) volatile and thermally stable compounds
(*b*) non volatile and thermally unstable compounds
(*c*) high molecular weight compounds
(*d*) the separation of only inorganic ions

17.9 With Van Deemeter's equation we can easily evaluate in HPLC the:
(*a*) height of theoretical plate (*b*) particle size of stationary support
(*c*) rate of flow of eluant (*d*) nature of diffusion of column

17.10 The elution with single eluent is termed as
(*a*) isocratic elution (*b*) gradient elution
(*c*) selectric elution (*d*) stepwise elution

17.11 The kind of pressure used in typical HPLC separation mode is:
(*a*) 10 psi (*b*) 150 psi
(*c*) >1000 psi (*d*) 50000 psi

17.12 The role of precolumn in HPLC is to function as:
(*a*) emergency column
(*b*) for removal of impurities from the solvent
(*c*) to raise flow rate of an eluant
(*d*) to have rehearsal of separation before actual separations

17.13 The pellicular support in HPLC consists of:
(*a*) glass beads coated with resins (*b*) large size support to glass bead
(*c*) small size support to beads (*d*) beads made out of diatomaceous earth

17.14 The kind of detector commonly used in HPLC is:
(*a*) photometric detector (*b*) refractive index detector
(*c*) amperiometric detector (*d*) conductometric detector

17.15 The kind of stationary support used in HPLC is:
(*a*) ion exchange resin (*b*) silica and alumina bonded support
(*c*) diatomaceous earth (*d*) inorganic exchangers

17.16 Nucleic acids are separated in HPLC with column containing:
(*a*) porous layer of beads (*b*) bonded silica gel
(*c*) anion exchange resin (*d*) glass beads

17.17 The kind of detector used in separation of isomers of glucose with acetonitrile as eluant is:
(*a*) refractive index detector (*b*) photometric detectors
(*c*) fluorimetric detectors (*d*) polarographic detectors

17.18 A compound with high mass, nonvolatile and thermally unstable can be best separated by technique of:

(*a*) GLC (*b*) HPLC
(*c*) ion exchange chromatography (*d*) SFC

17.19 The HPLC technique facilitates the separation of:
(*a*) single component sample (*b*) multiple samples
(*c*) any number of samples at a time (*d*) no restriction on number of samples

17.20 HPLC-MS hyphenated technique:
(*a*) facilitates the elucidation of the structure of compound
(*b*) operates at microgram concentration
(*c*) reduces the time of analysis
(*d*) facilitates rapid separation

17.21 The most commonly used detector for separation of coloured species is:
(*a*) refractive index detector (*b*) thermal conductivity detector
(*c*) uv absorbance detector (*d*) fluorosence detector

17.22 The isocratic elution involves the use of as many as:
(*a*) one solvent (*b*) two solvents
(*C*) four solvents (*d*) no restriction on the number of solvents

17.23 The most commonly used pump for separation in HPLC is:
(*a*) reciprocating pump (*b*) displacement pump
(*c*) pneumatric pump (*d*) constant pressure pump

ANSWERS

17.1 c	17.2 a	17.3 a	17.4 d	17.5 c	17.6 d
17.7 c	17.8 b	17.9 a	17.10 a	17.11 c	17.12 d
17.13 a	17.14 b	17.15 b	17.16 c	17.17 a	17.18 d
17.19 d	17.20 a	17.21 c	17.22 d	17.23 d.	

REFERENCES

17.1 D.A. Skoog, F. J. Holler and T. A. Nieman, *Principles of Instrumental Analysis,* 5th Ed., Harcourt Brace Pub., p. 725 (1998).

17.2 F.A. Settle, *Handbook of Instrumental Technique for Analytical Chemistry,* Pearson Indian Education, Publisher, p. 147 (2004).

17.3 H.H. Willard, L.L. Meerit, J.A. Dean, F.A. Settle, *Instrumental Methods of Analysis,* 6th Indian Ed., CBS Publishers, p. 529 (1986).

17.4 R.D. Braun, *Introduction to Instrumental Analysis,* McGraw-Hill & Co., p. 890, (1987), Indian Edition (2004).

17.5 S.M. Khopkar, *Basic Concepts of Analytical Chemistry,* 3rd Ed., New Age International Publishers (P) Ltd. (2007).

Chapter 18

Supercritical Fluid Chromatography and Extraction

INTRODUCTION

The discovery of Supercritical Fluid Chromatography (SFC) is of recent origin with numerous advantages. As the name implies it deals with fluid, which exist either in form of liquid or gaseous state. One can easily define supercritical fluid as the substance like fluid which is existing above the critical temperature, when it neither exist as liquid nor it can be gasified. However if we compare viscosity of fluid it is similar to gas. This technique facilitates separation of both volatile and non-volatile compounds. It can also analyse impurities and is useful for the isolation and purification of the compounds. It can easily accomplish the separation of closely related compounds. This is a non destructive method. It permits qualitative as well as quantitative analysis. The fractionation of complex mixture is possible by SFC. The significant advantage of SFC is extraordinary mobility of the phase with very large diffusion coefficient, which in turn permits separation in short time.

Carbon dioxide is most favoured gas in SFC method. Of course one can also use acetonitrile or methanol as supercritical fluids. Also ammonia, n-butane, n-propane are also called as supercritical fluids. One can also use water as the supercritical fluid, though it has high critical pressure. Acetonitrile is having highest critical temperature as opposed to CO_2 which has lowest critical temperature which in turn suits the purpose of separations. Amongst inert gases Argon was also explored as the supercritical fluid on reducing pressure to 70-80 atmosphere in SFC. For CO_2 the elution time is remarkably decreased. One sometimes needs a modifier as methanol is best modifier for CO_2 which helps in the process of separation.

It is necessary to dissolve the sample in liquid or supercritical fluid chromatography but it can be also solid for supercritical fluid extraction (SFE). About 0.1–19 μl sample is used with approximate concentration of 1-20 mgs. The sample preparation involves steps like dilution, preconcentration, filtration, extraction via liquids or fluids. The time for analysis varies between 2-60 mts for SFC and 10-60 mts for SFE. The time varies with nature of sample. For instance in analysis of the environmental samples like chlorinated hydrocarbons both methods viz. SFC and SFE can be used together for rapid monitoring.

The technique has numerous applications in environmental monitoring for food processing on commercial scale or analysis of organic or biological products. The separation of metals by SFC is not as effective as is SFE method. SFC can measure active drugs, degradation products in pharmaceutical dosage forms. The hazardous insecticides and pesticides can be also analysed. The surfactants, polymers additives, hydrocarbons can be best separated by SFC.

However there are few limitations. For complex mixtures, the resolution is difficult. SFC permits only separation of one sample at a time. The time of analysis depends upon the nature of sample. By increasing the pressure the solubility of sample remarkably increases. With rise in the temperature the diffusivity also increases but viscosity decreases specially at critical point. Though CO_2 is cheap, for SFE we need very pure CO_2. The absorbance and flame ionization detectors are commonly used in metal separations. For SFE sample preparation is not needed as the greater diffusivity and low viscosity of fluid facilitates good separations. Even SFC is preferred over HPLC as former has better resolution. It is rapid and can be extended to thermally liable compounds. As regards column, the packed and capillary columns are used, in SFC for better separations fatty acid, lipids were best analysed by SFC. For separation of steroids N_2O is used as mobile phase under special conditions. The phenomena of the coextraction is absent in SFE separations. In pharmaceutical industry, SFE is beneficial for quality control and separation of complex mixtures.

The best results are obtained by using combination of SFC and SFE together with analysers similar to GC-MS hyphenated techniques. SFC can accept any kind of detector like FID, ECD. FT-IR, instrument can be coupled with SFC to get quantitative information. Finally SFE can be directly coupled with GC, HPLC, FT-IR and MS methods for better resolution. Several significant separations have been accomplished by hyphenated method involving the use of supercritical fluid extraction.

PROBLEMS

Multiple Choice Questions

Tick mark the correct answer from the following multiple choices:

18.1 SFC is generally used for the separation of

(*a*) thermally non liable compounds (*b*) pure compounds
(*c*) impurities (*d*) fractionation of complex mixture

18.2 SFC needs sample in liquid state but SFE can use also

(*a*) gaseous samples (*b*) solid samples
(*c*) only in liquid samples (*d*) liquid semi-solid or solid samples

18.3 In SFC the level of analysis of the sample is at

(*a*) gram concentration (*b*) semi micro level
(*c*) micro level concentration (*d*) sub micro level for analysis

18.4 The exact level of sample needed for analysis is dependent on

(*a*) kind of detectors used (*b*) type of solvent used
(*c*) physical state of the sample is (*d*) chemical structure of the sample

18.5 Without MS or IR instrument use with SFC is not benefited as

(*a*) sample identification and quantitative analysis is difficult
(*b*) separation is time consuming
(*c*) it is applicable at milligram concentrations
(*d*) the quality of the separation is not very good

18.6 The separations involving SFC or SFE
(*a*) is restricted to single sample separation
(*b*) is giving bad resolution with solvent
(*c*) needs longer time for analysis
(*d*) is possible to use only by trained personnel.

18.7 SFC is defined as a substance which is a fluid
(*a*) in vapour form (*b*) above critical temperature
(*c*) below its boiling point (*d*) below its critical pressure

18.8 Supercritical fluid (SCF) has one following properties common with gas
(*a*) viscosity being different from gas (*b*) density same as gas
(*c*) viscosity identical with gas (*d*) density identical with gas

18.9 The significant advantage of SFC technique is that we have mobile phase with
(*a*) large diffusion coefficient (*b*) high viscosity
(*c*) uniform concentration (*d*) better mobility

18.10 One of the following compounds is unsuitable for SFC separations
(*a*) CO_2 (*b*) methanol
(*c*) acetonitrile (*d*) dinitrogen oxide

18.11 Which of the following is excellent supercritical fluid?
(*a*) n-butane (*b*) n-propanol
(*c*) ammonia (*d*) carbon dioxide

18.12 Name supercritical fluid which has highest critical pressure?
(*a*) water (*b*) ammonia
(*c*) methanol (*d*) dinitrogen oxide

18.13 Which of the following fluid has the highest diffusion coefficient?
(*a*) NH_3 (*b*) CO_2
(*c*) n-propane (*d*) n-butane

18.14 Which of the following inert gas can be used as supercritical fluid?
(*a*) He (*b*) Ar
(*c*) Ky (*d*) Xe

18.15 The decrease in pressure on CO_2 as supercritical fluid leads to
(*a*) reduced elution time (*b*) enhanced elution time
(*c*) unclean separation (*d*) no significant effect on elution process

18.16 The kind of detector most commonly used in SFC technique is:
(*a*) ECD (*b*) FID
(*c*) TCD (*d*) photometric detector

18.17 One of the following compounds is used as modifier with CO_2 in SFC. It is:
(*a*) O_2 (*b*) N_2
(*c*) CH_3-OH (*d*) H_2

18.18 The process of derivatisation improves separations with detectors as:

(*a*) FID (*b*) ECD
(*c*) TCD (*d*) amperiometric detector

18.19 The SFC operates for molecules with molecular weight of:
(*a*) 10^3 Dalton (*b*) 10^2 Dalton
(*c*) 10^4 Dalton (*d*) 10^6 Dalton

18.20 The average plate height of the theoretical plate in SFC technique is of the order of:
(*a*) 0.2 mm (*b*) 0.039 mm
(*c*) 0.013 mm (*d*) 0.0013 mm

18.21 In environmental monitoring SFC is used for rapid analysis of:
(*a*) chlorinated hydrocarbons (*b*) nitrogenous compounds
(*c*) sulphur compound (SO_2) (*d*) chiral compounds

18.22 Polymers can be analysed by one of the following hyphenated technique as:
(*a*) SFC-MS (*b*) SFC-FTIR
(*c*) SFC-SFE (*d*) CFC-CZE

18.23 In comparison with SFC, the SFE technique has the advantage of:
(*a*) fast speed
(*b*) low level of separation
(*c*) wide application for separation of compounds
(*d*) being inexpensive method

18.24 Which one of the following technique is rapid for purpose of separation?
(*a*) GLC (*b*) HPLC
(*c*) SFC (*d*) Ion exchange

18.25 The kind of fluid most popularly used in SFC technique is:
(*a*) Water vapour (*b*) Argon gas
(*c*) Carbon dioxide (*d*) Methane vapour

18.26 Which one fluid provides highest critical pressure?
(*a*) NH_3 (*b*) CO_2
(*c*) N_2O (*d*) butane

18.27 SFC operates satisfactorily if molecular weight of the sample is:
(*a*) 10^6 (*b*) 10^2
(*c*) 10 (*d*) 10^9

18.28 The long time is required for the SFC separation of:
(*a*) pesticides (*b*) drug residues
(*c*) cholesterol (*d*) PAH in river sediments

ANSWERS

18.1 d	18.2 c	18.3 c	18.4 a	18.5 a	18.6 a
18.7 b	18.8 c	18.9 a	18.10 d	18.11 d	18.12 a
18.13 b	18.14 b	18.15 a	18.16 b	18.17 c	18.18 a
18.19 d	18.20 c	18.21 a	18.22 a	18.23 c	18.24 c
18.25 c	18.26 a	18.27 a	18.28 a		

REFERENCES

18.1 S.M. Khopkar, *Basic Concept of Analytical Chemistry,* 3rd Ed., New Age International Publishers (P) Ltd. pp. 186–187 (2007).

18.2 R.M. Smith, *Supercritical Fluid Chromatography*, CRC Press Boca Raton (1988).

18.3 F.V. Bright, M.F. P. McNally, *Supercritical Fluid Technology,* ACS Symposium No. 488 (1988).

18.4 F.A. Settle, *Handbook of Instrumental Techniques in Analytical Chemistry,* Pearson Education Publishers, p. 183 (2004).

18.5 D.A. Skoog, F.J. Holler, T.A. Nieman, *Principles of Instrumental Analysis,* Harcourt Bruce Publishers, 5th Ed., p. 768 (1998).

18.6 G.D. Chrishan, *Analytical Chemistry*, 5th Ed., John Wiley (1999).

Chapter 19

Exclusion Chromatography

INTRODUCTION

Gel permeation or exclusion chromatography was used very commonly in biochemical separations. Such separation is based upon exploiting the difference in property of molecular geometry of solutes. For instance a compound with high molecular weight ($MW = 80{,}000$) if passed along with compound with low molecular weight, packed with gels like Bio Gel, Sephadex, Styrogel, while the molecules with low molecular weight and small size would emerge latter from would enter into pores of gel via interstitial space while molecules with large size molecular weight compounds cannot penetrate through gel therefore they pass through void fraction and do not enter inside the gel and appear first out of the column. While low molecular weight compounds with small size emerge out of column at latter stage. Thus we exploit the difference in the molecular size leading to differential migration front of the compound through the gels. It is based upon primarily size differences of solutes.

While in ion exclusion technique which does not involve true ion exchange process, it was noted that if mixture of ionic and non-ionic compound is passed on the column containing resin ionic maternal emerges out first while non-ionic material emerges out latter on the column. The ionic compound can penetrate through resin (provided it is in same form as the ionic compound) while non-ionic compound emerges latter as it goes round resin in void fraction and appears latter from the column. By washing column with water both ionic and non-ionic compounds can be separated.

Now in columnar behaviour we observe the following relation:

$$V_b = \left(V_o + V_i + V_v\right) = V_o V_S, \text{ if } V_i = mS_r/\rho_j$$

$$V_i = \frac{S_r \rho_r}{\rho s\left(1 + S_r/\rho s\right)} = \left(V_o + K_d Vi\right) \text{ if } V_i/V_o = \left(1 + K_d\, V_i/V_o\right)$$

$$K_{av} = \frac{\left(V_a - V_o\right)}{\left(V_b - V_o\right)} \text{ if } K_{av} K_{as} = K_d V_i$$

If V_i = inner volume, V_r = gel matrix volume, V_s = total volume of stationary phase

m = weight of gel, S_r = solvent up take by gel, ρ_s = density of solvent

K_d = distribution coefficient, ρ_r = density of gel, V = volume, b = bed, o = exterior of the gel

This can be also explained alternatively in following expression, viz.:

$$V = V_m\left(1+\frac{KV_s}{V_m}\right) = V_m + KV_s.$$

If V_m V_s are volume of mobile and stationary phases resp. K = capacity factor almost analogues to distribution ratio (K_D).

This equation facilitates us to calculate total pore Volume (V_s), distribution coefficient ($K^1 = K_d$), molecular weight (M) of gel used, the (W) retention volume of unknown compound. Such calculations are shown in the problems given in subsequent part of this chapter.

The exclusion chromatography is most suitable method for the separation of major from minor components. It is inexpensive. It is used for the desalting, determination of molecular weight. Now if V_b = Volume occupied by the inert material, V_i = volume of mobile phase inside the beads, V_o = volume of mobile phase outside beads, for small molecule size compounds, elution volume is ($V_i + V_o$) while for large molecular compounds which cannot enter the beads it is V_o if adsorption occurs then the retention volume is > ($V_o + V_i$). Now if V_R = retention volume $V_R = V_o + K_v$; where K is the fraction of ion to a given solute. The best separations are carried out with slow flow rate (< 3.5 ml/mt). The separation of ionic from non-ionic materials can be best accomplished by ion exclusion technique e.g., sugar and common salt on Dowex 50WX8 (Na^+ form resin).

In ion retardation technique we use both cationic as well as the anionic resins. In this compound is absorbed but with weaker force, they can be washed with water. In ion retardation ions are retraded as they flow down the column. It can permit the separation of electrolyte from non-electrolyte or another electrolyte, e.g., NH_4OH and NaOH or $FeSO_4$ and $ZnSO_4$.

The detailed information on size exclusion chromatography is described in various books in the references listed at the end of this chapter.

SOLVED PROBLEMS

19.1 In a typical bioanalytical work a column of Sephadex - 75 with size of 2.5 × 36 cm^2 gave following results:

Ribo nuclease A(M.W. = 1.3×10^4; V_a = 109.5 ml)
Aldolase (M.W. 1.58×10^5; V_a = 55 ml)
Blue dextrane (M.W. 2×10^6; V_a = 531 ml)
Ovalbumin (M.W. 4.5×10^3; V_a = 71.8 ml)

Calculate the molecular weight of protein [X] with separation volume of 76.5 ml?

Ans: We plot graph of K_d versus log (molwt) it gives an idea of mol. weight but K_d is evaluated from V_o, V_c, and V_i where V_o is measured with large molecules like blue dextrane from (M.W. = 2×10^6) V_g is negligible while calculating V_i and V_t; we arrive at molecular weight of unknown compound.

19.2 Examine the following data to calculate molecular weight of unknown compound if

Lysozyme (M.W. = 14,000, RM relative mobility = 0.90)
Trypsine (M.W. = 23,000, RM = 0.73)
Ovalbumin (M.W. = 43,000, RM = 0.54)
Catalase (M.W. = 62,000, RM = 0.38)

The unknown compound has relative mobility of 0.68?

Ans: In polyarylamide gel electrophoresis is carried. SDS denature proteins to coil conformation, so electromobility depends upon molecular weight. The relative electrophoresis mobility to that of low molecular weight is linear and it helps to find out the molecular weight of the unknown protein molecule.

19.3 In molecular exclusion chromatography, column was made out of old 50 ml. burret. Can you forecast the elution volumes for this column?

Ans: Since the burret has maximum capacity of 50 ml, the largest elution volume should be obviously less than 50 ml.

19.4 What was the void volume (V_m), total pore volume (V_i), distribution ratio (K_d) if molecular species (M.W. = 8×10^5) was passed through a exclusion column with exclusion limit of 4×10^5 containing a stationary phase? The retention volume was 25 ml. A low molecular weight compound had retention volume 152 ml?

Ans: The universal equation used is

$$V = V_m\left(1 + \frac{KV_s}{V_m}\right) = V_m + KV_s$$

where in V_s, V_m are volume of stationary and mobile phase inside the column,
K^1, = capacity factor = (KV_s/V_m)
Now MW = 80,000, $K = 0$, $V = 25$ ml. For low molecular weight compound
$V = 152$ ml. $K = 1$. Substitute the value in equation given above
$V = (V_m + KV_s)$, *i.e.*, $152 = 25 + 1V_s$) so $V_s = 127$ ml. We calculate K_d
$102 = 25 + L(127)$ so $K_d = 0.61$.

19.5 What was the volume, total peak volume and distribution ratio if an organic compound (M. Wt. 7×10^5) was passed through an exclusion column with exclusion limit of 3.5×10^5, on a stationary phase. If the retention volume was 501 ml. A low molecular weight compound with sample component had retention volume of 204 ml and sample component had a retention volume of 150 ml?

Ans: We have $V_m + KV_s = V$
If M.W. = 7×10^5, $K = 0$, $V = V_m = 50$ ml. For low molecular weight compound
$V = (V_m + KV_s)$, *i.e.*, 150 ml = $(50 + V_s)$ as $K = 1$
$V_s = (150 - 50) = 100$ ml.
So also we calculate K (capacity factor) as
$204 = 50 + K(106) = 50 + K(100)$
$154 = K(100)$ So $K = \frac{154}{100} = 0.154$.

19.6 What was the retention volume (V_R) if external solution (V_o) is 16 ml and internal solution is 5 ml and fraction of (V_i) acceptable to solute (K) is 12 ml?

Ans: Since $V_R = (V_o + KV_i)$
Now $V_o = 16$ ml, $V_i = 5$ ml and if $K = 12$ mm
$V_R = 16 + (12 \times 5) = 60 + 16 = 76$ ml.

19.7 What was the volume occupied by stationary support in exclusion chromatography if column had total volume (V_t) of 30 ml with volume of solvent in pores or voids (V_i) was 10 ml and free external volume was 15 ml (V_o)?

Ans: Since $V_t = (V_o + V_i + V_g)$
We have $V_t = 30$ ml; $V_o = 15$ $V_i = 10$ ml. So $V_g = x$
$30 = 15 + 10 + x$, *i.e.*, $V_g = 5$ ml.

19.8 What was the volume with internal stationary support like gel if column was packed with gel too large to enter gel pores ($K = 0$) and ($V_e = V_o$) the elution volume was 25 ml while volume of external solution was 15 ml with $K = 0$?

Ans: $V_e = (V_o + KV_i)$ Now $V_e = 25$, $V_o = 15$, $K = 0$
So $25 = 15 + V_i$, *i.e.* $(25 - 15) = V_i = 10$ ml.

19.9 What was the distribution ratio in exclusion chromatography where in external volume was 20 ml, and interstitial volume was 10 ml (V_i) with the elution volume of 50 ml?

Ans: Since $K_d = \left(\frac{V_e - V_o}{V_1}\right)$ if $V_e = 50, V_o = 20, V_i = 10$

$$K_d = \left(\frac{50 - 20}{10}\right) = \frac{30}{10} \text{ i.e., } K_d = 3.0.$$

PROBLEMS FOR PRACTICE

19.10 What was the inner volume of a column of Sephadex *G*, if the wet density (g/ml) was 1.19 on prepared column the bed volume was 30 ml, and the void volume was 9 ml?

[**Ans:** $V_t = 15$ ml]

19.11 Evaluate the magnitude of V_e/V_o and K_{av} for system if a column with bed volume of 1.6 × 96 cm had the elution volumes was for raffinose (84.8), glucose (102), KCl (124) with void volume 69 ml. [**Ans:** 1.23 $K_{ab} = 0.092$.
KCl = 1.81 $K_{av} = 0.32$]

19.12 Calculate V_e/V_o and K_{av} if in desalting of compound on column (1.3 × 36 cm) had the peak at 36.5 ml for organic compound and 42.5 ml for common salt from one 1 ml of sample.

[**Ans:** 1.8 K_{av} 0.13 for organic compounds 1.37 K_{av} 0.27 for common salt]

19.13 What was the molecular weight of protein as the K_{av} for a bed volume of 30 ml on support with one ml aliquots, if zone broadening was six times sample volume (assume one component was globular protein (M.W. = 20,000)?

[**Ans:** $K_{av} = 0.29$, for $\Delta V_e = 6$ ml, M.W. = 80,000]

19.14 What kind of refinement in separation would occur if bed volume was changed from 30 to 100 ml in above problem?

[**Ans:** $\Delta K_{ar} = 0.088$ from MW = 12,000 to 28,000 can be separated]

19.15 Assuming Sephadex *G* 200 was replaced by Sephadex *G* - 75 do you expect a change in separation? [**Ans:** $V_p = 30$, $\Delta K_{av} = 0.32$, M.W. = 6,000]

19.16 Point out a peak position and amount of stationary support required if the sample volume was 10 ml (M.W. = 40,000) to be desalted on Sephadex *G* 25?

[**Ans:** Bed volume 4 times sample, *i.e.*, 40 ml eq. to 8 gm of dry gel $V_o + V_1 = 35$ ml elution volume, $V_s = 24$ ml, $(\Delta V_e - V_s)$ peak is 24 hours for 11 ml of the sample.]

19.17 Find out the molecular weight of the unknown compound designated as x?

	Cytorchrome	Myoglobin	Ovalbumin	Urease	Unknown (x)
Mol. wt	1.3×10^4	10.9×10^3	4.6×10^4	4.88×10^5	x
V_1 ml.	140	133	107	45	125

[Ans: M.W. = 2300]

19.18 Calculate the molecular weight of a compound if the retention volume of polyethylene oxide was determined with diethyl formamide as the mobile phase?

(Assume retention volume of 13.5 ml)

Mol. wt: 1.1×10^3; 5.1×10^3, 2×10^4, 7.7×10^4, 7.0×10^5

V_1 ml: 17.2 ml, 16.3 ml, 15.1 ml, 14.1 ml, 12.7 ml. **[Ans:** M.W. = 2.1×10^5]

19.19 What was the retention volume of blue dextrin if it was separated from glucose by exclusion chromatography when the distribution coefficient was 0.81?
(Assume retention volume of sucrose = 195 ml and retention volume of blue dextrin = 39 ml)

[Ans: 170 ml]

19.20 What was the total pore volume and distribution ratio of the sample if the retention volume of protein (M.W. 1.2×10^6) had a exclusion limit of 8×10^4 was 15 ml. the retention volume of napthalene (M.W. 128) was 124 ml. If a unknown sample had retention volume 87 ml.

[Ans: 10 gml, $K_d = 0.66$]

REFERENCES

19.1 S.M. Khopkar, *Basic Concepts of Analytical Chemistry,* 3rd edn., New Age International Publishers (P) Ltd. (2007).

19.2 J.A. Dean, *Chemical Separation Methods,* Van Nostrand Reinhold Co. pp. 304-5 (1970).

19.3 E.W. Berg, *Physical and Chemical Methods of Separation,* McGraw-Hill and Co. (1963).

19.4 T. Kremman and G.L. Boross, *Gel Chromatography,* John Wiley (1979).

19.5 H. Determann, *Gel Chromatography,* Spring Verlag (1968).

19.6 E. Leaderer and M. Leaderer, *Chromatography—Theory and Practice,* 2nd edn. Van Nostrand (1957).

19.7 D.G. Peters, J.M. Hayes and G.M. Hieftj, *Chemical Separation and Measurement* Saunders, Wiley-Interscience (1974).

19.8 B.L. Krager and L.R. Snyder, C. Horwath, *An Introduction to Separation Science,* Wiley-Interscience (1980).

19.9 G.D. Christen, *Analytical Chemistry*, 5th edn., Wiley-Interscience (1997).

19.10 R.D. Braun, *Introduction to Instrumental Analysis*, McGraw-Hill and Co., p. 888 (1987).

Chapter 20

Electro Chromatography

INTRODUCTION

The term electrophoresis and electro chromatography are frequently used as the synonymous terms but there is minute difference between them. When we carry out separation by taking advantage of the ionic mobility we term process as electrophoresis. In such separations mobile phase is absent. The technique involving free boundary or continuous electrophoretic separation is called as the electrophoresis. However when we use stabilised media like filter paper for the purpose of separation acting as stationary support with mobile and stationary phase we term it as electrochromatography.

The fundamental principles is involved in any electro chromatographic separation is to exploit the difference in migration of the ions on stabilised media on account of difference in ionic mobility of the ions under the influence of applied electric potential. The ions with high mobility move faster while those ions with low mobility move very slowly there by creating differential migration front consequently accomplishing clean-cut separation of ions on the stabilised media like filter paper.

The migration of charged ion in paper depends upon the magnitude and sign of the net charge of ion, surface, charges, applied, voltage, electrolyte concentration; ion strength, pH, temperature, viscosity, solute absorbility, and other properties of the migrating medium. The moving in unidirection also affect mobility of those ions which are migrating in the opposite direction. For all practical purposes the path traversed by a charged particle is given by equation $d = \frac{ute}{qk} = \frac{utv}{L}$ where V = volts applied, L = length of channel, K = conductivity, e = current flowing, u = mobility of ion at a time t_1 q = cross-sectional area and d = path traversed

In addition to this another expression usually generally employed to compute volume of buffer used for electrolysis is:

$$V = \frac{0.00373 \times i \times t}{C}$$ if i = current (amps), t = time in hours, C = concentration (M).

V = volume in ml.

The apparent mobility of two charged compounds is estimated as

$$\mu^1 = \frac{\mu K_a}{[H^+] + K_a}$$ if μ^1 = apparent mobility, μ = actual ionic mobility.

K_a = ionisation constant

We can also evaluate ionic mobility (*M*) from an expression

$$\mu = \frac{d}{t(E/S)} \text{ if } d = \text{distance travelled cm, } t = \text{time in seconds}$$

E/S = Voltage (*E*) or current passed per seconds (*S*):

In case we know pK_1 and pK_2 and if limiting mobility's are also known viz. (μ_1/μ_2) we can easily calculate the pH for the separation of these two species from binary mixture as

$$[H^+] = (K_1K_2)^{1/2}\left[\frac{(\mu_1/\mu_2)^{1/2} - (K_1/K_2)^{1/2}}{1-(\mu_1/\mu_2)^{1/2}(K_1/K_2)^{1/2}}\right] \text{ From the magnitude of } -\log[H^+] = \text{pH}$$

Finally in some problems dealing with calculation of the concentration, involving over potential we use Nernst equation as $E = E_o - \frac{RT}{nF}\log\frac{(\text{oxid})}{(\text{Reduc})}$ as $RT/F = 0.0591$, so we have

$$E = E_o - \frac{0.0591}{n}\log\frac{(\text{oxid})}{(\text{Reduc})}$$ in this expression E_o = standard electrode potential,

n = nos. of electrons transferred

Apart from curtain or normal electro chromatography the techniques involving reverse osmosis or electrodialysis also uses same principle of ion migration. Only difference in reverse osmosis is we use semi-permeable membrane while in electrolysis we use the cationic and anionic membrane to accomplish separation. In reverse osmosis, water is extracted from bulk of waste water, while in electrodialysis ions are separated depending upon the kind of charge they carry, e.g., cations migrate to anode through anion exchange resin while the anions migrate towards cathode through cation exchange resin. During migration ions move towards electrode having an opposite charge.

SOLVED PROBLEMS

20.1 What was the volume of buffer required for electrophoresis if the concentration was 0.05 M and time needed for separation was 3 hours when 4 mA of current was flowing?

Ans: The volume was calculated from following empirical equation as

$$V = \frac{0.00373 \times i \times t}{C}$$ if i = current in amperes (mA), t = time in hrs

C = concentration in molar so we have

$$V = \frac{0.00373 \times i \times t}{C} = \frac{0.00373 \times 4 \times 3}{0.05} = 0.9 \text{ liter } i.e., 900 \text{ ml.}$$

20.2 Calculate the pH for the separation of the binary mixture of the two organic acids containing nitrogen if there pK_2 was = 9.87, pK_2 = 9.21 with limiting mobilities as (2.67×10^{-4}) and $(2.06 \times 10^{-4}$ resp.)

Ans: We calculate pH from the equation

$= \mu^1K_a/[H+] + K_a$ if μ's apparent mobility, K_a = ionisation constant so

$$[H^+]=(K_1\ K_2)^{1/2}\left[\frac{(\mu_1/\mu_2)^{1/2}-(K_1/K_2)^{1/2}}{1-(\mu_1/\mu_2)^{1/2}\ (K_1K_2)^{1/2}}\right]$$

$\mu_1 = 2.67 \times 10^{-4}$; $\mu_2 = 2.06 \times 10^{-4}$, $pK = -\log K, = 9.87$
$pK_2 = -\log K_2 = 9.21$ on substituting the values in above equation

$$[H^+]=(6.10)^{-10}\left(1.35\times10^{-10}\right)^{1/2}\left[\frac{(2.06/2.67)^{1/2}\left(\dfrac{6.11\times10^{-10}}{1.35\times10^{-10}}\right)^{1/2}}{\left(\dfrac{(2.06\times2.67)^{1/2}\left(6.16\times10^{-10}\right)^{1/2}}{1.35\times10^{-10}}\right)}\right]$$

Hence pH = 9.38 ~ 9.40 = 0.02.

20.3 Find apparent mobility of two aminoacids (I) and (II) buffer of pH 9.4. Was it possible to estimate distance of migration if electrophores was conducted for 5.75 hr with V = 180 v per 53 cm and current of 200 m/A?

Ans: Since $\mu^1 = \mu K_a/([H+]\ K_a)$ we substitute the values

$$\mu^1=\frac{\left(2.06\times10^{-4}\right)\left(6.16\times10^{-10}\right)}{\left(4\times10^{-10}\right)+\left(6.16\times10^{-10}\right)}=125\times10^{-4}\ \text{cms/v/sec for } A_1$$

$$\mu^1=\frac{\left(2.67\times10^{-4}\right)\left(1.35\times10^{-10}\right)}{\left(4\times15^{-10}\right)6+\left(1.35\times10^{-10}\right)}=0.67\times10^{-4}\ \text{cms/v/sec for } A_1$$

Now as per equation

$d = \mu t\ (E/S)$ substitute these values if $\mu = 1.25 \times 10^{-4}$, $E = 180/53$, $S = 5.7 \times 60 \times 60$

$d = (1.25 \times 10^{-4})\ (5.75 \times 60 \times 60)\ (180/53) = 8.8$ cm for A_1

and for second aminoacid; $\mu = 0.67 \times 10^{-4}$, S = 5.76 × 60 × 60; E = 180/53

$d = (0.67 \times 10^{-4})\ (5.75 \times 60 \times 60)\ (180/53) = 4.7$ cm for A_2

20.4 Find out the volume of buffer necessary for working if 0.10 M of sodium salt of weak acid, had current of 8 mA by inducing voltage gradient and operation is continued for 360 minutes.

Ans: Since $V=\dfrac{0.00373\times it}{C}=\dfrac{0.00373\times8\times6}{0.10}$; $V=\dfrac{0.00373\times8\times6}{0.10}=900$ ml. (as $i=8$, $t=6$ hrs.)

20.5 Find out the volume of 0.2 N buffer solution required for an overnight run of 900 minutes at current of 3.5 mA.

Ans: $C=0.2, i=3.5, t=15$ hr $\left(\textit{i.e.,}\ \dfrac{900}{60}=15\ \text{hr.}\right)$

$$V=\frac{0.00373\times i\times t}{c}=\frac{0.00373\times3.5\times15}{0.2}=100\ \text{ml.}$$

20.6 Calculate the concentration, over potential for electrode where copper was deposited at the electrode with concn. as 10^{-4} while one in the solution is 10^{-2} M.

Ans: By Nernst equation

$$E = E_o - \frac{RT}{nF}\log\frac{(\text{oxid})}{(\text{red})} \text{ if } RT/F = 0.0591 \text{ if } E^0_{\text{Cu}} = 0.345, n = 2$$

$$E = 0.345 + \frac{0.0591}{2}\log\left[10^{-2}\right] = 0.286 \text{ V}$$

concentration over potential is

$$E = 0.345 + \frac{0.0591}{2}\log 10^{-4} = 0.227 \text{ V}$$

$$\text{Overpotential} = (E - E_{\text{ref}}) = (0.227 - 0.286) = -0.059 \text{ V}.$$

20.7 It was possible to evaluate the ionic mobility of an acetate anion if the rate of electroosmosis was evaluated from movements of the molecules which were neutral at pH of the buffer. When the pK_a of a caffeine was 0.15 it had zero charge at pH above 3.0 assuming mobility of caffein was same as the rate of osmosis. At any pH > 3.0 picrate ion (pK_a = 0.89) had one charge and mobility was dependent on pH. Further a data obtained showed a migration time of 100 mts at 10 V/cm on 40 cm bed. The acetate ion travelled till was 12.7 ems.?

Ans: Ionic mobility (μ) is expressed as

$$\mu = \frac{d}{t(E/S)}$$ if d = 12.5 distance travelled, t = 100 mt = 100 × 60 second and if (E/S) = 18

we have,

$$\mu = \frac{d}{t(E/S)} = \frac{12.5}{(100 \times 60) \times (18)} = 1.6 \times 10^{-4} \text{ cm}^2/\text{V/Second}.$$

20.8 Calculate the best pH for the separation of lysine and histidine glycine and sevine if ionic mobility of lysine was 9.7, histidine was 7.6 and glycine was 6.0 and serine was 2.64?

Ans: (At pH 7.6, 9.3) The separation depends only on mobility difference as pK_a value were virtually identical. pK_a of lysine (2.20), histidine (1.8), glycine (2.44) and serine (2.21) were closer to each another.

PROBLEMS FOR PRACTICE

20.9 What was the expected electrophoretic pattern of a mixture of alkyl amines named as *X, Y, Z,* whose mobilities at zero ions strength were 5.99, 4.85, 4.3 × 10^{-4} cm V^{-1} resp?

[**Ans:** Since all species exist as cations, the mobility would decrease in order]

20.10 Where were the zones located in relation to origin if on a silicagel impregnated with buffer of pH 6.6 the pattern of lysine and histidine was 7 hr. 100 v an bed 53 in length?

[**Ans:** Lysine moved 22 cm towards cathode, histidine moved only 7.8 cm in the same direction of cathode]

20.11 The experimental mobility value were obtained for bovine β-lipoprotein 2.75 × 10^{-5} at pH 2.67; 0.3 × 10^{-5} at pH 4.9 and −2.3 × 10^{-5} at pH 7.0. Calculate value of p/e?

[**Ans:** p/e = 5.2]

20.12 In desalting set up the positions of substance operating at 25 v/cm for 15 mt on a sheet of paper saturated with pyridine acetic acid (pH 6.0) was located in order of $KNaSO_4Cl\ H_2PO_4$. Can one locate the position of cations and anions and locate desalted area?

[**Ans:** Towards cathode were located Na^+, K^+, ions while for anode were PO_4^{-3}, SO_4^{2-} Cl^- anions. The desalted area was between the position of Na and H_2PO_4 from starting zone.]

20.13 Calculate the mobilities of electrososmosis in a buffer pH 4.6 when electrophoresis was carried out for 3.5 hr at 6.3 v/cm vanillin moved 49.5 m towards cathode while in the direction of the anode ferrulic acid moved 10.5 m and vanillnic acid 28 mm. If acetone moved 50.9 mm towards positive pole?

[**Ans:** Corrected values in units of 10^{-5} cm/v/sec were vanillin 0.16, ferrulic acid 7.24 and vanillin acid 9.3]

20.14 What is the apparent mobility of glycine and serine in a phosphate buffer of pH 9.2?

[**Ans:** Glycine 0.95×10^{-4}, serine 1.43×10^{-4} cm/v/sec.]

20.15 What volume of 0.2 N buffer solution would be required for an overnight run of 15 hour at current of 3.5 nA? [**Ans:** About 1 liter]

20.16 Calculate the relative mobilities of dimer of unknown protein with molecular weight from the following data:

Protein	Lysozyme	Trypsin	Ovalbumin	Catalase	Bovine	Unknown
Mol. wt:	1.4×10^4	2.33×10^4	4.3×10^4	6.2×10^4	6.8×10^4	X
Rel. Mobility	9×10^{-1}	7.3×10^{-1}	5.4×10^{-1}	3.8×10^{-1}	4.0×10^{-1}	6.8×10^{-1}

[**Ans:** M.W.> 20,000]

20.17 What was the time needed to deposite 400 mg of copper by electrolysing it in an acidic solution of copper sulphate with a current of 4 A with 50% current efficiency?

[**Ans:** $t = 607$ seconds]

REFERENCES

20.1 S.M. Khopkar, *Basic Concepts of Analytical Chemistry,* 3rd edn. New Age International Publishers (P) Ltd. (2007).

20.2 J.A. Dean, *Chemical Separation Methods,* Van Nostrand Reinhold and Co., pp. 330, 347-9 (1970).

20.3 M. Bier, *Electrophoresis*, Academic Press (1959).

20.4 M. Lederer, *Paper Electrophoresis*, Academic Press (1959).

20.5 R.J. Block, R. Lestrange and G. Zweig, *Paper Chromatography a Laboratory Manual,* Academic Press (1973).

20.6 L. Smith, *Chromatography and Electrophoresis,* Heinemann-Inter Science (1960).

20.7 E.W. Berg, *Physical and Chemical Methods of Separation*, McGraw-Hill and Co. (1958).

20.8 R.D. Braun, *Introduction to Instrumental Analysis,* McGraw-Hill and Co., p. 883 (1987).

Chapter 21

Electrophoresis

INTRODUCTION

Amongst the various methods of separations used in the chemical analysis, electrophoresis technique is, most established method for the purpose of separations. It was largely used for the separation of DNA and RNA in biochemical separations.

This technique was later used for the separation on the support like filter paper. Many separations were possible on account of differential migration rate of ions under the influence of the electric charge. Such difference originated due to the difference in the ionic mobility of the ions. Those of the ions belonging to s-block metals had better mobility and therefore they migrated faster in relation to other ions like heavy metals with less ionic mobility.

Electrophoresis cannot be classified as the true chromatographic separation technique. This is because all separations are just possible in one phase only, while in chromatography methods there are at least two phases as one is stationary phase while other is mobile phase. However in chromatographic method of separation is due to differential migration rate of the charged ions as it facilitates the individual separations. Such difference originates from the difference in the ionic potential and ionic mobility. The technique works more efficiently on small samples. Slab electrophoresis is special class of electrophoretic technique involving the use of the thin layer of porous semi-solid gel buffered with solution as the stationary support. While planning electrophoretic separation one needs the knowledge of isoelectric point. It is defined as the pH at which no net migration of charged particles occur. It is abbreviated as pI. In the electrochromatography, better separations are possible due to difference in electroosmotic pressure of each ion. In order to get better separation by electrophoresis the ratio of the charge of ion to size of ion must be as large as possible because it facilitates the migration of ions under influence of large direct current. The rate of the separation in electrophoresis is quite slow in comparison to electrochromatography technique of the separation.

However this problem of rate was circumvented with the recent discovery of capillary electrophoresis; because miseller electro-kinetic technique permitted the separation of even neutral molecules. It was possible on account of reliable and the excellent detectors available during the analysis. Further the capillary electrophoresis (CE) technique permitted the separation at low concentration of the sample (viz, 10^{-15} fematogram concentration).

As a matter of fact excellent methods for the separation are available in chemical separation science. However the technique of electrophoresis is still popular with chemist working in biological sciences for the purpose of separation of biological products and complex materials.

We summarize phenomena of the electrophoresis as the movement of charged particles in conducting solution due to application of an electric field. The rates and the direction of the migration depends upon the sign and the magnitude of the charge or charge to size ratio. The electroosmotic flow (EOF) causes movement of the electrolyte through the medium. Highly charged ions move faster than the low charged ions. On passage of electric charge Joule's effect on heat generation is also observed leading to zone spreading. The addition of non-conductive stabilizing media like gel reduces the convective diffusion. As the compounds migrate through the detector at different times a electroplerogram is generated in which separated components appear as peaks with different migration rates. If solution migrates with velocity V_{ep} we have $V_{ep} = \mu_{ep}$ E. if μ_{ep} is electrophoretic mobility. E = applied voltage in good setup good sensitivity can be easily obtained. The solutes are separated if $\mu_{ep} = \dfrac{q}{\sigma\pi\eta r}$ if q = charge of ionic solutes, η = viscocity of an electrolyte and r = radius of the solute. $A(q/r)$ with greater charge to size ratio offers better separation. Neutral molecules can not be easily separated but cations and anions can be easily separated. The order for separation of the anions is: $Br^- < Cl^- < SO_4^{2-} < NO_2^- < NO_3^- < MoO_4^{2-} < SCN^- < F^- < PO_4^{3-}$ etc.

PROBLEMS

Multiple Choice Questions

Tick mark correct answer from following multiple choices:

21.1 Electrophoresis is separation method based upon
(*a*) difference in the rate of migration of charged species under influence of current
(*b*) the difference in ionic potential of elements
(*c*) electron affinity of the ions
(*d*) electro-neutrality of the ions

21.2 The most striking feature of electrophoresis separation is
(*a*) the separation of DNA and RNA
(*b*) metal and non-metal separations
(*c*) charged cation or anion separations
(*d*) separation of substance with low molecular weight

21.3 Electrophoretic separation is carried out
(*a*) on very small sample of solution
(*b*) employing low DC potential in buffer solution
(*c*) on small size charged ions only
(*d*) samples with smaller ratio of charge to size

21.4 Slab electrophoresis is usually carried on
(*a*) glass plates
(*b*) filter paper
(*c*) thin layer of porous semi-solid gel with buffer solution
(*d*) in small capillary tube

21.5 Slab electrophoresis is widely used for the separation of
(*a*) macro molecules (*b*) polymers
(*c*) biochemical and biological products (*d*) microlevel concentrations

21.6 The phenomena of electrophoresis proceeds due to difference in
(*a*) size of ions (*b*) charge of ions
(*c*) eletrophoretic mobility (*d*) ionic and non-ionic nature of the species

21.7 The difference in electrophoresis and the electrochromatography originates from the fact that
(*a*) one phase is involved in electrophoretic separation
(*b*) the stationary support is different in two methods
(*c*) rate of migration of ions is slower in chromatography
(*d*) multi element separation is not fissible in chromatography

21.8 Isoelectric point is defined as (pI) the
(*a*) pH at which no net migration occurs
(*b*) is useful in analysis of aminoacids
(*c*) ionisation constant is not dependent in system
(*d*) applicable only in analysis of proteins

21.9 Electropherogram is graph of
(*a*) absorbance vs migration time (*b*) transmittance vs rate of migration
(*c*) amount of ion vs migration time (*d*) potential vs size of carbon

21.10 Electrochromatography works better due to electroosmotic pressure to facilitate
(*a*) separation of uncharged ions (*b*) positively charged ions
(*c*) negatively charged ions (*d*) due to difference in concentration in mobile and stationary phase

21.11 The first application for separation by slab electrophoresis was in
(*a*) biochemistry of proteins, (*b*) inorganic separation
(*c*) organic products (*d*) radioactive material

21.12 The kind of electric supply used in electrophoresis is
(*a*) AC (*b*) DC
(*c*) AC/DC (*d*) modulated current

21.13 The charge to size ratio in good separation should be
(*a*) as large as possible (*b*) as small as fissible
(*c*) no definite relation (*d*) exceeding small

21.14 Slab electrophoresis is designated as such because one uses:
(*a*) porous material (*b*) glass tubes
(*c*) filter paper (*d*) column

21.15 The ions would migrate on slab electrophoresis continuously till:
(*a*) stationary support is good (*b*) stationary phase is compatible
(*c*) DC potential is applied (*d*) AC potential is used

21.16 One needs to use DC potential, which is reasonably
(*a*) low (*b*) high
(*c*) any magnitude (*d*) very small

21.17 Electroosmotic pressure doesn't facilitate the separation of
(*a*) cations (*b*) neutral molecules
(*c*) anions (*d*) amphoteric products

21.18 Misceller Electro Kinetic (MEK) capillary electrophoresis permits separation of
(*a*) neutral molecules (*b*) charged molecules
(*c*) uncharged molecules (*d*) coloured molecules

21.19 Numerous application are encountered for slab electrophoresis in separations of
(*a*) organics (*b*) cations
(*c*) anions (*d*) biological products

21.20 Capillary electrophoresis is most useful in
(*a*) biochemistry (*b*) organic separations
(*c*) inorganic separations (*d*) electrochemical research

21.21 Electrophoresis technique is not comparable with chromatography as
(*a*) separation are slow (*b*) doesn't permits separation of similar ions
(*c*) separation of ionic species (*d*) separation only at milligram level

21.22 Capillary electrophoresis proved versatile in relation to slab electrophoresis due to
(*a*) availability of excellent detectors (*b*) availability of good capillary
(*c*) possibility of eluant to travel in capillary (*d*) presence of impurities in sample

21.23 Capillary electrophoresis became popular as it facilitated analysis at
(*a*) very low concentration (*b*) high concentration
(*c*) microgram concentration (*d*) femagram concentration

21.24 In comparison to slab electrophoresis separations, the capillary electrophoresis is carried out in few
(*a*) hours (*b*) minutes
(*c*) seconds (*d*) days

21.25 Electrophoresis in general is reasonable tool for separation of the charged:
(*a*) molecules (*b*) ions
(*c*) cationic species (*d*) anions in complexes

ANSWERS

21.1 b	21.2 a	21.3 a	21.4 c	21.5 c	21.6 c	21.7 a
21.8 a	21.9 a	21.10 a	21.11 a	21.12 b	21.13 a	21.14 a
21.15 c	21.16 b	21.17 b	21.18 a	21.19 d	21.20 a	21.21 a
21.22 a	21.23 d	21.24 c	21.25 b.			

REFERENCES

21.1 F.A. Settle, *Handbook of Instrumental Techniques in Analytical Chemistry,* Pearson Education Publishers, p. 765 (2004).

21.2 E.W. Ewing, *Instrumental Methods of Analysis,* McGraw-Hill Publication, p. 505 (1958).

21.3 D.A. Skoog, F.I. Holler and T.A. Nieman, *Principles of Instrumental Analysis,* 5th edn., Harcourt Brace Pub, p. 779 (1995).

21.4 S.M. Khopkar, *Basic Concepts of Analytical Chemistry,* 3rd edn., New Age International Publishers (P) Ltd. (2007).

21.5 S.M. Khopkar, *Instrumental Methods in Bioanalytical Chemistry*, New Age International Publishers (P) Ltd. (In press).

Chapter 22

Capillary Electrophoresis

INTRODUCTION

As described earlier, normal electrophoresis technique was mainly used for the separation of biological products like amino acids, proteins, however it was not used for the analysis of inorganic ions. Only with the invention of the technique of capillary electrophoresis, it was possible to analyse inorganic ions by this new method. It was further rendered simple on account of the availability of efficient detectors for the measurement of absorbance, conductance. On account of the intrinsic sensitivity of these detection techniques, one could go down to the lowest level of analysis. Such lowest level of analysis was extended from 10^{-15}-10^{-18} gm.

For the analysis of volatile compounds capillary electrophoresis was rarely used. For such analysis the preferred method was gas liquid chromatography. One cannot directly analyse any solid samples by any of the chromatographic method. The sample has to be dissolved in suitable electrolyte before the analysis. Although the technique is not as fast as HPLC, it provides reliable results. This variety was used for the analysis of the compounds with low molecular weights specially, which might be polar in nature. It is necessary to apply the potential as high as 30 kv in order to get the reproducible results. Further one prefers tubes made out of silica for easy handling of the samples. The multicomponent mixture which contains several ions show no promising results if they are analysed by capillary electrophoresis. For detection of smallest possible ions one has to resort to the use of laser induced fluorescence method. For the analysis of neutral molecules one prefers the misceller electro-kinetic capillary chromatography where analysis time is also short.

In comparison with slab electrophoresis this technique has good resolution and is relatively rapid due to the availability of sensitive detectors. The separations are efficient on account of the presence of numerous theoretical plates in the separating column. Inspite of the availability of numerous detectors the most commonly used detector is based on spectroscopic principle. The ions with very small ionic size are effectively analysed by capillary electrophoresis.

In the technique of capillary iso-electrophoresis all the ions in band move with the identical velocity. If one is confronted with the problem of microscale analysis, it is always better to prefer the technique of capillary gel electrophoresis. Several chemical properties can be predicted from electropherogram. It is graph of absorbance against time of migration. The detectors to be used are conductometry, amperiometry or pulse polarography but one never uses potentiometric methods for obvious reasons. We expect brisk activity in the field of biochemical analysis with the technique of capillary electrophoresis.

PROBLEMS

Multiple Choice Questions

Tick mark correct answer from the following multiple choices given

22.1 Capillary electrophoresis cannot be used for the separation of:
(*a*) polar and non-polar compounds (*c*) inorganic cations and anions
(*b*) ionic and non-ionic materials (*d*) volatile components

22.2 The solid sample can be analysed by capillary electrophoresis if
(*a*) it is dissolved in liquid or with an electrolyte injected in capillary
(*b*) such analysis depends upon kind of detector used
(*c*) where the level of analysis is critical
(*d*) the time needed for separation is rather large

22.3 The capillary electrophoresis permits analysis of the samples at the level of
(*a*) 1-50 ηg (nano) (*c*) 1-10 mg (milli)
(*b*) 1-100μg (micro) (*d*) 1-10 ρg (pico)

22.4 Capillary electrophoresis technique is not sensitive in comparison to
(*a*) HPLC (*c*) RPPC
(*b*) LLPC (*d*) GLC

22.5 Low molecular weight volatile compounds are better analysed by
(*a*) SFC (*c*) HPLC
(*b*) SFE (*d*) GLC

22.6 Capillary electrophoresis permits analysis of:
(*a*) non-polar low molecular weight compounds
(*b*) non-polar high molecular weight polymers
(*c*) evaluation of isoelectric point of biomolecule
(*d*) dry solid samples

22.7 The extent of voltage employed in capillary electrophoresis is of the order of
(*a*) 30 kv (*c*) 10 kv
(*b*) 20 kv (*d*) 5 kv

22.8 The kind of material used for construction of capillary support is made out of
(*a*) Glass (*c*) Fused silica
(*b*) Stainless steel (*d*) Titanium metal

22.9 One of the following parameter does not affect electroosmotic pressure in separation, it is
(*a*) increase in voltage (*c*) change in ionic concentration
(*b*) change in ph of an electrolyte (*d*) with sample containing too many ions

22.10 The detector with the lowest detection limit is
(*a*) uv-visible detector (*b*) fluorescence detector
(*c*) laser induced fluorescence detector (*d*) amperiometric detector

22.11 Miceller electrokinetic capillary chromatography (MECC) is used to separate one of the following compounds viz.
(*a*) Neutral molecules (*b*) Charged molecules
(*c*) Proteins (*d*) Detergents

22.12 One of the prominently known drawback is not encountered in capillary electrophoresis separation for
(*a*) mixture containing many component
(*b*) use of high voltage during separation
(*c*) long separation time due to long columns
(*d*) need to use small volumes of the samples

22.13 Capillary electrophoresis facilitates quantitative analysis of
(*a*) neutral molecules (*b*) inorganic cations as well as anions
(*c*) only cations (*d*) only anions

22.14 In comparison with slab electrophoresis; capillary electrophoresis
(*a*) has high speed & good resolution
(*b*) has low speed but offers reasonable separation
(*c*) is applicable at milligram concentration
(*d*) it needs large volume of sample

22.15 Capillary electrophoresis includes
(*a*) detectors which facilitating quantitative analysis
(*b*) chromogenic ligands for identification
(*c*) porous support to develop fronts
(*d*) use of large volume of samples

22.16 The total number of plates in capillary electrophoresis are approximately
(*a*) 5,000-10,000 (*b*) 10,000-20,000
(*c*) 20,000-50,000 (*d*) 100,000-200,000

22.17 The preferred method for quantitative analysis of metals in capillary electrophoresis is
(*a*) absorption spectrophotometry at low level of analysis
(*b*) emission spectroscopy at trace concentration
(*c*) mass spectrometery if concentration is high
(*d*) conductivity measurements for large samples

22.18 The small size ions are better separated by technique of
(*a*) slab electrophoresis (*b*) capillary gel electrophoresis
(*c*) capillary zone electrophoresis (*d*) electrochromatography

22.19 In capillary isotachophoresis all bands travel during separation
(*a*) at same velocity (*b*) at different velocity
(*c*) no particular order (*d*) do not travel at all

22.20 The micro scale electrophoretic separations are carried out by
(*a*) slab electrophoresis (*b*) capillary zone electrophoresis
(*c*) capillary isotachophoresis (*d*) capillary gel electrophoresis

22.21 Electrochromatography is hybrid technique consisting of
(*a*) slab electrophoresis and HPLC (*b*) gel electrophoresis and GLC
(*c*) capillary electrophoresis and HPLC (*d*) electrophoresis and liquid chromatography

22.22 Electropherogram is graphical representation of
(*a*) absorbance vs time of migration (*b*) absorbance vs concentration
(*c*) flow rate vs sample size (*d*) intensity of signal vs volume of sample

22.23 One of the following technique is not used for the detection in capillary technique, it is
(*a*) conductometry detector (*c*) potentiometry detector
(*b*) amperiometry detector (*d*) pulse polarography detector

22.24 The smallest concentration at which capillary electrophoresis operates is
(*a*) microgram (*b*) nanogram
(*c*) picogram (*d*) femagram

22.25 All the analyte bands migrate at same velocity in
(*a*) capillary isotachophoresis (*b*) capillary zone electrophoresis
(*c*) capillary gel electrophoresis (*d*) capillary isoelectric focusing

ANSWERS

22.1 d	22.2 a	22.3 d	22.4 a	22.5 d	22.6 a
22.7 a	22.8 c	22.9 d	22.10 c	22.11 a	22.12 c
22.13 b	22.14 a	22.15 a	22.16 d	22.17 a	22.18 c
22.19 a	22.20 d	20.21 c	20.22 a	22.23 c	22.24 d
22.25 a.					

REFERENCES

22.1 F.A. Settle (Ed.) *Handbook of Instrumental Techniques in Analytical Chemistry,* Pearson Education Publishers, p. 165 (2004).

22.2 D.A. Skoog, F.J. Holler and T.A. Nieman, *Principle of Instrumental Analysis*, 5th edn., Harcourt Brace Publishers, p. 778 (1998).

22.3 R.D. Braun, *Instrumental Methods of Analysis*, McGraw-Hill & Co. Indian Edition (2004).

22.4 S.M. Khopkar, *Basic Concepts of Analytical Chemistry*, 3rd edn., New Age International Publishers (P) Ltd. (2007).

22.5 S.M. Khopkar, Instrumental Methods in Bioanalytical Chemistry, New Age International Publishers (P) Ltd. (In press).

Chapter 23

Analytical Spectroscopy

INTRODUCTION

The analytical absorption and emission spectroscopy has become very popular tool for quantitative analysis since ages. When an atom or molecule is excited, the electrons from ground level migrate to higher energy level for absorption of radiation. However such an excited electron cannot remain in excited state indefinitely, so it has to release the acquired additional energy; and return back to ground state. Such release of energy is possible by either absorption or emission of radiation at a particular wavelength. This is also possible with either molecules or atoms. In the common, absorption spectroscopy, the energy is just absorbed by sample at particular wavelength. Such absorption is possible in ultraviolet, visible or in infrared region. This leads to the origin of molecular absorption spectroscopy like, UV visible and IR. In an event such absorption occurs, after excitation of atoms, we call it atomic absorption spectroscopy (AAS). Instead of absorption, the energy can be emitted in form of radiation in molecular phenomena it is called molecular emission or as molecular luminescence while in atom it causes atomic emission although atomic fluorescence is rare. This gives rise to atomic emission spectroscopy (AES) or flame emission spectroscopy, (FES) or ICP-AES Plasma spectroscopy.

If P_o is intensity of incident radiation, P_a = intensity of the absorbed radiation, P_t = intensity of transmitted radiation and P_r = intensity of reflected light, we have $P_o = (P_a + P_t + P_r)$ for practical purpose $P_r < 4\%$ is neglected so $P_o = (P_a + P_t)$ Beer's and Lambert law is formally defined as one correlating concentration with the absorbance since $T = \left(\frac{P_t}{P_o}\right) = 10^{-abc}$ if T = transmittance, b = optical path and c = concentration

$\hat{a}$ = absorpivity, we have then,

$\log T = \log\left(\frac{P_t}{P_o}\right) = -abc$ by invertendo we get an expression as

$\log\frac{I}{T} = \log\frac{P_o}{P_t} = abc = A$ while $\frac{I}{T} = T^{-1}$ is called opacity.

$-\log T = abc = A$ if A is called absorbance of the solution.

We have $A = abc = -\log T = \log \frac{1}{T}$

$A = abc$ is Beer's Law we observe in actual practice that

$A \alpha b$ if c is constant, $A \alpha c$ vary. if b is constant. If $b = 1$, $c = 1$ then

$A = a$ it is absorbance of solution of one gram mole/liter solution if optical path = 1 cm. Beer's law is valid for dilute solutions only.

Since we have $E = h\nu$ if energy $h = 6.625 \times 10^{-34}$, ν = frequency.

If C = Velocity of light then we have *i.e.* $x = \nu = 1/\lambda$ if we consider

$E = h\left(\frac{C}{x}\right) = hC\nu$ refractive index (n) then we have.

$E = \left(\frac{h}{n}\right)\left(\frac{C}{x}\right)\left(\text{because } n = \frac{C}{\nu} \text{ or } \lambda\nu = C/n\right)$ and n = refractive index

What are causes of deviation of Beer's Lambert law (1852)? We must keep temp, pressure constant. Instrumental error is caused if we have fluctuating source, bad wavelength selector (non-chromatic) poor slit apperture control. The chemical deviation like pH, presence of complexing agent, refractive index also causes error in analysis showing wrong value for absorbance or percentage transmittance.

The reliability of spectral method depends upon sensitivity and relative concentration error in analysis. The concentration corresponding to 0.010 absorbance is called 'Sandeli's sensitivity' expressed as mg/ml/cm^2. The optimum absorbance which can be used for analysis with minimum error is evaluated from Ringbom's plot of (log $C = T$) log concentration versus transmittance.

$A = abc$ so $c = \frac{A}{ab}$ as $A = -\log T$

$c = \frac{-\log T}{ab}$. Differentiating both sides with respect to T we have

$\frac{dc}{dT} = \frac{-0.4343}{T(ab)} = \frac{0.4343c}{T(\log T)}$.

$\frac{dc/c}{dT} = \frac{-0.4343}{T(\log T)}$. Now if we have error proportional to T (log$T$) if we set function to zero differentiate w.r.t. (T) we have

$T = \frac{1}{a} = 0.368;\ A = 0.4343,$ this shows when the absorbance of solution A = 0.4343 or percentage transmittance T = 36.8% we have the minimum relative error. A plot of log c against T is straight line graph with slope = θ

We have

$$\frac{dc/c}{dt} = \frac{2.303}{T/d\log c} = \frac{2.303}{\text{Slope of Plot}}$$

Such plot of –Log c vs T is called as Ringbom plot. This gives an idea of absorbance value with minimum error in analysis.

Finally let us consider a situation of a coloured solution which is having high refractive index (n) we know relation (n) with wavelength as $n = C/v$ if C = Velocity of light, ν = frequency if λ = wavelength, $\lambda v = C/n$ so in terms of energy of radiation we have equation $E = \frac{hC}{n\lambda} = \frac{dCv}{n}$. This shows '$n$' will affect energy and hence absorbance so in order to correct absorbance we should multiply each of the absorbance value by function $\left(\frac{an}{(n^2+2)^2}\right)$ if a = absorbivity in order to get corrected absorbance

$$A = \left[\frac{an}{(n^2+2)^2}\right]$$ to give true value of absorbance it is called as Kortum Seiler equation.

Thus we note that in 1864, the law of spectroscopy was discovered simultaneously by Beer, Lambert and Bouger. It forms the fundamental law of any spectroscopic technique. The term (A) has definite value, it is always in fraction. The value $A = 0.001$ has no meaning excepting it is used to express absolute sensitivity similarly $A > 1.0$ is also absurd as beyond 0.850 we do not expect linearity of Beer's Plot. In UV, visible spectroscopy we express it in absorbance and wavelength while in IR spectroscopy we express in % T transmittance w.r.t. (ν) frequency. The term a which is called absorbivity also called as 'molar absorbivity' if c = gms moles/liter has definite meaning. If any value of a is 1×10^3 to 10^5 the method is considered to be a good method. One cannot go beyond $a = 10^5$ even with dyes as chromogenic ligands as instrumental accuracy for measurement of (A) is limited to 3-places of decimal for absorbance.

In this chapter numerous problems have been given to illustrate the Beers law and concept of sensitivity and kind of errors encountered in spectral analysis.

SOLVED PROBLEMS

23.1 The absorbivity of a coloured complex was 3.20×10^3 at 240 nm. Calculate the absorbance of a 5×10^{-5} M solution in 50 mm cell when it was measured at 240 nm?

Ans: As per Beer's law since $A = abc$

If $a = 3.20 \times 10^3$, $b = 5$ cm $c = 5 \times 10^{-5}$ M we substitute these values.

$A = (3.20 \times 10^3)\ (5)\ (3 \times 10^{-5}\ \text{M}) = 0.800.$

23.2 Calculate the molar absorbivity of a compound with molecular weight = 144 if 1×10^{-5} g/ml solution exhibited an absorbance of 0.400 with the optical path of unity.

Ans: In this example

$$c = \frac{1\times10^{-5}}{144} \qquad A = 0.400,\ b = 1,\ a = \text{unknown so}$$

$$A = abc \text{ or } \frac{A}{bc} = a, \text{ i.e., } \frac{0.400}{(1\times10^{-5})\times1} \text{ i.e., } 6.9\times10^{-5} = c$$

So $0.400 = a \times 1 \times 6.9 \times 10$

$a = 576 \times 10^7$ gm/mol/cm^2

23.3 What was the concentration of the uranium in 100 ml volume if 25 ml if an aliquat had an absorbance of 0.428 at 430 nm. If from external one ml solution containing 0.050 mg of uranium was added and made up to 25 ml it showed absorbance of 0.517?

Ans: The corrected absorbance for second solution was

$$0.517 \times \frac{26}{25} = 0.538 \text{ for 25 ml of solution}$$

i.e., absorbance by one ml of 0.050 uranium was (0.538 – 0.428) = 0.195.
0.195/25 ml = 0.078/ml was in 100 ml. *i.e.*, 0.00078/ml in 2 ml.
one ml contained 0.00078/2 = 0.000399 mg/100 ml = 3.9×10^{-4}/100 ml.

23.4 Calculate the absorbance as well as transmittance of 3.4×10^{-4} M coloured thiocyanate complex of iron which had molar absorbivity of 3.2×10^2 at 580 nm?

Ans: $A = abc$ so $A = (3.2 \times 10^2)\ (1)\ (3.4 \times 10^{-4})$ so $A = 1.09$, $a = 3.4 \times 10^{-4}$, $b = 1$, $c = 3.4 \times 10^{-4}$.
$A = (2 - \log\% T)$ since $A = 1.09$.
$1.09 = (2 - \log\% T)$ or $-\log\% T = -0.091$ so $T\% = 8.1\%$.

23.5 What is the detection limit of beryllium complex with Eriochrome cyanine *R* if it had maximum absorbance at 530 mm due to species $[Be(H_2O)_6]$ and molar absorbivity of 1×10^{-1}?

Ans: The detection limit was the smallest amount of chromophoric metal species showing an absorbance of 0.010 = *A*, if cell path was 10 mm; if $a = 1 \times 10^{-1}$ cm we have $A = abc$,
$A = 0.010$, $b = 1$, c = unknown,
$0.010 = (1 \times 10^{-1})\ (1)\ (c)$ or

$$c = \frac{0.010}{0.10} = 1 \times 10^{-3} \text{ gm mole/liter/cm}^2$$

This is smallest detectable concentration of berylyum complex.

23.6 Calculate the sensitivity of molybdenum thiocynate complex at 620 mm if optical path was 10 mm, $a = 1 \times 10^3$; c = unknown, $A = 0.010$ sensitivity limit?

Ans: so $A = abc$ $0.010 = (1 \times 10^3)\ (1)\ (c)$ on simplification
$c = 1 \times 10^{-5}$ M is smallest measurable concentration or detection limit.

23.7 If $[Fe(H_2O)_6]^{3+}$ complex at 440 mm showed an absorbance with minimum detectable limit of 0.100 with absorbivity as zero. State whether you can measure the hydrated species by spectrophotometer?

Ans: $A = abc$, $A = 0.100$, $a = 0$, $b = 1$, c = unknown
$0.100 = (0)\ (1)\ C$ so $c = 10^{-2}$ M. Yes it cannot be easily measured by spectrophotometry.

23.8 Convert 200 nm to (1) cm (2) cm^{-1} (3) cv (4) cg, (5) cal and (6) Hz.

Ans: (1) 2000°A = 200 nm or 2×10^3 A° so A° → cm Factor –(10^{-8} cm/A°)
$2 \times 10^3 \times 10^{-8} = 2 \times 10^{-5}$ cm

(2) cm-cm^{-1} use $\left(\text{factor} \frac{1}{\text{cm}}\right)$ so $\frac{1}{2 \times 10^{-5}} = 5 \times 10^4 \text{ cm}^{-1}$

(3) cm^{-1} → ev use factor 8068.3 cm^{-1}/cv

$$\frac{5 \times 10^4}{8068.3} = 6.2$$

(4) cm erg use $\left(\text{factor } E = \frac{hc}{\lambda}\right)$ so $E = \frac{3\times10^{10}}{2\times10^{-5}} = 9.940\times10^{-12}$ erg/mol.

(5) ev → cal use (factor 23,063 cal mol^{-1} cv^{-1})
So 6200 ev × 23,063 = 1.430 × 10^5 cals/mole

(6) cm → Hz use (Factor ν = 3 × 10^{10})
2 × 10^{-5} cm × ν = 3 × 10^{10} so ν = 1.5 × 10^{15} Hz.

23.9 **What is the relative concentration error for the coloured complex having an absorbance of 0.200 with concentration of 2 × 10^{-2} M?**

Ans: The relative concentration error was evaluated as:

$$c = \frac{A}{ab} = \frac{-\log T}{ab} \text{ or } \frac{dc}{dT} = \frac{0.4343}{T(ab)} \text{ i.e., } \frac{dc/c}{dT} = \frac{0.4343}{T(\log T)}$$

So if we should plot graph of log *c* against *T* (transmittances) to find out relative concentration error.

23.10 **Convert *A* = 0.523 to transmittance and 75% transmittance to absorbance scale.**

Ans: (1) Since *A* = –log *T*
0.523 = –log *T*, $T = 10^{-abc}$ = 75; *T* = 75 = 10^{-abc}, log 75 = *abc* log 10 = 0.602

23.11 **Calculate the (*i*) molar absorbivity; (*ii*) absorbance and (*iii*) concentrations with following data:**

A = 0.474, *c* = 6.51 × 10^{-4}, *b* = 1 mm

(*i*) molar absorbivity if *b* = 2.5 cm, *c* = 3.40 × 10^{-5} M % T = 88.4

(*ii*) absorbance if *a* = 2.67 × 10^4, *b* = 1.5, *c* = ppm (325) % T 31.2

(*iii*) Concentration if *A* = 0.126, *b* = 1.50, *a* = 2.67 × 10^4 *c* = *x*

Ans: (*i*) *A* = *abc* 0.474 = a × 0.1 × 6.51 × 10^{-4} = 7.28 × 10^3
% *T* = 88.4 given, *b* = 2.5, *c* = 3.40 × 10^{-5}; *A* = –log *T*
A = –log *T* = *abc*, a × 2.5 × 3.40 = –log*T* *a* = 63%

(*ii*) absorbance *A* = *abc*; *a* = 2.67 × 10^4, *b* = 1.5, *c* = 325 g/w, %*T* = 31.2
–log *T* = 2.67 × 10^4 × 1.5 × 325 = –log 31.2 so *A* = 0.506

(*iii*) *A* = *abc* 0.126 = 2.67 × 10^4 × 1.50 × *c*

$$c = \frac{0.126}{2.67\times1.5\times10^4} = 3.15\times10^{-6}\text{M.}$$

23.12 **What is the absorbance of solution if the molar absorbivity is**

(*i*) *a* = 7 × 10^2, *c* = 1.42 × 10^{-3}, *b* = 1, and

(*ii*) *a* = 7.63 × 10^3, *c* = 5 × 10^{-5}, *b* = 2.0

Ans: (*i*) *A* = *abc*; *A* = (7 × 10^2 (1) (1.42 × 10^{-3}) so *A* = 1.016

(*ii*) *A* = *abc*, *A* = (7.63 × 10^3) (2) (5 × 10^{-5}) so *A* = 0.763.

23.13 **Calculate the absorbance of coloured solution if concentration is 1.5 × 10^{-6} M which has molar absorbivity of 1.45 × 10^4 with optical path of 10 cm?**

Ans: *A* = *abc* *a* = 1.45 × 10^4, *b* = 10, *c* = 5 × 10^{-6} M
So *A* = (1.45 × 10^4) (10) (5 × 10^{-6} M) = 0.763

23.14 **Estimate probable optical path of solution with concentration of 8.32 × 10^{-5} M with molar absorbivity as 8.74 × 10^3 with absorbance of 0.897.**

Ans: $A = abc$ or $\dfrac{A}{ac} = b$

$$b = \frac{0.897}{(8.32\times10^{-5})\times(8.74\times10^{3})} = 11.86 \text{ mm}$$

23.15 Evaluate the concentration of a solution if it had an absorbance of 0.555 with molar absorbivity of 2.22×10^4 and optical path 1.5 cm.

Ans: $A = abc$, $A = 0.555$, $a = 2.22 \times 10^4$, $b = 1.5$.

$$\text{So } \frac{A}{ab} = c = \frac{0.555}{(2.22\times10^{4})(1.5)} = 3.15\times10^{-4} \text{ M.}$$

PROBLEMS FOR PRACTICE

23.16 Convert (*a*) absorbance to % transmittance 0.210, 0.650 (*b*) convert % T to absorbance; 32%, 5.4%. [**Ans:** (*a*) 61.7, 22.4% (*b*) 0.500, 1.3]

23.17 What is the absorbance of solution of concentration 1.03×10^{-3} M with molar absorbivity of 7.2×10^2 and optical path of 10 mm? [**Ans:** 0.740]

23.18 Calculate the molar absorbivity of solution of concentration 1×10^{-4} M with absorbance of 0.710 and optical path of 10 mm. [**Ans:** 7.1×10^3 liters/mol/cm^2]

23.19 When chromium (III) is oxidised to chromium (VI) as chromate charge transfer band is at 373 nm, with molar absorbivity of 1.5×10^4. State whether one can measure 10^{-5} M solution as chromate? [**Ans:** yes]

23.20 If the path length was changed from 5 cm to 10 cm, what will be % transmittance if observed value is 24.7%? [**Ans:** $T = 0.0284$ or 28.4% *T*]

23.21 Find out the frequency in reciprocal seconds in hertz corresponding to wavelength of 222 nm and 530 nm. [**Ans:** (*i*) 1.35×10^{15} Hz (*ii*) 5.66×10^{14} Hz]

23.22 Find out the frequency in wave number for following wavelengths as 0.222 nm; 530 nm; and 0.030 cm.

[**Ans:** 4.50×10^4 cm^{-1}; 1.89×10^4 cm^{-1}; 5.86×10^6 cm^{-1}]

23.23 Calculate the molar absorbivity of solution containing 8.43 ppm substance having transmission of 37.5% with 1.5 cm cell? [**Ans:** $a = 33.7$]

23.24 What is the absorbance of 3.61×10^{-5} M solution at 580 nm if cell width was 0.5 cm with molar absorbivity = 7×10^3 lit/cm/mol? [**Ans:** $A = 0.127$]

23.25 Calculate the concentration of a coloured complex if it showed an absorbance of 0.250 with cell path 100 mm with molar absorbivity of 5.89×10^3? [**Ans:** $c = 4.24 \times 10^{-5}$ M]

23.26 What was the optical path if a coloured complex had molar absorbivity of 1480 with absorbance of 0.750 with concentration of 1.40×10^{-3} M? [**Ans:** 7 mm]

23.27 Calculate the concentration of solution of coloured complex which had absorbivity of 6.74×10^3/lit/cm/mol if optical path was 2.5 cm and % $T = 7.77\%$. [**Ans:** $c = 6.5 \times 10^{-5}$ M]

23.28 Calculate the % transmittance if its concentration was trippled when solution of molybdenum thiocyanate complex had a concentration with transmittance of 80%? [**Ans:** 26.6]

23.29 Calculate the double transmittance and dilution required if given solution had transmittance 20% of light a certain wavelength if the absorbance was half value.

[Ans: (I) 1/2 and (II) 4/7]

23.30 A column of a solution of potassium dichromate had a transmittance of 34% for light at wavelength of 480 mm. What was the transmittance of light through a column was adjusted to 1/5 of original length?

REFERENCES

23.1 S.M. Khopkar, *Basic Concepts of Analytical Chemistry*, 3rd Ed. New Age International Publishers (P) Ltd. (2007).

23.2 D.A. Skoog and D.M. West, *Analytical Chemistry,* 3rd edn., Holt, Reinhart and Winston, pp. 470, 476-7 (1979).

23.3 D.J. Pietrzyke and C.W. Frank, *Analytical Chemistry—An Introduction,* Academic Press, pp. 111, 103, 121, 127 (1979).

23.4 J.S. Fritz and G.H. Schenk, *Quantitative Analytical Chemistry,* Allyn and Baccon Inc. pp. 434-5, 459 (1979).

23.5 D.G. Peter, J.M. Hayes and G.M. Hieftze, *A Brief Introduction to Modern Chemical Analysis,* Saunders and Co., pp. 407, 620 (1980).

23.6 E.B. Sandell, *Colorimetric Determination of Trace Metals,* 3rd edn., Wiley-Interscience (1962).

23.7 G. Charlot, *Colorimetric Determination of Elements,* Elsevier Pub. Co. (1964).

23.8 Z. Marchenko, *Spectrophotometric Determination of Metals,* Eills Horwood and Co. (1980).

23.9 F.D. Snell, *Photometric and Fluorimetric Methods of Analysis of Metals*, Wiley-Interscience (1980).

23.10 A. Babko and A. Pilipenko, *Photometric Analysis*, Mir Publishing House, Moscow (1971).

Chapter 24

Ultraviolet Visible Spectrophotometry

INTRODUCTION

The ultraviolet absorption spectrometry is mostly used for the elucidation of chemical structure of the organic compounds or structre of metal-ligand complexes. However visible spectrophotometry is largely used for the quantitative analysis of metals which form coloured co-ordination complex with chormophoric ligands.

The absorption in uv-region is due to involvements of π σ and n-electrons. The electrons jump from ground state or orbital to anti-bonding orbital, *e.g.*, $\sigma \rightarrow \sigma^*$, $\pi \rightarrow \pi^*$, $\pi \rightarrow \sigma^*$ or $n \rightarrow \pi^*$ transitions wherein "σ^* and π^* are anti-bonding orbital", in CH_4 λmax at 125 mm due to $\sigma \rightarrow \sigma^*$ transitions while series of saturated compounds with unshared electrons exhibit $n \rightarrow \sigma^*$ transition $\pi \rightarrow \pi^*$ transitions with maxima absorption between 200-700 nm.

Apart from these, transitions in d and f orbitals electron gives rise to UV visible absorption maxima, *e.g.*, lanthanides and actinides. In fact d-orbital degenrate to t_{2g} and e_g, orbitals giving rise to (Δ) *i.e.*, crystal field spliting into t_{2g} and e_g orbitals. This further lead to formation of spectrochemical series as *e.g.*, $I^- < Br^- < Cl^- < OH^- < C_2O_4^{2-} < H_2O < NH_3 < SCN^- < en < NO_3^- < CN^-$. Another bright possibility of absorption of radiation is due to what is known 'charge transfer spectras', *e.g.*, $[Fe^{2+}\ Fe^{3+}\ Fe(CN)_6]^{3-}$ or $[Fe(o\text{-}phen)_3]^{3-}$ or $[Fe(SCN)_3]^{3-}$ it leads to charge transfer phenomena where in the components are both electron donar and electron acceptor. The complete absorbance at longer wavelength gives red shift or bathochromic shift of wavelength when metal behaves as electron acceptor. If the absorption shifts to shorter wavelength this is called blue shift or hyposochromic shift. In metal complexes common shift is to longer wavelength is called as bathochromic or red shift. The complexation of metals with chromophoric ligands leads to bathochromic or red shift of maxima. These kind of compounds are designed as bathochromic or hyperchromic compounds.

Supposing we have mixture of two coloured solutions like $K_2Cr_2O_7$ and $KMnO_4$ and if we desire to analyse them with spectrophotometer without undertaking separation then we can resort to technique of simultaneous spectrophotometry. Thus for compound (1) if λ_1 is maxima with molar absorbivity as a'_{λ_1}. If $A\lambda_1$ is a absorbance of mixture at first wavelength and $A\lambda_2$ is the absorption of a mixture at λ_2, and if C_1 is concentration of (1) and C_2 is concentration of (2) second compound and if the concentration to be found for mixture is unknown. We know $A = abc$ if optical path is kept constant $b = 1$ we have $A = ac$ or $A/c = a$. We evaluate at λ_1 for first complex a molar absorbivity so also at λ_2 we get two absorption values for second individual compounds and we measure absorbance of mixture at $\lambda_1\lambda_2$

We have

$$A\lambda_1 = (a_1c_1)\lambda_1 + (a_2c_2)\lambda_1 \quad ...(i)$$

$$A\lambda_2 = (a_1c_1)\lambda_2 + (a_2c_2)\lambda_2 \quad ...(ii)$$

which on simplifications gives values of C_1 and C_2 as

$$C_1 = \frac{(a_2)\lambda_1 A\lambda_1 - (a_2)\lambda_1 A\lambda_2}{(a_1)\lambda_1(a_2)\lambda_2 - (a_1)\lambda_2(a_2)\lambda_2} (ii) \text{ and } C_2 = \frac{(a_2)\lambda_1 A\lambda_1 - (a_2)\lambda_1 A\lambda_2}{(a_1)\lambda_1(a_2)\lambda_2 - (a_2)\lambda_1(a_1)\lambda_2} (iii)$$

of which three at λ_1 and three are at λ_2 both for mixture and individual solutions and solve simultaneous equations.

In case of differential spectrophotometry, there are no intricate calculations. In high absorbancy method we use a blank which is just less concentrated than strong solution. While in low absorbance method we make adjustment of sensitivity with absorbance scale at 0.700 instead of 0 and we measure the decrease or change in absorbance value from difference.

In photometric titrations, we plot graph of absorbance against volume of titrant. Since solution undergoes dilution we give correction to volume by multiplying resulting reading each time with $\frac{(V+v)}{v}$ if V = original volume; v = increment of vol. added.

Also we can evaluate pK_a of ligand by spectral methods involving spectrophotometry as follows:

$$HR = H^+ + R^- \text{ so } K_a = \frac{[H^+][R^-]}{[HR]} \text{ Taking logs of both sides}$$

So $$\log K_a = \log[H^+] + \log[R^-] - \log[HR]$$

i.e., $$-\log[H^+] = -\log K_a + \log[R^-] - \log[HR]$$

$\mathrm{pH} = pK_a + \log\frac{[R^-]}{[HR]}$ in other words, we rewrite it as $\mathrm{pH} = pK_a + \log\frac{(\text{alkaline form})}{(\text{acidic form})}$ of the ligand. In actual practice, we measure absorbance of ligand in alkaline and in acidic form and also of ligand buffered at fixed pH at these two wavelengths and then evaluate pK from molar absorption of alkaline anion and acid form of ligand.

To ascertain composition of coloured complex we usually use Yoe's mole ratio method, or Job's method of continuous variation. A plot of absorbance against $\left[\frac{M}{R}\right]$ or A against $\left(\frac{M}{M+R}\right)$ mole fraction is made and slopes gives value of 'n' in MR_n. This method is valid for monomeric complexes but for the polymeric or extracted complex reliable results are not available.

Finally the evaluation of sensitivity can be experimentally done by measurement of concentraction of absorbing species with absorbance of 0.01 or 0.001. Former is called Sandell's sensitivity. One can also measure sensitivity and molar absorbivity from following equation $S = n\left(\frac{M}{a}\right)$ if S = sensitivity,

M = Mol. wt., n = nos. of atoms in element so $a = n\left(\frac{M}{S}\right)$ where 'a' is molar absorbivity. Further if we plot the graph of A against C (concentration) in Beer's Law plot the slope θ gives the magnitude of the product of molar absorption and optical path (ab). In order to get reproducible results, complexation reaction must be complete and the stable, complex must absorb in UV or visible region. The absorption spectrum of complex should not overlap with spectra of ligand.

While judging spectrophotometric method, one must check molar absorpivity, stability and sensitivity of complex, influence of pH; absorption maxima (λ_{max}), nature of the reaction, adherence to Beer's law and kind of interferences encountered while verifying Beer's law plot the slope is not always 45°. It may be 30° or 60° one with 60° is best showing high absorbance with low metal concn. If cuvetts are coloured plot will show interception on y axis with constant absorbance.

The basic requirement in simultaneous spectrophotometry is absorbance of two species must be additive and two components must absorb at two different wavelength which are separated wide apart. The compounds showing shift to longer wavelength are called as bathochromic compounds. While those showing decrease in the absorbance intensity are called hypsochromic compounds.

The significant advantage of using photometric titration is colour change can be easily measured. Further the method can be extended to strongly coloured compounds, end point is obtained by adding small increment of titrant. The end point can be obtained by extrapolation of result. Indicator colour change are very quickly detected.

The spectrophotometric method permits one to easily calculate concentration from Beert plot or evaluate molar absorbivity. By use of simultaneous spectrophotometry we can find out concentration of two species in mixture without carrying out the separation. The absorption of solution with very high or very low value can be measured by differential spectrophotometry. The pK of reagent can be spectrophotometrically evaluated.

SOLVED PROBLEMS

24.1 A 1×10^{-3} M. solution of potassium dichromate showed on absorbance of 0.200 at 450 nm and an absorbance of 0.050 at 530 nm. A 1.0×10^{-4} M solution of potassium permanganate showed zero absorbance at 450 nm and absorbance of 0.420 at 530 nm. Calculate the concentraction of mixture of two compounds present in a solution which showed an absorbance of 0.370 and 0.710 resp., at 450 and 530 nm assuming cell width was constant, (viz., 1 cm)?

Ans: Calculate the molar absorptivity (a) for $KMnO_4$ and $K_2Cr_2O_7$ at 530 and 450 nm so $A = abc$, $b = 1$, $A = ac$ or $A/c = a = \alpha$

$$\alpha_1 \text{ at } \lambda_1 \text{ for } K_2Cr_2O_7 \text{ at 450 nm } = \frac{0.200}{1\times10^{-3}} = 200$$

$$\alpha_2 \text{ at } \lambda_2 \text{ for } K_2Cr_2O_7 \text{ at 530 nm } = \frac{0.05}{1\times10^{-3}} = 50$$

α_1 at λ_2 for $KMnO_4$ at 450 nm $A = 0$ $a = 0$

$$\alpha_2 \text{ at } \lambda_2 \text{ for } KMnO_4 \text{ at 530 nm } = \frac{0.03}{1\times10^{-3}} = 50$$

Now we set simultaneous equation if λ_1 = 450 nm and λ_2 = 530 nm

$A\lambda_1 = (a_1c_1)\lambda_1 + (a_2c_2)\lambda_1$ so $A\lambda_1 = 0.370 = 200(c_1) + 0(c_2)$ at 450 nm (λ_1)

$A\lambda_2 = (a_1c_1)\lambda_2 + (a_2c_2)\lambda_2$ so $A\lambda_2 = 0.710 = 50(c_1) + 4200(c_2)$ at 530 nm (λ_2)

On solving simultaneous equation for c_1 and c_2

$$c_1 = 1.9 \times 10^{-3} \text{ M of } K_2Cr_2O_7$$

$$c_2 = 1.46 \times 10^{-4} \text{ M of } KMnO_4.$$

24.2 Uranium oxinate in chloroform showed an absorption maximum at 430 nm with molar absorptivity of 1.4×10^4 at 430 nm. By differential spectrophotometry the complex was measured. A 1.0×10^{-2} solution was used as blank instead of chloroform with this blank the transmittance was 30.2% T. Evaluate the concentration of unknown solution if optical path was 10 nm.

Ans: $A = abc$ since $a = 1 \times 10^4$, $b = 1$, $c = 1 \times 10^{-4}$ M

So $A = (1 \times 10^4)\ (1)\ (1 \times 10^{-4}) = 1.0$ or 10% T as $A = -\log T$ *i.e.*,

The scale has been expanded 10 times original value of 30.2% T which is actually 3.02% T.

This is equivalent to $3.02\% = -\log\left(\frac{302}{100}\right) = -\log 0.0302 = \text{absorbance}$

Now $A = abc$ if $A = -\log 0.0302$, $b = 1$, $a = 1 \times 10^4$ c = unknown

$-\log 0.03 = (1 \times 10^4)\ (1)\ (c)$ so $c = 1.52 \times 10^{-4}$ M.

24.3 β napthlyamine had following values for log of molar absorpivity (ε) at given wavelength?

β-napthol	3.40	0
β-napthol anion	3.70	3.47
	(at λ 285 nm)	(at λ346 nm)

A solution of buffer containing 1×10^{-4} M β-napthlyamine with pH 9.2 showed an absorbance of 0.373 at 285 nm and 0.098 at 346 nm. What is the pK value of β-napthlyamine?

Ans: We must calculate concentration of R-OH and RO-anion.

R · OH $a_{285} = 3.087 \times 10^3$ (as $\log a = 3.40$); $a = 0$ ($\log a = 0$) at $\lambda = 346$ nm.

RO^- $a_{285} = 5.010 \times 10^3$ (as $\log a = 3.70$); $a_{346} = 2.93 \times 10^3$ (as $\log a = 3.47$)

Now as per equation $A\lambda_1 = (a_1c_1)\lambda_1 + (a_2c_2)\ \lambda_2$

$0.373 = (RO^-)\ (5.01 \times 10^3)$ at 285 nm + (ROH) (3.087×10^3) at 285 nm

Since $RO^- = 3.3 \times 10^{-5}\text{M} = \frac{0.0981}{2.950 \times 10^3} = 3.30 \times 10^{-5}\text{M}$ as $c_1 = \left(\frac{A_1}{a_1}\right)$ at λ_1

(ROH) = (3.3 ×. 0^5) (3.01×10^3) + (ROH) (3.08 c 10^3) so we calculate ROH

$$\text{ROH} = \text{pH} - \log\left[\frac{\text{base}}{\text{acid}}\right]$$ substitute pH = 9.20, log (base) = log 3.33×10^{-5}

while log (acid) = log 6.67×10^{-5}

So $$pK_a - 9.20 - \log\left[\frac{3.33 \times 10^{-5}}{6.67 \times 10^{-3}}\right] = 9.20 - \log 0.05 = 9.50.$$

24.4 Calculate the concentration of the manganese solution if one gm sample of stainless steel was dissolved in mixture of dilute nitric and hydrochloric acid. Manganese in solution was oxidised to $KMnO_4$ and made to 100 ml for this absorbance was 0.700 at 530 mm with optical path 10 mm $KMnO_4$. A 1.52×10^{-4} M $KMnO_4$ known showed on absorbance of 0.350.

Ans: $\dfrac{0.700}{0.350} = \dfrac{c_1}{1.52 \times 10^{-4}}$ if c_1 = unknown as $\dfrac{c_1}{c_2} = \dfrac{a_1}{a_2}$

$$c_1 = \frac{(0.700)(1.52 \times 10^{-4})}{(0.350)} = 3.04 \times 10^{-4} \text{ M of [Mn]}.$$

$$\% \text{ Mn} = \frac{(3.04 \times 10^{-4})(\text{at wt. Mn}) \times \text{dil. factor} \times 100}{1.0} \text{ if dil. factor} = 0.1$$

$$\% \text{ Mn} = \frac{3.04 \times 10^{-4} \times 54.94 \times 1 \times \frac{100}{1000} \times 100}{1.0} = 0.17\%$$

24.5 What is the molar absorbivity and the concentration of solution (M.W. 54.95) with an absorbance of 0.700 if its concentration was of 1.52×10^{-4} M with optical path of 1 cm and absorbance as 0.350?

Ans: $a = \dfrac{A}{bc}$ if $A = 0.350$, $b = 1$, $c = 1.52 \times 10^{-4}$ so

$$a = \frac{0.350}{1 \times 1.52 \times 10^{-4}} = 2300 \text{ litres/mole}.$$

$$c = \frac{A}{ab} \text{ i.e., } \frac{0.700}{2300 \times 1.0} = 3.04 \times 10^{-4} \text{ M}.$$

24.6 Calculate the concentration of the unknown solution if two aliquots of 5 ml each were prepared from standard solution ($1.52^{-} \times 10^{-4}$ M) permanganate while the second solution was obtained from one gm of steel sample dissolved in mixture of acids and made to 100 ml. The second solution was diluted from 5 ml to 10 ml with distilled water while the first solution of 5 ml was diluted with 5 ml of 1×10^{-4} M of $KMnO_4$ was added. The second solution showed an absorbance of 0.350 and first showed absorbance as 0.465?

Ans: This was ascertained by equation viz.

$$C_{\text{unknown}} = \frac{A_{\text{unknown}} \times C_{\text{stand}}}{A_{(\text{unknown+stand})} - A_{\text{unknown}}} \text{ at fixed wavelengths of 550 nm.}$$

Thus $A_{\text{unknown}} = 0.350$, $C_{\text{stand}} = 0.500 \times 10^{-4}$ M, $A_{(\text{unknown + stand})} = 0.465$, A = unknown = 0.350
By substituting all the values we have

$$C_{\text{unknown}} = \frac{(0.350)(0.500 \times 10^{-4})}{(0.465 - 0.350)} = 1.52 \times 10^{-4} \text{M } KMnO_4.$$

24.7 Calculate the concentration of manganese in terms of % of the manganese if absorbance was 0.700 if one gram sample of manganese (1.52×10^{-4} M) showed an absorbance of 0.350 at 550 nm with optical path 1 cm?

Ans: $a = \dfrac{A}{bc} = \dfrac{0.350}{(1)(1.52 \times 10^{-4})} = 2.3 \times 10^{2}$ lit/mol/cm

$$\text{Now } c = \frac{A}{ba} = \frac{0.700}{2.3\times10^2 \times 1} = 3.04\times10^{-4}\,\text{M}$$

$$\%\,\text{Mn} = \frac{C_1 \times 54.94\times100\times1}{1.0} = 0.17\%\ \text{Mn.}$$

24.8 What will be amount of cobalt and nickel in a given mixture if a mixture of complexes of cobalt and nickel showed an absorbance of 0.816 at 365 nm and 0.0820 at 700 nm. The standard solution of cobalt had molar absorbivity of 3529 (at 365 nm) and 428.9 (at 700 nm) while that of nickel showed absorbivity as 3228 (at 365 nm) and no absorbivity at 700 nm (the optical path was 10 mm)?

Ans:

For	Cobalt	Absorbance	Nickel	Unknown soln
λ_1 = 365 nm	3.529 × 10^3	0.816	3.288 × 10^3	0.816
λ_2 = 700 nm	428.9	0.082	0	0.0820

Since $A\lambda_1 = (a_1c_1)\lambda_1 + (a_2c_2)\lambda_1$ so $0.816 = 3.529c_1 + 3.228\times10^3 c_2$

$A\lambda_2 = (a_1c_1)\lambda_2 + (a_2c_2)\lambda_2$ so $0.0820 = 428.9c_1 + Oc_2$

By solving simultaneous equation, *i.e.*, for c_1 and c_2 *i.e.*, of cobalt and nickel resp.
c_1 = 1.91 × 10^{-4} M for cobalt c_2 = 4.4 × 10^{-3} M for Nickel.

24.9 What would be absorbance of two solutions at two different wavelength if the equilibrium (*K*) for conjugate acid base pair was

$$H_{in} + H_2O \rightleftharpoons H_3O + In \qquad K = \frac{[In][H_3O]}{[Hin][H_2O]}$$

The absorption of species and its anion was as follows:

at 430 nm	a × 1720	0	10 × 10^{-4}
at 600 nm	a × 0	1960	2.5 × 10^{-4}
	HIn	*In*⁻	unknown

Ans: The graphical respresentation will give best results
(*i*) 0.286 at 430 nm 0.670 at 600 nm
(*ii*) 0.020 at 430 nm and 0.468 at 600 nm.

24.10 What should be the magnitude of the molar absorbivity if 1.2 × 10^{-5} M solution had an asorbance of 0.210 with optical path of 10 mm at 430 nm?

Ans: $A = abc$, $A = 0.210$, $b = 1$, $c = 1.2 \times 10^{-5}$ M
On substituting these values we have

$$0.210 = a \times 1 \times (1.2 \times 10^{-5}) \text{ or } a = \frac{0.210}{1.2\times10^{-5}} = 1.75\times10^4$$

24.11 Calculate the concentration of a solution which had molar absorbivity of 1.32 × 10^4 at 430 nm with absorbance of 0.410.

Ans: $A = abc$ or $c = \frac{A}{ab}$ if $A = 0.410$, $b = 1$, $a = 1.32\times10^4$

$$\text{On substituting } c = \frac{0.410}{(1.32\times10^4)(1)} = 3.1\times10^{-4}\,\text{M.}$$

24.12 How will you calculate the molar absorbivity and concentration of unknown solution which had an absorbance of 0.930 when it had absorbance of 0.650 at 330 nm for 0.01 gm/L of solution? (M.W. = 272.8).

Ans: $\frac{c_1}{c_2} = \frac{A_1}{A_2}$ if $c_1 = 0.010$, $c_2 =$ unknown, $A_1 = 0.650$, $A_2 = 0.930$

So $\frac{0.010}{c_2} = \frac{0.650}{0.930}$ *i.e.*, $c_2 = \frac{0.01 \times 0.930}{0.650} = 0.0143$ g/L.

Now molar absorbivity is given as $A = abc$ if A = 0.930, $b = 1$, $c = 0.0143$ g/0.930 = a × 1 × (0.0143 divided by 0.930 = $a \times 1 \times (0.0143) \times 272.8$

Now 0.0143 g/L × 272.8 gm mole/L.

$a = 1.77 \times 10^4$/L cm/mole.

24.13 A certain compound of complex by Job's method of continuous variation gives ratio of mole fraction (absorbance) as 0.80 (0.231), 9.22(0.632), 0.45(0.531), 0.65(0.334) 0.91(0.087) What is its composition?

Ans: We plot a graph of mole fraction against absorbance. The slope of graphs is 3.0 so complex formed is MR_3.

24.14 Calculate the concentration of unknown solution in ppm if concentration versus (absorbance) readings are: 1(0.070), 2(0.150), 3(0.200), 5(0.350) 10(0.68) unknown (0.430).

Ans: You have to plot graph of absorbance (*x*-axis) against concentration (*y*-axis). It is st. line graph. Mark a point with absorbance of 0.430 on graph. The corresponding concentration in ppm for unknown is 6 ppm.

24.15 Calculate the concentration of ammonia in the unknown solution when analysed by Nessler's colorimeter. The height of solution was 40 while height of standard was 36. (The standard solution contained 3 mg/liter.)?

Ans: In visual colorimetery we have

$h_1/h_2 = c_1/c_2$ if h_1 was height of analyt, h_2 = height of standard.

$c_1 = (x)$ con. of unknown solution, and c_2 = concentration of standard then solution h_1/h_2 = 40/36 = $0/c_2 = 3/x$ as $h_1 = 40$, $h_2 = 36$ $c_1 = x_1$ $c_2 = 3$

So $x = 0.90$ mg/L

24.16 Colorimetric determination of nitrate was carried out by brucine method. 100 ml of sample was diluted to one liter. It contained 1.5g $NaNO_3$. When the ratio of unknown/known solution = 44/40. What is the concentration of nitrate in polluted sample of water?

Ans: Let the content of nitrate per 100 ml of water to be analysed be equal to '*x*' gram then

If c_x = unknown a_x = unknown, a_{stand} = for standard absorbivity c_{stand} = con. of standard solution

Since $c_x\ a_{stand} = c_{stand}\ a_x$

So we have

$$c_x\ a_{stand} = c_{std}\ a_x \text{ so } c_s\ a_x = \frac{1.5 \times 44x}{(NaNO_3) \times 1000 \times 1000} = c_x a_{st} = \frac{(x) \times 40}{1000} = 0.0119/L$$

The solution contained 11 mg/Liter of nitrite.

24.17 Ascertain the sensitivity of determination of manganese and molar absorbivity of $KMnO_4$ from following data:

c_{Mn} (µg/ml)	:	0.5	1.0	1.5	2.0
Absorbance	:	0.220	0.430	0.650	0.870

Ans: Since sensitivity = –.5 µg/ml/cm² corresponding to SD = 0.22 where *s* = sensitivity, *SD* = stand deviation.

$$\text{The sensitivity is equal to } \frac{0.5\times0.005}{0.22}=0.11\ \text{g/ml/cm}^2\ \text{sensitivity}$$

$$\text{molar absorbivity} = \frac{0.430\times1000\times54.9}{1.0} = 23{,}600,\ \textit{i.e.},\ 2.3\times10^4.$$

24.18 Find out the concentration of chromium (VI) and Manganese (VII) from the following data at 550 nm for Cr (VI) = 0 = a_1 at λ_1 for Mn (VII) at 550 nm = 2100 = $a_2\lambda_1$ 430 nm for Cr (VI) = 0.220 = a_1 at λ_2 for Mn (VII) at 430 nm = 500 = $a_2\lambda_2$.
If absorbance $A\lambda_1$ = 0.420 at 430 nm and $A\lambda_2$ = 0.710 at 550 nm for mixture when *b* = 2?

Ans: If *x* = con. of Mn and *y* = con. of Cr (VI) then we have

$$x=\frac{0.710}{2100\times2}=1.69\times10^{-4}\,\text{M concentration of manganese}$$

$$y=\frac{0.420-\left(0.710\times\dfrac{500}{2100}\right)}{220\times2}=5.7\times10^{-4}\,\text{M concentration of chromium (VI)}$$

So con. of Mn = 1.69×10^{-4}, con. of Cr (VI) = 6.3×10^{-4}M.

24.19 Find out the concentration of uranium (VI) and uranium (IV) in concentrated phosphoric acid with following data: absorbance at 410 nm was 2.0 for U (IV): while at 410 nm for U (VI) was 11.0 *a*, at 630 nm. It was 330 for U (IV) and for U (VI). It was Zero. The corresponding absorbance were 0.300 at 410 nm and 0.800 at 630 nm?

Ans: We calculate the concentration if
x = conc. of U (IV) and *y* = conc. of U (VI) so we have
0.800 ÷ 0.330 = 0.0024

$$\text{U(VI)}_{\text{un}}=\left(0.50-\frac{0.80\times2}{330}\right)\div11=0.041.$$

Where 2/330 was the factor for conversion of the absorbance of U (VI) at 430 nm to absorbance of U (VI) at 410 nm. So corresponding concentration were:
U (VI) = 0.0024 M, U (VI) = 0.40 M.

24.20 In the analysis of potassium by chromogenic ligand such as picric acid the following readings were obtained in visual colorimetry:

con. (µg/ml)	(*c*) :	*x*	10	20
Readings	(*n*) :	35.5	20.5	39.5

Find out the concentration of (*x*) in unknown solution?

Ans: Since the concentration *c* is proportional to the reading (*n*) the proportionality ratio was *n* equal to 10/20.5 or 20/39.5 *i.e.*, 0.49 in first case and 0.51 in second case. The mean value of 0.49 and 0.51 was 0.50 so for unknown solution reading was 35.5 total (35.5 × 0.50) = 17.8 µg. ml.

PROBLEMS FOR PRACTICE

24.21 What is the absorbance of solution containing 3 ppm if transmittance was 65% in 10 mm cuvett? **[Ans:** 0.190]

24.22 Calculate the molar concentration and absorbivity of a certain compound (10 mg/l) which had an absorbance of 0.650 at 310 nm, while unknown compound (M.W. = 272.8) had on an absorbance of 0.930. **[Ans:** 14.3 mg/L; $\varepsilon = 1.77 \times 10^4$]

24.23 Calculate the molar absorbivity of complex of iron with o-phenanthroline complex if ferric solution (4.7 mg/L) when treated with reagent (made up to 100 ml) gave absorbance of 0.467 at 510 nm. **[Ans:** $\varepsilon = 1.11 \times 10^4$ L/mo/cm.]

24.24 Find out concentration of $KMnO_4$ and $K_2Cr_2O_7$ from the unknown mixture with following data for simultaneous spectrophotometry?

λ(nm)	$KMnO_4$	$K_2Cr_2O_7$	Unknown solution
440 nm	0.035 (3.77×10^{-4})	0.308(8.33×10^{-4} M)	$(A_1)\lambda_1 = 0.385$
545 nm	0.886 (3.77×10^{-4})	0.099 (8.33×10^{-4} M)	$(A_2)\lambda_2 = 0.653$

[Ans: $K_2Cr_2O_7 = 9.73 \times 10^{-4}$ M, $KMnO_4 = 2.73 \times 10^4$ M]

24.25 Iron (III) formed complex with bipyridyl anion. The Job's method for continuous variation showed the ratio of mole fraction (absorbance) as 0.80 (0.231), 0.22 (0.500); 0.45 (0.531); 0.56 (0.422); 0.91 (0.087). Ascertain the composition of the coloured complex by Job's method of continuous variation? **[Ans:** Fe (2-2'bipyridyl)$_3$].

24.26 Calculate the concentration of unknown solution of gallium and indium when the following data was obtained in simultaneous spectrophotometry.

λ_{max}	Gallium	Indium	Unknown
$\lambda_1 = 510$ nm	$a = 36400$	$a = 5520$	$A\ \lambda_1 = 0.467$
$\lambda_2 = 656$ nm	$a = 1240$	$a = 17{,}500$	$A\ \lambda_2 = 0.347$

An unknown 0.3769 g was dissolved in 50 ml and 25 ml taken for complexation in 50 ml solution. **[Ans:** Ga = 1.56×10^{-2}% *In* = 2.98×10^{-2}%]

24.27 Find out the magnitude of *pK* value of an organic compound if $a = 1.0 \times 10^{-3}$ M solution buffered at pH 3.91 showed an absorbance of 0.440 (at 280 nm) and 0.080 (at 340 nm). If molar absorbivity was 6.607×10^3 (at 280 nm) 2.089×10^3 (at 340 nm) and while anion had no absorbivity at (340 nm) and $\varepsilon = 302 \times 10^3$ at (280 nm). **[Ans:** *pK* = 4.11]

24.28 Following data was obtained for simultaneous spectrophotometry by Murthy and Khopkar for complexes of cobalt and nickel with thiodibenzoyl methane: [Ref. Talanta **21**, 246 (1976)]

λ (nm)	Cobalt (0.0001 M)	Nickel (0.0002M)	Unknown
460 nm	0.150	0.684	0.721
500 nm	0.642	0.088	0.604

Calculate the concentration of cobalt and nickel?

[Ans: Co = 8.21×10^{-3} Ni = 1.75×10^{-4}]

24.29 A unknown solution of a compound had an absorption maxima at 400 nm with $a = 1.585 \times 10^4$. A solution 4.1×10^{-5} was measured by differential spectrometry. The two standard solutions were compared. If solution with unknown concentration had 82.5% transmittance, what is its concentration? **[Ans:** 4.02×10^{-5} M]

24.30 Calculate the concentration of the unknown solution of potassium dichromate if the transmittance measured was 75% *T*. When one ml of 0.015 N $K_2Cr_2O_7$ was added to 5 ml cell the reading changed to 65% T? **[Ans:** 0.186 mg/ml]

24.31 Calculate the sensitivity of complex of lead with diethyldithiocarbamate at 440 nm with optical path 50 mm, volume of solution being 25 ml and $a = 1.28 \times 10^4$. The reading for absorbance was 0.005. **[Ans:** 0.125 μg/cm²/ml]

24.32 The $Cr(HAA)_3$ complex was having strength of 3.16×10^4. If the gram of the sample was dissolved in 5 ml solution, when optical path was measured at 500 mm what is the smallest concentration of chromium which could be measured?
[Ans: Minimum reading = 0.025; 7.15×10^{-6}%.]

24.33 Utilising Yoe's mole ratio method ascertain the composition of complex of MR_n with ligand HR if HR(A): 0.20 (0.065), 0.40 (0.130), 0.60 (0.20), 0.8 (0.27), 1.0 (0.315) 1.5 (0.365)?
[Ans: 1:1 complex]

24.34 Calculate the dissociation constant for methyl red measured at 531 nm when pH (absorbance): was 2.30 (1.375), 3 (1.364), 4 (1.274), 5 (0.766), 6.30 (0.081), 8 (0.002)?
[Ans: $K_a = 7.98 \times 10^{-6}$]

24.35 A 3.75×10^{-4} cobalt in chloro complexes in acetone gave following results: chloride concern (1×10^4) (absorb) at 640 nm 1.99 (0.118) 4.95 (0.292) 6.83 (0.403) 8.3 (0.422), 9.84 (0.369), 1×10^4 640 nm. From mole ratio plot of absorbance ascertain the composition of the coloured species? **[Ans:** $CoCl_2$ and $CoCl_3$ present]

24.36 An indicator had absorption maxima at 410 nm with $a = 347$. The base formed λ_{max} at 640 nm and $a = 100$. The acid form had zero absorption at 640 nm and base had no absorption at 410 nm. To water small quantity of indicator was added and absorbances were measured as 0.118 (410 nm) and 0.267 (640 nm) if $pK_a = 3.90$. Find out the pH of the aqueous solution?
[Ans: pH = 4.80]

24.37 The % transmittance for optical path of solution was one mm as *T* 24.7%, with optical path of 50 mm, what is % transmittance? **[Ans:** % *T* = 97.2.]

24.38 What is the % *T* corresponding to following absorbance?
(*i*) 0.215, (*ii*) 0.585 (*iii*) 0.057 **[Ans:** (*i*) 61.0%, (*ii*) 26%, (*iii*) 87.7%]

24.39 What is the absolute absorbance value corresponding to following % transmittance?
(*i*) 22.0% (*ii*) 100% (*iii*) 4.2% **[Ans:** (*i*) 0.658; (*ii*) 0, (*iii*) 1.37]

24.40 Calculate the relative concentration error with different transmittance if a spectrophotometer had 0.5% absolute error in % transmittance for transmittance values as
(*i*) 0.803 (*ii*) 0.631 (*iii*) 0.195 (*iv*) 1.449.
[Ans: (*i*) 0.0284, (*ii*) 0.0172, (*iii*) 0.0174, (*iv*) 0.0421]

24.41 Find out the length of the optical path with least error in measurement with error in spectrophotometry sample had 0.237 transmittance, with absolute error of 0.3% T with 10 mm cuvett?
[Ans: 0.694 cm; 0.88% T]

24.42 Calculate the value of molar absorbivity and the % transmittance for Iron (III) for 5 ml solution (47 mg/L) when complexed with 1-10 phenathroline had absorbance of 0.476 at 510 nm?
[Ans: 34.1% T, $a = 1.11 \times 10^4$ L/mole/cm]

24.43 If 0.150 M sodium picrate in sodium hydroxide had absorbance of 0.419. Calculate the dissociation constant if 0.30 M solution had an absorbance of 0.581? **[Ans:** $K_a = 0.470$]

24.44 What was the magnitude of molar absorbivity if a certain compound had a M.W. = 92 if 0.015% solution by weight had 27% T with 10 mm cells?

24.45 Calculate the concentration of potassium permanganate in a photometric titration if 10 ml of it was titrated against 0.01 M oxalic and with absorbance values indicated in brackets: Vol. of titrant (A): 0(1.43), 1(1.30), 2(0.60), 3(0.27), 4(0.02), 4.75 (0)?

[Ans: 8.78×10^{-3} M]

24.46 A solution containing M^+ and L^- combine to form coloured complex with adherence to Beer's law with spectral data as follows:

λnm	*Molar absorbivity for M (metal)*	*Molar absorbivity for L (ligand)*
400	893	0
440	995	0
500	795	185
540	574	428
600	167	1178
700	0	1937

(1) If a solution containing both *M* and *L* had an absorbance of 0.312 with 10 mm cuvett at 540 nm; Indicate the concentration of *L* in solution when $M = 2.50 \times 10^4$.

(2) A solution had both *M* and *L* with absorbance of 0.696 at 440 and 600 nm with same cuvett. What was the concentration of *M* as well as *L*?

[Ans: (1) $3.94 \times 10^{-4} = [L]$. (2) 7.29×10^{-4} [*M*] and 4.88×10^{-4} M [L]]

24.47 What should be the wavelength of absorption maxima of the certain compound [$CH_3CH_2CH = CBr = C(CH_3) = 0$] using Woodword-Freser Rule? **[Ans:** λ_{max} = 252 nm]

24.48 What should be molar absorbivity of the organic compound (MW = 250) if at fixed wavelength had certain absorbance (concentration) as : 0.109 (0.3×10^{-4} M) 0.210 (1×10^{-4} M) 0.380 (1.5×10^{-4} M)? **[Ans:** ε = 436 L/mol/cm]

24.49 How will you calculate the concentration of two compounds if the absorbance of a mixture was 0.945 at 580 nm and 0.297 at 395 nm. With same cuvett width the molar absorption of first compound at 500 nm was 984 and 548 at 395 nm while second compound had molar absorbivity was 455 at 500 nm and 857 at 395 nm? **[Ans:** $C_1 = 9.44 \times 10^{-5}$, $C_2 = 2.93 \times 10^{-3}$M]

24.50 What should be the magnitude of molar absorbivity if a solution was prepared by dissolving 0.1298 gm of compound (M.W. = 214.3) in one liter of solution. The absorbance of the new solution at 255 mm was 0.824? **[Ans:** ε = 1430]

24.51 Calculate the composition of coloured complex by Job's method of continuous variation by typical spectrophotometric determination of the composition of the coloured complex following readings for ligand concentration were against absorbance for the ratio of M/R was 0.20 (0.11); 0.40 (0.28); 0.60 (0.338), 0.80 (0.470), 1.0 (0.563), 1.2 (0.679), 1.4 (0.784), 1.6 (0.678). 1.8 (0.252), 2.0 (0). **[Ans:** M:L::1:3]

REFERENCES

24.1 S.M. Khopkar, *Basic Concept of Analytical Chemistry,* 3rd edn., New Age International Publishers (P) Ltd., (2007).

24.2 D.J. Pietrzyke and C.W. Frank, *Analytical Chemistry—An Introduction,* Academic Press, pp. 145-6; 167-8; 620-64 (1979).

24.3 D.A. Skoog and D.M. West, *Analytical Chemistry,* 3rd edn., Holt, Reinhart Winston, pp. 478-9 (1979).

24.4 D.G. Peters, J.M. Hayes and G.M. Hieftze, *A Brief Introduction to Modern Chemical Analysis,* Saunders and Co., pp. 438-9; 666-8 (1980).

24.5 D.F. Boltz, J.A. Howell, *Colorimetric Determination of Non Metals,* John Wiley (1978).

24.6 R.M. Silvestein, G.C. Bassler, *Spectrometric Identification of Organic Compound*, 5th edn., Wiley-Interscience (1990).

24.7 D.G. Peters, J.M. Hayes and G.M. Hieftze, *Chemical Separation and Measurement,* Saunder Golden Series, pp. 666, 168, 438, 439 (1975).

24.8 G.D. Christian, *Analytical Chemistry*, 5th edn., John Wiley and Co. (1997).

24.9 D.A. Skoog and J. Leoroy, *Principles and Instrumental Methods of Analysis*, Holt, Reinhard and Winston (1996).

24.10 E.B. Sandell, *Colorimetric Determination of Traces of Metals*, 3rd edn., Wiley-Interscience (1962).

24.11 A Muskin, A. Kharaphkovsky, S. Shaikino and S. Efros, *Problem in Quantitative Analysis,* Mir Publishing Co., pp. 527, 532, 563, 544, 543, 555 (1984).

24.12 R.D. Braun, *Introduction to Instrumental Analysis,* McGraw-Hill, pp. 284, 292, 268, 297 (1985).

Chapter 25

Infrared Spectroscopy

INTRODUCTION

This is the technique in which we have spectra due to the molecular vibrations or rotations. Such vibrations may be wagging or twisting or rocking or scissoring or symmetrical, asymmetrical stretching or bending. In such situation the wave number of transmittance is evaluated from the formula as:

$$\nu = \frac{1}{2\pi}\sqrt{k\left(\frac{m_1 + m_2}{m_1 m_2}\right)} = \frac{1}{2\pi}\sqrt{\frac{k}{\mu}} \quad \text{if } m_1\ m_2 \text{ are mass of atoms} \quad \left[\because \mu = \frac{m_1 m_2}{m_1 + m_2}\right]$$

k = force constant; μ = reduced mass of atoms, ν = frequency

Now $\Delta\nu = (E_2 - E_1)$ then we calculate $E = \Delta\nu\ (h\nu)$ The radiation can be expessed in terms of wave number (σ) as

$$\sigma = \frac{1}{2\pi c}\sqrt{\frac{k}{\mu}} = \frac{1}{2\pi c}\sqrt{k\left(\frac{m_1 + m_2}{m_1\, m_2}\right)} \quad \text{if } c = \text{Velocity light}$$

It is thus possible to theoretically calculate from the transmittance frequency (ν) the other parameters like $m_1\ m_2$, k_1. The infrared spectroscopy is utilised to elucidate the structure of the organic compounds. The Colthup table gives list of frequency maxima, from knowledge of empirical formula boiling points of compound, one can assign a particular structure to the organic compound. In several problems we convert the wave number to wavelength and vice versa. The formula used is simple namely $v = \frac{nc}{\lambda}$ if ν = frequency, λ = wavelength, c = velocity of light and n = refractive index. As in UV visible spectroscopy, the composition of the complex can be easily ascertained by Job's mole fraction method or Yoe's mole ratio method.

It is also possible to calculate the total number of vibration bands from (N = Nos. of atoms present in molecule), we get it as (3N – 6) if N = Nos. of atoms present. We get total number of vibrational modes. We can also calculate the critical angle between prism with known value of the refractive index of prism and if refractive index of the sample is known. For instance if n_1 = refractive index of prism and n_2 = refractive index of the sample and supposing $\hat{\theta}$ = critical angle of prism, we get from relationship is $\theta = \sin\left(\frac{n_2}{n_1}\right)^{-1}$ derived from formula given below, viz., n_1 Sin $\theta = n_2$. At Sin 90° we can thus get value of θ. Similarly it is possible for one to evaluate the absorbance of the peak at known wave number if we know % transmittance at base line and % transmittance at peak for the sample thus if $T_b = T_0$ = % transmittance of base line and % Tp = % transmittance of sample then we have

$$A = \log\left[\frac{\%T_b}{\%T_p}\right] = \log\frac{\%T \text{ at baseline } (T_b)}{\%T \text{ at peak of sample } (T_p)}$$

This is called "base line method". It facilitates us to calculate absorbance from calibration curve and hence concentration. Finally from number of figures observed between particular frequency region one can easily calculate the cell width or path of the light beam. For instance if number of fringes are 'n' and frequency region is ν_1 (high) and ν_2 (low) we evaluate (b) optical path or cell width with relation which can be extended to calculate b as $b = \frac{5n}{(\nu_1 - \nu_2)}$ number is known as $b = \frac{n\lambda_1\lambda_2}{a(\lambda_1 - \lambda_2)}$. The IR spectroscopy is rarely used for the quantitative analysis of substances. It is used for identification and for elucidation of the chemical structure. If one has to use it for the quantitative analysis then one must use base line method. In such base line method we evaluate (P_O) intensity of incident radiation (P_t) intensity of the transmitted radiation which is measure of point of maximum absorption. If Log (P_O/P_t) is plotted against concentration [c], we get calibration curve, from this curve, we know % T of unknown compound and find out the concentration when two components are to be measured. One can resort to method of the simultaneous spectrophotometry which was discussed in earlier chapter.

SOLVED PROBLEMS

25.1 Zinc cyanide complex has following IR spectra at 2100 cm^{-1}: The absorbance with ratio of metal to ligand (M/R) is: 0.150 (1:2) 0.300 (1:1) 0.450 (3.2) 0.590 (2:1) 0.600 (5:2) 0.600 (3:1) 0.600 (7.2). What should be the composition of the metal complex?

Ans: We plot a graph of absorbance (y-axis) against the metal to ligand ratio $\frac{M}{R}$ (x-axis) to get two st. lines intersecting at M/R = 2:1 with absorbance of 0.590. Indicating the nature of the said complex as $Zn(CN)_2$ or Zn:CN::1:2.

25.2 The concentration of an antibiotic sample was to be analysed by IR technique at 2200 cm^{-1}. The unknown sample showed an absorbance of 0.180. Following set of readings were obtained for calibration curve as con. in ppm (absorbance): 0(0.40), 0.8(0.110), 1.4(0.160), 2.2(0.200) and 3.0(0.290). Calculate the concentration of antibiotic in unknown sample?

Ans: We have to plot graph of concentration against absorbance. The plot is st. line graph with slope of 40°. The concentration corresponding to 0.180 was 1.7 so 1.7 ppm present of unknown sample.

25.3 Identify the organic compound with zero absorbance in UV/visible region however in infrared spectra of unknown compound is showing following peaks: 3300 cm^{-1}, 3010 cm^{-1}, 2000 cm^{-1}, 1500 cm^{-1}, 1000 cm^{-1}.

Ans: From Colthup's Table one can draw some conclusions. All aldehyde, ketone, acid, nitrate and halogen were not present. Since molecule was not aromatic or unsaturated because of no absorption in UV region and IR band. The molecular species might have CH or CH_3 and NH_2 or OH groups which was H-bonded. If it was to contain NH group then 2.6-3.3, 5.7-6.3, 7.7-11.4, or 11.4-14.5 should have peaks while peaks in 2.6-3.3, 6.8-7.3, 7.4-7.8 and 9.5-10.0 indicated the good possibility of primary alcohol or saturated alcohol. For alcohol the peaks were at 3500-3000 cm^{-1}, partially at 1400-1500 cm^{-1}.

25.4 Can you draw a molecular orbital diagram for acetaldehyde and formaldehyde if acryldehyde was similar to formaldehyde except that C-C double bond was conjugated with > C = O group?

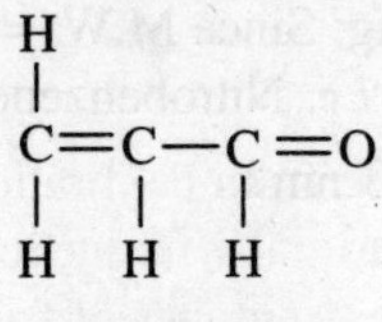

(acryldehyde)

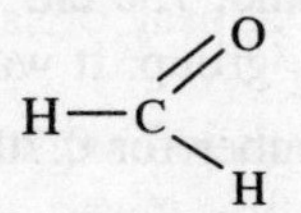

(Formaldehyde)
($n \rightarrow \pi^*$)

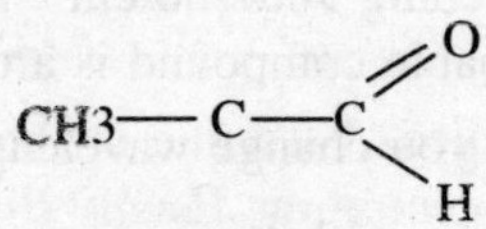

(Acetaldehyde)
($\pi \rightarrow \pi^*$)

Ans: In acrolein there was one π bonding orbital and one π^* anti-bonding orbital for each double bond but for formaldehyde there was additional non-bonding orbital. Both $n \rightarrow \pi^*$ and $\pi \rightarrow \pi^*$ occurred at lower energy at longer wavelength in acryldehyde than in formaldehyde.

25.5 Can you identify an organic compound if it had sharp absorption peaks at 710, 1300, 1510 and 2700 cm^{-1}. The compound was saturated alkaline?

Ans: The four peaks were due to C-H, C-C as 2940 cm^{-1} from C-H stretching at 1460 cm^{-1} from scissors like bending of H-C-H group at 1380 cm^{-1} from symmetrical distortion of end methyl group at 720 cm^{-1}. From rocking of methylene group aligned linearly. The peak at 2930 cm^{-1} and 2960 cm^{-1} were assigned to C-H stretching in methyl and methylene group. As the unsaturation was not there we have > 3000 cm^{-1} stretching. No aromatic compound as 200-670 cm^{-1}. The unknown compound was nujol oil CH_3 $(CH_2)_3CH_3$ a common solvent.

25.6 Can you identify an organic compound if it had molecular weight of (70+2) with peaks at 3000 cm^{-1}, 2300 cm^{-1}, 1500 cm^{-1}, 1100 cm^{-1}?

Ans. Band at 3000 cm^{-1} and 2300 cm^{-1} with one at 1500 cm^{-1} and 1300 cm^{-1} indicated presence of methyl and methylene groups. A sharp peak at 2300 cm^{-1} showed presence of nitrite group. The peak at 1500 cm^{-1} showed presence of isopropyl doublet, since there was no band at 1200 cm^{-1} isopropyl group was not possible M.W. 70. If we deduct atomic weight of nitrite carbon and nitrogen we have 44. The compound is likely to be *n*-butylnitrite, *i.e.*, CH_3 CH_2 CH_2 $C \equiv N$.

25.7 Identify an unknown compound with empirical formula of $C_6H_{16}O$ if it had peaks at 3000 cm^{-1}, 1500 cm^{-1}, 1050 cm^{-1}, 900 cm^{-1} and 7500 m^{-1}?

Ans: A sharp peak at 300 cm^{-1} indicates alcohol also another sharp peak at 1500 cm^{-1} states no amine bond was seen. The peaks at 3000 cm^{-1} 1500 cm^{-1} and 1300 cm^{-1} indicate presence of CH_3 and CH_2 groups. No splitting was observed at 130 cm^{-1}. Compound should be isopropyl or *t*-butyl group but empirical formula was C_6H_1 O. Therefore the likelihood of the compound was primary octylalcohol, *i.e.*, 1-Octanol.

25.8 Identify the unknown organic compound if it had IR peaks at 3000 cm^{-1}, 1600 cm^{-1}, 1050 cm^{-1}, 900 cm^{-1}, 750 cm^{-1} and 680 cm^{-1}?

Ans: Peaks at 3000 cm^{-1} and 1600 cm^{-1} indicated presence of CH_3 group, a peak at 1600 cm^{-1} showed > CO group, one at 750 – 680 cm^{-1} showed CH group. CH stretching at 3000 cm^{-1}, 1600 cm^{-1}, CO stretching, C-C stretching at 1050 cm^{-1} CH bonding at 750 cm^{-1}. Hence compound was $CH_3COCH(CH_3)_2$.

25.9 Identify the unknown organic compound if it had molecular weight of 120 and if IR peaks were at 3000 cm^{-1}, 2000 cm^{-1}, 1500 cm^{-1}, 1350 cm^{-1}, 1100 cm^{-1}, 1000 cm^{-1}, 950 cm^{-1}, 850 cm^{-1} and 700 cm^{-1}?

Ans: Peak at 3000 cm^{-1} CH. stretching peak at 2000 cm^{-1} CN stretching, peak at 1100 cm^{-1} CN stretching 900-700 cm^{-1} aromatic compound, 750 cm^{-1} CN bending: Since M.W. = 120. The probable compound is aromatic with NO_2 group. It was $C_6H_5NO_2$ *i.e.* Nitrobenzene.

25.10 Can you change wavelengths to wave number for 0.70, 2.8, and 33 nm?

Ans: Since $\nu = \frac{1}{\lambda} = \frac{n\overline{\nu}}{c}$ if c = velocity of light

So $\overline{\nu} = \frac{1}{0.70 \times 1/10^4} = 1.4 \times 10^4\ cm^{-1}$ for 0.70 nm

$\overline{\nu} = \frac{10^4}{2.8} = 3.6 \times 10^3\ cm^{-1}$ for 2.8 nm

$\overline{\nu} = \frac{10^4}{33} = 3.6 \times 10^3\ cm^{-1}$ for 3.0 nm

25.11 Can you calculate the total vibrations band for hexane?

Ans: Hexane (C_6H_{14}) had N = 20 and it was not linear so we had $(3N - 6) = 3\ (20) - 6 = 54$ vibrational modes.

25.12 (*a*) What should be the critical angle between prism with refractive index of n = 2.03 and sample with refractive index of 1.34?

(*b*) What organic compound it is likely to be present if its b.p. was 31°C and empirical formula was C_5H_8O when major peaks were located at wavelength of 2880 - 2980 cm^{-1}, 1750 cm^{-1}, 1460 cm^{-1}, 1155 cm^{-1}, 960 cm^{-1} and 835 cm^{-1}?

Ans: The critical angle can be calculated if θ = critical angle and n_1 = 2.03 for glass, n_2 = 1.34 for sample then $\theta = \sin\left(\frac{n_2}{n_1}\right)^{-1}$ so we substitute in this equation refractive indices.

$$\hat{\theta} = \sin\left(\frac{1.34}{2.03}\right)^{-1} = 41.3°.$$

Note peaks at higher wave numbers at 1155 cm^{-1} no compound 1750 cm^{-1} was carbonyl > C = O peak. From Tables we see that 1460 cm^{-1} represent $-CH_2$-or-CH_3, 1410 cm^{-1} for > CO - CH_2, 960 cm^{-1} for C = C-H and 835 cm^{-1} for CH = C. There were several possible structures. The compound with b.p 110.5°C was $CH_3CH = CH_2CHO$ or $(CH_3)\,C = CHO$. The curve fitting shows structure it as cyclopentanone.

25.13 What should be the absorbance of the peak at 1470 cm^{-1} if the % transmittance was 27.4% (presume that % T on graphic interpolated base line was 81)?

Ans: $A = \log \dfrac{\%T_b}{\%T_p}$ if T_b = base line % transmittance and if T_p is % Transmittance peak of sample usually the spectrum of pure solvent was used to get % of base line if needed

$$A = \log \frac{81.0}{27.4} = 0.471 \text{ (as } T_b = 81,\ T_p = 27.4)$$

25.14 What is the path length of the cuvetts if the number of fringes were 51 between wave number of 3500 – 900 cm^{-1}?

Ans: As $b = \dfrac{5n}{(v_1 - v_2)}$ if b = path length, n = nose of fringes, $v_1 = v_2$ frequency

Since $n = 51$, $v_1 = 3500$, $v_2 = 900$ on substituting these values we have

$$b = \frac{5(51)}{(3500 - 900)} = 0.098 \text{ nm}.$$

25.15 Calculate the optimum path length of the cuvett if the number of fringes were $(n) = 45$ between wave number 3000 – 600 cm^{-1}. (If n = 12 in second case between 3500 - 500cm^{-1})

Ans: (*a*) Since $b = \dfrac{5n}{(v_1 - v_2)}$ if $n = 45$, $v_1 = 3000$, $v_2 = 600$

We have $b = \dfrac{5(45)}{(3000 - 600)} = \dfrac{225}{2400} = 0.097 \text{ mm}$. was path length.

(*b*) In second case $n = 12$, $v_1 = 3500$, $v_2 = 500$

$$b = \frac{5n}{(v_1 - v_2)} = \frac{5 \times 12}{(3500 - 500)} = \frac{60}{300} = \frac{1}{50} = 0.042 \text{ mm}. \text{ was path length.}$$

PROBLEMS FOR PRACTICE

25.16 Assuming we have knowledge of wave number (cm^{-1}) and wavelength (nm) of the first and second harmonic vibrations (i.e., overtones) assign values to functional group > C = O (Formaldehyde); C-C (benzene); and C = N in CH_3CN.

[**Ans:** First overtone at 3488 cm^{-1} and 2867 nm, second overtone at 2930 cm^{-1}, 3413 nm and third overtone at 4286 cm^{-1} and 2333 nm.]

25.17 Calculate the wave number of a band and energy of separation at two levels if transition between two vibrational energy levels of a molecule gave an IR peak around 6.43 mm.

[**Ans:** (*a*) 1555 cm^{-1} (*b*) 3.09×10^{-12} erg.]

25.18 Find out the wavelength (nm) and wave number (cm^{-1}) of IR for following frequencies and wavenumber?

(*i*) 1.096×10^{14} Hz (*ii*) 1.127×10^{14} Hz (*iii*) 7.236×10^{13} Hz

[**Ans:** (*i*) 2735 nm, 3656 cm^{-1} (*ii*) 2660 nm, 3756 cm^{-1} (*iii*) 4143 nm, 2413 cm^{-1}.]

25.19 Can you explain mole phenomena of broad band if 2-4 pentadione had a broad absorption band at 3000 cm^{-1} and at 1800 cm^{-1}? Can you state why 3000 cm^{-1} band was seen when both oxygen were doubly bounded to carbon compound?

[**Ans:** The end form will have a broad O-H band due to the intramolecular H-banding.]

25.20 Is it possible to indicate wavelength for maximum absorbance and molar absorbivity of nitrobenzene, propane and hexanone? [**Ans:** yes]

25.21 Can you easily predict the Raman Shift for first compound if following modes of vibration were Raman active? Identify one in which IR is active. (The groups were –C = N and > CH).

[**Ans:** Of these five groups > C = O, -C = N and > C = H are Raman active, normal modes of vibration were IR active when shifts were 1744, 1465, 2143, 1128 and 3051 cm^{-1} resp.]

25.22 Is it possible to convert following vibrations given in frequencies in wave numbers: 2.459×10^{14} Hz; 4.785×10^{3} Hz 3.279×10^{13} Hz and 6.003×10^{13} Hz?

[**Ans:** 8203 cm^{-1}; 1595 cm^{-1}, 1094 cm^{-1}, 2002 cm^{-1}]

25.23 Compare the actual resonance vibration group frequency (in Hertz) for diatomic molecules if force constant (k) was given for C = N; and C = O?

[**Ans:** 426×10^{13} Hz and 5.230×10^{13} Hz]

25.24 Predict the normal modes of vibrations for a water molecule.

[**Ans:** $\nu_1 = 1.09 \times 10^{14}$, $\nu_2 = 4.8 \times 10^{43}$, $\nu_3 = 1.1 \times 10^{14}$]

25.25 Can you convert water molecule frequency (ν_2) 4.8×10^{13} Hz into wavelength assuming $C = \mu\nu$ and $C = 2.998 \times 10^{8}$ m/sec.? [**Ans:** 6.27 nm]

25.26 Will it be possible for you to calculate the optical path length if an empty cell showed the pattern of interference having 10 peaks with maxima at 1918 to 1150 cm^{-1} if $\frac{n}{2}\left(\frac{1}{\nu_1 - \nu_2}\right)$ where '*b*' is the optical path? [**Ans:** 6.5×10^{-3} cm]

25.27 What is path length '*b*' of each cell and indicate interference fringe design in above problem? [**Ans:** 0.0015 cm.]

25.28 What is the critical angle between a prism with a refractive index of $n_1 = 2.12$ and a sample with refractive index $n_2 = 1.28$?

[**Ans:** $\theta = \sin (n_2/n_1)^{-1}$. If $n = 6$, $\nu = 1200$, $\nu_2 = 1000$ $b = 0015$ cm if θ = angle]

25.29 Find out the critical angle between a prism with a refractive index of 1.95 and a sample with a refractive index of 1.31? [**Ans:** 42.2°]

25.30 Convert the following to units as shown:

(*a*) 6.0, 12.5 and 2.7 μm to wave numbers

(*b*) 3165, 1825 and 625 cm^{-1} wave number to wavelength (μm)

[**Ans:** (*a*) 1700, 800, 3700 cm^{-1} (*b*) 3.160, 5.479, 16.0 nm]

REFERENCES

25.1 S.M. Khopkar, *Basic Concepts of Analytical Chemistry,* 3rd edn., New Age International Publishers (P) Ltd. (2007).

25.2 D.J. Pietrzyke and C.W. Frank, *Analytical Chemistry: An Introduction,* Academic Press, p. 133 (1979).

25.3 J.S. Fritz and G.H. Schenk, *Quantitative Analytical Chemistry,* Allyn and Bacon Inc. pp. 433, 457-458, 462 (1979).

25.4 D.G. Peters, J.M. Hayes and G.M. Hieftze, *Chemical Separations and Measurements,* Saunder Golden Series, p. 747 (1974).

25.5 D.G. Peters, J.M. Hayes and G.M. Hieftze, *A Brief Introduction to Modern Chemical Analysis,* Saunder and Co. (1980).

25.6 J.R. Dyer, *Application of Absorption Spectroscopy of Organic Compounds,* Prentice Hall (1965).

25.7 R.J. Bell, *Introductory Fourier Transform IR Spectroscopy,* Academic Press (1972).

25.8 L.J. Bellomy, *The Infrared Spectra of Complex Molecules,* 2nd edn. Wiley (1958).

25.9 G.W. Ewing, *Instrumental Methods for Chemical Analysis*, 5th edn. McGraw-Hill and Co. (1980).

25.10 G.D. Christian, *Analytical Chemistry,* 5th edn., John Wiley and Co. (1997).

25.11 R.D. Braun, *Introduction to Instrumental Analysis,* McGraw-Hill and Co. pp. 346-409 (1987).

25.12 R.M. Silverstan, G.C. Basster, *Spectrophotometric Identification of the Organic Compounds,* 5th edn., John Wiley and Co. (1990).

Chapter 26

Raman Spectroscopy

INTRODUCTION

According to the simplest description of Raman Spectroscopy, is when a monochromatic light is scattered by molecules. A small fraction of scattered light is seen with different changed frequency than the frequency of irradiating light, such phenomena is called 'Raman effect'. Such effect is produced by molecular vibration and rotation of molecules. Raman lines appear on either side of exciting radiation both at lower and higher frequencies in comparison with incident radiation. Those lines produced with high frequency are called Stoke's lines while those lines produced with lower frequency are called antistokes lines. Such lines have small intensity compared to the stokes line. These are in compounds like carbon tetrachloride, carbon disulphide and dichloromonomethane. They possess the intense Raman lines. Such lines show frequency difference such change is called "Raman Shift". The intensity of line is proportional to concentration. The intensities varies if the scattered radiation is polarised. Raman spectra are valuable for both qualitative and quantitative analysis. This is because each compound has its unique spectrum. The ratio of stokes and anti-stokes line varies with temperature. Prof. C.V. Raman along with Dr. K.S. Krishnan discovered this effect in 1928 and in 1931 Raman was awarded a Nobel Prize for his discovery.

There are various kinds of sources to generate incident radiation. They involve argon ion, krypton ion, He-Ne, Nd.YAG laser sources of which last laser source is very effective in wavelength region of 1064 nm. Raman scattering spectrum and IR absorption spectrum resembles each another but certain species are active in former but not in latter techniques. Fluorescence however shows interference in Raman spectroscopy The Raman frequency shift and shift in IR absorption peak are identical. This technique is most effective for the characterisation of inorganic species. The Resonance Raman Spectroscopy, is surface enhanced Raman method and Non-linear Raman spectroscopy are other variations in the technique. The significant advantage is it facilitates analysis of aqueous solution. Raman spectroscopy is useful for characterisation of the organic compounds and determination of functional groups. It is useful for the identification of inorganic compounds and find composition of surface determination. It provides vital information on excited state structure. Apart from this in fact there are numerous applications in chemistry. It can be used to study reaction kinetics. Molecular impurities or additives can be detected by it. Resins, polymers and plastic are easily analysed. It is non-destructive analysis.

All solids, liquids or gases can be analysed by Raman technique from 10 mg to 50 ηg. As described earlier, aqueous solutions of complexes can be analysed by Raman using glass cuvettes and not by IR spectroscopy. Fluorescence interferes in Raman Spectral Characterisation. The time needed for analysis varies depending upon the sample preparation. Average sample is 15–20 ml in adequate for complex analysis. The coloured samples pose a problem as they absorb the incident radiation. The sensitivity is ± 0.5%. As lines are weak as compared to IR. Therefore this technique needs samples with large volume and high concentration. Further Raman effect involves scattering while IR involves absorption of radiation. A strong bend in IR will be weak in Raman and vice versa. A thermal heating by laser source must be avoided during measurements, so PMT is used in instrument.

FT Raman spectroscopy if is free from effect of fluorescence then dyes can be also analysed. The geochemist use Raman microprobe wherein use of microscopic objective is used to focus light. While in Remote Raman Analysis one which involves use of LIDAR. Here Rayleigh scattering from an inter laser beam is collected by telescope. A laser emitting at 308 and 353 nm is used for analysis of SO_2 in atmosphere. Finally, it can be used in quantitative analysis as we have an expression.

A sample with 1-4 ppm concentration can be analysed. We plot a graph of Z versus C to know unknown concentration if Z = intensity of Raman band.

The analysis of PCl_3, aviation turbine fuel and qualitative analysis of the pharmaceutical products like anti-histamic tablets is done by Raman method. With discovery of Nd-YAG laser the technique has become very popular with chemist.

PROBLEMS

Multiple Choice Questions

Tick (√) mark the correct answer from following alternatives given.

26.1 The Stokes line in comparison to antistokes lines are
(*a*) stronger in intensity (*b*) weaker in intensity
(*c*) same in intensity (*d*) have no particular relation

26.2 Raman effect is produced by the following process in spectroscopy
(*a*) Molecular Vibration (*b*) Atomic absorption
(*c*) Atomic emission (*d*) Atomic scattering of radiation

26.3 In Raman radiation if the intensity is compared to absorption radiation are
(*a*) weak in intensity (*b*) strong in intensity
(*c*) of equal intensity (*d*) no difference in intensities of radiation

26.4 Raman lines appear when intensities of incident radiation are in
(*a*) lower frequency (*b*) higher frequency
(*c*) lower or higher frequencies (*d*) no regular pattern.

26.5 The wavelength of Rayleigh scattered radiation is related to wavelength of incident radiation being
(*a*) greater (*b*) lower
(*c*) equal
(*d*) no relation with wavelength of incident radiation.

26.6 In nephelometry is another technique involving scattering of the intensities of incident and emitted radiation is
(*a*) analogus (*b*) lower
(*c*) higher (*d*) lower or higher

26.7 The vibrations which are active in Raman spectroscopy are weak and not active in:
(*a*) Infrared spectroscopy (*b*) Visible spectroscopy
(*c*) Ultraviolet spectroscopy (*d*) NMR spectroscopy.

26.8 The source of radiation used in Raman spectroscopy is
(*a*) Tunable mercury-neon laser lamp (*b*) Tungsten lamp
(*c*) Deutrium lamp (*d*) Hollow cathode lamp

26.9 The significant merit of Raman spectroscopy over infrared spectroscopy is facility of
(*a*) handling aqueous samples
(*b*) facility for powder samples
(*c*) use of non-aqueous sample
(*d*) analysis of gaseous sample.

26.10 Most commonly used laser source in far infrared region is
(*a*) Helium-neon (*b*) Krypton ion
(*c*) Argon ion (*d*) Nd/YAG.

26.11 Most significant limitations in Raman spectroscopy is
(*a*) Aqueous sample cannot be analysed
(*b*) Intensity of Raman line is very strong
(*c*) Fluorescent impurities interfere in spectra
(*d*) Concentration of samples analysed are in milligram levels.

26.12 The ΔE energy in antistokes line is:
(*a*) Negative (*b*) Positive
(*c*) Same as Stoke's lines (*d*) Too large to measure.

26.13 The wavelength used in the laser sources for Raman spectroscopy is in
(*a*) Ultraviolet region (*b*) Infrared region
(*c*) Microwave region (*d*) Visible region

26.14 Can more information on functional groups in the organic compounds can be obtained by Raman Spectroscopy?
(*a*) Yes (*b*) No
(*c*) Sometimes (*d*) Never.

26.15 The difference in wavelength between incident and scattered visible radiation corresponds to wavelength in
(*a*) Ultraviolet region (*b*) Infrared region
(*c*) Visible region (*d*) Microwave region

26.16 The sample holder in Raman spectroscopy is made of:
(*a*) Glass or quartz (*b*) KBr pellet
(*c*) Nujol mull oil (*d*) Mylar material

26.17 The Raman spectroscopy was discovered by scientist named as
(*a*) Mosley (*b*) Hoffman
(*c*) Raman (*d*) Roengtgen

26.18 An excellent Raman spectra is obtained by irradiating the sample by
(*a*) Laser source (*b*) UV source
(*c*) IR source (*d*) by W-lamp

26.19 The magnitude of Raman shift and wavelength of the excited radiation are
(*a*) Interdependent (*b*) Independent
(*c*) Partly dependent (*d*) Partly interdependent

26.20 The samples can be analysed in aqueous sample because the water is
(*a*) Weak absorber of Raman radiation
(*b*) Strong absorber of infrared radiation
(*c*) Strong absorber of Raman radiation
(*d*) Weak absorber of IR radiation

26.21 For analysis of biological sample, Raman technique is used as
(*a*) Large sample can be used
(*b*) Sizeable interference can be tolerated
(*c*) Interference by water is absent
(*d*) Fluorescent materials do not interfere.

ANSWERS

26.1 a	26.2 a	26.3 a	26.4 c	26.5 c	26.6 d
26.7 a	26.8 a	26.9 a	26.10 d	26.11 c	26.12 b
26.13 d	26.14 a	26.15 b	26.16 a	26.17 c	26.18 a
26.19 b	26.20 a	26.21 c.			

SOLVED PROBLEMS

26.22 Calculate the force constant (K) for Raman line and identify the normal mode of vibrations is Raman active?
(Given C—C bond in benzene, $K = 12.3 \times 10^5$ dynes/cm.)
(C—C bond in ethane, $K = 4.5 \times 10^5$ dynes/cm)

Sol.: Hook laws state is $F = -Kx$ if x = distance, F = force, K = constant. We give fundamental vibration frequency by equation as:

$$V = \frac{1}{2\pi}\sqrt{K\frac{(m_1+m_2)}{m_1m_2}} = \frac{1}{2\pi}\sqrt{\frac{K}{\mu}}$$ as K = force constant, m_1, m_2 = are mass of 1 & 2.

V = vibration frequency.

$$\mu = \left(\frac{m_1+m_2}{m_1m_2}\right)$$ is called reduced mass so substituting values

$$\mu = \frac{24}{12\times 12} = \frac{24}{144} = \mu.$$

We substitute the values

$$r = \frac{1}{2\times 3.14}\sqrt{\frac{12.3\times 10^5}{17.6}} = 17.44 \text{ cm}^{-1}$$ for second also here $K = 12.3\times 10^5$

$$r = \frac{1}{2\pi}\sqrt{\frac{K}{\mu}} = \frac{1}{3.314\times 2}\sqrt{\frac{4.5\times 10^5}{17.6}} = 1465 \text{ cm}^{-1}.$$

Here $K = 4.5\times 10^5$.

PROBLEMS FOR PRACTICE

26.23 In Raman spectroscopy can you estimate relative sensitivity at 275 nm as against 776 nm? **[Ans.** Yes]

26.24 Do you call a species Raman active if weak Rayleigh line appears in absence of Raman line where such Raman line was very weak? **[Ans.** No]

REFERENCES

26.1 J.H. Harley and S.E. Wiberley, *Instrumental Analysis,* John Wiley & Sons. New York p. 96 (1954).

26.2 D.A. Skoog, F.J. Holler and T.A. Nieman, *Principles of Instrumental Analysis*, Harcourt Brace Publication p. 429 (1998).

26.3 H.H. Willard, L.L. Merritt, J.A. Dean and F.A. Settle, *Instrumental Methods of Analysis,* 6th Ed. CBS Publisher India p. 217 (1986).

26.4 R.D. Braun, *Introduction to Instrumental Analysis*, McGraw-Hill Book Co. p. 432 (1987).

26.5 S.F. Settle, *Handbook of Instrumental Technique for Analytical Chemistry,* Pearson Publication India. p. 285 (2004).

26.6 S.M. Khopkar, *Basic concepts of Analytical Chemistry*, 3rd ed., New Age International Publishers (P) Ltd. (2007).

Chapter 27

Atomic Absorption Spectroscopy

INTRODUCTION

Atomic absorption spectroscopy has become most powerful tool of analysis in the last two decades. The principle reason for the popularity is that if a sample contains interfering ions, these ions need not be separated before analysis. In fact, it is possible to analyse one metal or element in the presence of another element by selectively choosing an appropriate wavelength for the measurement. The other reasons are at a time several metals can be analysed by it quantitatively at nanogram concentrations. It is possible to eliminate spectral interferences by choice of suitable wave length. The method is very sensitive for most of the metals for instance aluminium, gold boron, beryllium, bismuth, calcium, cobalt, chromium and gallium can be analysed to lower detection limit of 0.0001%. While metal like copper is analysed to limit of 0.00008%. While magnesium can be analysed easily. The elements which show poor sensitivity are hafnium (0.02%), potassium (0.3%) niobium (0.01%) strontium (0.002%) and uranium (0.38%).

In AAS, we have two modes of operation, first is flamed model AAS while another is non-flamed or graphite furnace model. In AAS with the flamed model we can go up to temp. of (2800° - 3000°C) with the acetylene-nitrous oxide or acetylene oxygen flame, while in graphite furnace model permits analysis up to any temp. upto 15000°C. We use AAS for quantitative analysis where in problems dealing with analysis of set of readings for concentration and (absorbance) are given and for unknown sample the absorbance is measured. On plotting graph of concentration (*x*-axis) against absorbance (*y*-axis), we get st. line graph passing through the origin from which concentration of unknown sample can be computed.

In standard-addition method, we use the relationship as follows:

$A_x = K_c$ but $A_T = K(C_x + C_s)$. If C_x = analytical concentration, C_s = concentration of the added standard, A_x and A_T are absorbance before and after addition of standard of same elements which is being analysed. If

$$C_x = C_s\left(\frac{A_x}{A_T - A_x}\right)$$ is concentration of the unknown sample which can be also calculated directly from this relationship.

The calculation of wavelength of spectral line resulting from the transition of electrons from lower energy level (e.g., 4*p*) to higher energy level orbital (4*s*) is calculated from the relation as follows:

$E = h\nu = \frac{hc}{\lambda} = (E_2 - E_1)$ if $(E_2 - E_1)$ is excitation potential for a level usually hc = 12,400 so, we have relationship where '*h*' refers to Plank's constant and '*c*' is velocity of light

$$E = \frac{hc}{\lambda} = \frac{12.400}{\lambda}.$$

We can calculate E or λ if one of them is known.

The amplification is n^d, if n = average electron per dynode

d = number of dynodes used

We get amplification to the power of $10^4 - 10^5$ using several dynodes.

Finally the observed wavelength (λ) of the radiation emitted from an atom in vacuum can be located if the speed of the moving electron from source to the detector is known. We use the following relationship viz.

$$\gamma_0 = \left(1 + \frac{u}{v}\right) v_0,\ \text{if } v_0 = \text{observed frequency of radiation}$$

$v = c/n$ if c = velocity of light,

u = velocity of moving electron from the source to the detector. Thus if v, u are given one can easily calculate the value of (v_o) frequency of radiation.

For some elements like copper, the atomic absorption method has better sensitivity and precision as AAS is not dependent on experimental conditions. So no internal correction for minor difference in the flame is needed. Virtually there are negligible interferences of other ions during the analysis. The greatest advantage is multi-element analysis, *i.e.*, ability to analyse many elements simultaneously in a sample. Sometimes we encounter what is called as 'Doppler broadening' effect which is due to rapid motion of atomic particles in the flame plasma. It is observed for the absorption line, consequently those atoms travelling towards the source absorb radiation at shorter wavelength. The atomic absorption lines are narrow to give narrow bandwidth.

Various kinds of interferences are encountered in AAS analysis. The spectral interference can be eliminated by correction method to wavelength. We have also chemical interferences involving formation of compounds of low volatility, presence of dissociation equilibrium, and ionisation of flame. All these chemical interference can be easily eliminated by using standard addition method. One can also use the releasing agents also called as 'spectroscopic buffers' to prevent formation of refractory oxides of calcium. The interference by ionisation is eliminated by choosing different wavelength (λ) and not using first resonance line for the measurements. The difference in viscosity of standard and sample should be minimum so that no problem is created for automisation. Zeeman background correction takes advantage of the polarisation of spectral line in magnetic field (*i.e.*, Zeeman effect). Finally, the Zeeman instrument overcomes the demerit of continuous background correction.

AAS can analyse as many as 48 elements. The sensitivity is very high, this permits multi-elements and multiple analysis of the sample in the experiment. One needs hollow cathode of same element which is being analysed. One uses hydride generator for atomisation of selenium, tellurium and antimony. For analysis of mercury we use cold vapour AAS.

For detailed discussion, reader is advised to read the authoritative reference cited at end of this chapter.

SOLVED PROBLEMS

27.1 In atomic absorption spectroscopic measurements with cold vapour technique for mercury from waste water at 253.7 nm (with air acetylene flame) gave following reading of concn. of mercury (absorbance) shown in brackets.
0(0), 2(0.053), 3.8(0.140), 5.8(0.160), 8(0.220), 9.6(0.260), 11.2(310) unknown (0.170)? Calculate the concentration of mercury in the unknown sample of water?

Ans: If we plot a graph of amount mercury (x-axis) and the absorbance (y-axis). We get st. line graph passing through the origin. Then corresponding to 0.170 absorbance the concentration is 6.4 ppm of mercury on the st. line graph.

27.2 The sample of water containing calcium was analysed by atomic absorption technique at 422.7 nm with nitrous oxide acetylene flame when following readings for absorbance were observed: concn. of calcium in ppm (absorbance) at 422 nm was 0(0), 1(0.090), 2.2(0.194), 3.10(0.273), 3.9(0.340), 5(0.440), 6(0.0525). The unknown solution had an absorbance of 0.380. What is the concentration of calcium in the unknown sample?

Ans: Once again we plot Beer's law plot for concentration (x-axis) against absorbance (y) axis. We get st. line graph passing through origin. On plotting 0.380 on graph the corresponding concentration is 4.3 ppm.

27.3 Calculate the concentration of indium when measured at 303.9 nm from the atmosphere, when 1000 m^3 of air was sucked by high volume sampler. The suspended particulate matter was dissolved in mineral acid and made up to 100 ml. The absorbances reading at 303.9 nm were as follows: amount of indium (μg/ml) versus [absorbance]
2(0.150), 4.0(0.310), 6.0(0.470), 8.0(0.600), 10(0.770) unknown (0.580)?

Ans. The absorbance value of 0.580 from graph of [In] against absorbance is 7.50 μg/ml. This is in the total solution of 100 ml so amount is 7.50 μg/ml × 100 = 750 μg/ml. When 1000 m^3 of air was sampled.

So, we have $\dfrac{\text{Total ion}}{\text{Air Vol.}} = \dfrac{750\,\mu}{1000} = 0.750\ \mu\text{g In/cm}^3$.

27.4 Calculate the concentration of rubidium from unknown sample when a 1.2456 gram of salt was diluted to 100 ml and analysed using atomic absorbance technique at 780 nm. The absorbance reading against amount of rubidium were mg/l (absorbance) were:
0.5 (0.240), 1.0 (0.490), 2 (1.030), 2.5 (1.200), 4.0 (1.900) unknown (1.210).

Ans: In graph of [Rb] and absorbance, corresponding to concentration for 1.210 was roughly 2.5 mg/l resp. since one ml of sample was used for analysis total concn. = 2.5 mg/l × 100 = 250 mg/l present in 1.2456 gm of sample which is 0.20% rubidium.

27.5 What is the concentration of zinc in the unknown sample when 2.979 of zinc was dissolved in dilute acid and made up to 100 ml and analysed by atomic absorption spectroscopy. The readings were:
Zn μg/ml (absorbance): 1(0.110), 3(0.300), 5(0.540), 6(0.670), 20(1.130) unknown sample (0.370)?

Ans: From the plot of [Zn] against [absorbance] for 0.370 absorbance corresponding concentration was around 400 μg/ml of zinc.

27.6 Calculate the concentration of cadmium if a sample of manganese (0.9421 g) containing cadmium was analysed by AAS. At 228.8 nm by standard addition method, Manganese was dissolved in mineral acid and made up to 100 ml. Then 25 ml of diluted solution containing 4.50 ppm of cadmium solution was added which had an absorbance of 0.220 with air propane flame while an absorbance of sample of standard was 0.310 zinc?

Ans: In standard addition method

$A_X = K_C$ or $A_T = K(C_S + C_X)$, if C_X = anlyt concn. C_s = concentration of the added standard A_x and A_T were two measured absorbance of analyt sample and standard resp.

$$C_x = C_s\left(\frac{A_x}{A_T - A_x}\right)$$

Now in a given problem A_T = 0.310, A_X = 0.220, C_s = 4.50 ppm, C_x = unknown, when vol = 25 ml.

So $$C_x = \frac{(4.5)\times 0.220}{(0.310-0.220)} = \frac{4.5\times 0.220}{0.090} = 25\times 4 \text{ ppm}$$

$$25\times 4 = \frac{4.5\times 0.220}{0.9421}\times 4 = 263 \text{ ppm of cadmium per 100 ml.}$$

27.7 What is the wavelength of the spectral line resulting for the transition from 4*p* back to 4*s* ground state provided if the excitation potential for 4*p* level is 2.95 ev?

Ans: The energy difference between excited state (E_2) and the ground state (E_1) is 2.95 ev.

We obtain if $hc = 12{,}400\ \Delta E = 2.95 = (E_2 - E_1) = \frac{12{,}400}{\lambda} = \frac{hc}{\lambda}$ if h = Planck's Constant

c = velocity of light travelling.

So we have hc = 12,400 ($E_2 - E_1$) = 2.95 ev. So λ = 4200 A° or 420 nm.

27.8 What is the wavelength for transition if for flame excited lithium atom, two transitions occur at 3*p* → 2*s* and 4*d* → 2*p* with energy difference of (3.85 → 0 ev) and (4.60 – 1.90 ev) resp.?

Ans: ($E_2 - E_1$) = (3.85 – 0) and for (3*p* → 2*s*) = 3.85 ev is ΔE

($E_2^1 - E_1^1$) = (4.60 – 1.90) and for (4*d* → 2*p*) = 2.70 ev is ΔE

Now if hc = 12,400, then we have $\Delta E = (E_2 - E_1) = \frac{hc}{\lambda} = \frac{12{,}400}{\lambda}$

$$(E_2 - E_1) = \Delta E = 3.85\,\text{ev} = \frac{12{,}400}{\lambda} = 3220°\text{A} = 322 \text{ nm}$$

$$(E_2^1 - E_1^1) = \Delta^1 E = 2.70 \text{ ev} = \frac{12{,}400}{\lambda} = 4590°\text{A} = 459 \text{ nm.}$$

27.9 Calculate the total amount of selenium analysed by hydride generator in AAS at 196 nm when following readings were obtained: selenium (μg)/(absorbance) 0 (0.002), 0.30(0.090), 0.60(0.175), 1(0.263) and 2(0.440). The two unknown samples had an absorbance of 0.040 and 0.305 resp.?

Ans: We plot graph of concentration of selenium (μg) against absorbance. We get st. line graph passing through origin. Now concentration corresponding to an absorbance of 0.040 and 0.305 were respectively 0.12 μg and 1.16 μg/ml, so concentration per ml was 0.012 μg and 0.116 μg.

27.10 Can you estimate amplification factor of photomultiplier tube if it had 8-dynodes and each dynode emitted on an average of 4.4 electrons for every impinging electron?

Ans: The amplification factor is calculated from the relations on amplifications = n^d

n = average electrons per dynode = 4.4

d = nos. dynodes = 8.0

We have n^d $(4.4)^8 = 1.4 \times 10^5$.

27.11 What is the concentration of zinc in unknown sample with a absorbance of 0.520 at 213 nm from biological sample by standard addition method if:

Zinc μg/ml (absorbance): 0(0.280), 2(0.440), 4(0.600), 6(0.757), 8(0.912)?

Ans: Plot the graph of zinc concn. (μg/ml) versus absorbance we get intercept on y-axis (0.280) due to added standard addition method so 0.520 absorbance corresponds to 3 μg/ml of zinc.

27.12 Calculate the wavelength (λ) of radiation emitted from a sodium in vacuum if velocity of detector was 8×10^5 m/s and if the wavelength of radiation emitted sodium was 588.95?

Ans: If $\lambda v = C/n$ frequency of observed for radiation (v_o) when source was moving towards detector was related to the frequency (v) of the radiation with no movement of sodium by equation as

$$\nu_0 = \frac{1 + u/v}{\left[1 - (u/v)^2\right]^{1/2}} \text{ (Given } r = c/n,\ u = \text{velocity)}$$

In this problem, $u = 8 \times 10^5$ and $r = 588.95$ substituting the values,

$$\nu_0 = 1 + \left(\frac{8\times 10^5}{588.95}\right) \div \left(1 - \frac{8\times 10^5}{588.95}\right)^{\frac{1}{2}} \times 588.95 = 5.1035 \times 10^{14} \text{ Hz in frequencies.}$$

27.13 Calculate the original concentration of gallium if after extraction of gallium when standard addition method was used in AAS at 294.4 nm. Two samples each of 50 ml were taken and to the first sample 300 μl (containing 50 mg/l) of gallium was added. The organic phases containing gallium was then analysed by AAS at 294 nm when unknown sample had absorbance of 0.325 while one sample spiked with known amount of gallium was 0.670?

Ans: 300 μl had 50 mg/L = 5 mg/100 or 0.05 mg/10 = 0.05 mg/ml, *i.e.*, 50 μg/ml.

So 0.5 μg/μl × 3 = 1.5 μgl/300 μl = 0.325

(Unknown + standard) = 0.670 (0.670 – 0.325) = 0.345 for unknown sample.

The unknown sample had (0.283 mg/l)

$$\text{We use formula } C_x = C_s\left(\frac{A_x}{A_s - A_x}\right) \times 0.0283 \text{ as } A_s = 0.670, A_s = 0.325,$$

$$C_s = 0.15\,\mu\text{g/ml}\,283\ \mu\text{g/l Gallium.}$$

$$C_x = \frac{0.015 \times 0.325}{(0.670 - 0.325)} = 0.0283\,\mu\text{g/ml or } 283\ \mu\text{g/l.}$$

PROBLEMS FOR PRACTICE

27.14 A sample of Type metal containing antimony was dissolved in aquariga and was made up to 500 ml and with distilled water. It was analysed by AAS at 217.6 nm with air acetylene flame. The readings were: antimony con. (absorbance) 0.(1.0), 5(0.061), 10(0.122), 15(0.185), 20(0.248), 23(370). Unknown sample had absorbance of (0.124).

Calculate concentration antimony in unknown sample? **[Ans:10]**

27.15 What is the percentage of manganese in the unknown sample of soil if it was analysed by AAS for micronutrients at 279.5 nm with air acetylene flame. The readings were as 0(100%), 0.20 (88.2%), 0.40(77.6%), 0.60(68.2%), 0.80(60.1%), 1(52.7%), 1.2(46.5%) 2.4 (unknown)?

[**Ans:** 93.01%]

27.16 A sample of sodium at 589 nm gave a reading of 37% T. The 2nd and 3rd solution containing same quantity of unknown solution plus 40 and 80 μg/ml of standard sodium gave results as 65% T and 93% T resp. Calculate the concentration of sodium (μg/ml) in original unknown sample? [**Ans:** 20 μg/ml]

27.17 Calculate the concentration of cadmium in unknown sample if it was analysed by AAS at 228.8 nm. The following data was obtained for cadmium μg/m (absorbance):
1(0.110), 3(0.300), 5(540), 6(0.670), 7(0.790), 20(1.13) and unknown (0.670)?

[**Ans:** 6 μg/ml]

27.18 Calculate the concentration of strontium in unknown sample of drug with following readings at 460.7 nm is AAS, with strontium, mg/ml (Intensity reading) was 0.50(23), 0.75(35), 1(45), 1.25(58), 1.50(70) unknown (45)? [**Ans:** one mg/ml]

27.19 Potassium was analysed at 766.5 nm with atomic absorption methods. Find out concentration in the unknown sample if the readings were as follows: [*K*] ppm at 765.5 nm. 74.3(100), 55.7(87), 37(69), 18.5(46), 7.4(22) and 0(3). If one gm of oxide was dissolved in 100 ml gave reading of 78?

[**Ans:** 77.4 ppm or 0.104% K]

27.20 What is the concentration of rubidium and caesium in unknown mixture if 1000 ppm of potassium was used as the internal standard in analysis? The readings for standard samples were taken for caesium at (852.1 nm) and for rubidium at (780 nm). The readings were as follows: amount (intensity).

Rb	0.1 (0.11)	0.5 (0.52)	1 (1.2)	5 (5.9)	10 (10.5)	*x* (1.4)
Cs	0.1 (0.15)	0.5 (0.68)	1 (1.3)	5 (7.7)	10 (81)	*x* (0.73)
K	1000 (0.86)	1000 (80)	1000 (91)	1000 (91)	1000 (81)	*x* (95)

[**Ans:** Rb 1.13 ppm Cs = 0.44 ppm.]

27.21 What is the concentration of indium and gallium in unknown sample if the following results were obtained in analysis by AAS when measured for gallium at 294.4 nm and indium at 303.9 nm.

In (μg/ml)	53.25 (0.514),	4 (0.073)	6.27 (0.612)	8.5 (0.101)	15.2 (0.232)
Ga (μg/ml)	19.50 (0.356),	4 (0.392)	2.1 (0.220)	8.5 (0.155)	1.052 (0.232)

The unknown sample had reading of 0.247 for gallium and 0.371 for indium?

[**Ans:** 0.0019 indium; 0.0068 gallium.]

27.22 The s-block metals dissociate in flame as $M = M^+ + \overline{e}$. Now if unknown $p = 10^{-6}$ atom find out fraction of lithium atoms ionised at 2000 K and 3500 K?

[**Ans:** 0.00121 at 2000 K, 0.83 as 3500 K]

27.23 Calculate the amount of metal from unknown sample with intensity ratio of 0.599 at 334 nm and 332.7 nm was resp. 0.126 for 0.0045% of metal and 1.29 for 0.070% in metal at corresponding wavelength? [**Ans:** 0.028%]

27.24 Find out the ratio in vacuum of the intensity of 589 nm sodium line at 3000 K to the same line at 2000 K under similar conditions? [**Ans:** 58.6]

27.25 What should be extent of rise in vacuum for potassium atom in the excited state that led to 766.5 nm line when the temperature was changed from 1700-4500°C? **[Ans:** 26,400%]

27.26 What should be the amplification of photomultiplier cell which contained 6-dynode with an average of 4.7 electrons emitted? **[Ans:** 1.1×10^4]

27.27 The 8-dynode of phototube had an amplification of 4.9×10^5, what was the average number of electrons of each dynode? **[Ans:** 5.2]

27.28 Calculate the amplification of photomultiplier tube which had in all 9-dynodes with average of 4.3 electrons emitted? **[Ans:** 5 × 105.]

27.29 Calculate the number of electrons emitted by each dynode if the amplification factor for a 9-dynode photomultiplier tube was 5.8×10^5? **[Ans:** 4.4]

27.30 A photomultiplier tube had in all 8 dynodes with average of 2.0 electrons emitted. Can you calculate the amplification factor of this photomultiplier tube? **[Ans:** Yes]

REFERENCES

27.1 S.M. Khopkar, *Basic Concepts of Analytical Chemistry*, 3rd edn., New Age International Publishers (P) Ltd. (2007).

27.2 D.J. Pietrzyke and C.W. Frank, *Analytical Chemistry: An Introduction,* Academic Press pp. 181, 184-5 (1979).

27.3 J.S. Fritz and G.H. Schenk, *Quantitative Analytical Chemistry,* Allyn and Bacon Inc., pp. 463, 492 (1979).

27.4 D.G. Peters, J.M. Hayes and G.M. Hieftze, *A Brief Introduction to Modern Chemical Analysis*, Saunders and Co. (1980).

27.5 G.F. Kirkbright, *Atomic Absorption Spectroscopy and Related Methods,* Marcell and Dekker (1977).

27.6 M.D. Amos, *Basic Atomic Absorption Spectroscopy—A Modern Introduction,* Varion Vectron (1975).

27.7 W.T. Elwell and J.A.F. Gidley, *Atomic Absorption Spectrophotometry*, Pergaman Press Ltd. (1975).

27.8 J. Ramirez Muinoz, *Atomic Absorption Spectroscopy*, Elsevier Pub. Co. (1968).

27.9 J.B. Robinson, *Atomic Absorption Spectroscopy*, Marcell Dekkar and Co. (1966).

27.10 H.H. Willard, L.L. Meeritt and J.A. Dean, *Instrumental Methods of Analysis*, 5th edn., D. Van Nostrand and Co., (1980).

27.11 R.D. Braun, *Introduction to Instrumental Analysis,* McGraw-Hill Co., pp. 175-217 (1987).

27.12 D.A. Skoog and D.M. West, *Principles of Instrumental Analysis,* 2nd edn., Saunder Holt Co. pp. 315-34 (1980).

Chapter 28

Atomic and Flame Emission Spectroscopy

INTRODUCTION

The spectrographic analysis using arc/spark was known to physicist for long time, however it was not used extensively for a quantitative analysis of metal. Only with the discovery of better source of excitation like flame or plasma, the technique became popular with the chemist for quantitative analysis.

The basic idea involved in this technique is that the sample to be analysed is excited in arc, spark, flame or plasma torch. This process proceeds in few stages. The hydrated salt gets dehydrated, then it undergoes evaporation and ionization thereby separating cations from anion, the ionised material further dissociate to form atoms and then ion and which emits electrons. This electron is then promoted to higher level state by absorption of an energy from the source. Further the electron in high energy level dissipates the energy in form of the emission and the electrons without energy returns back to ground state. The characteristic emission released by element in process of relaxation has definite wavelength. The extent of the intensity of emission is directly proportioned to the amount of metal ion present in sample.

In problems involving atomic emission spectroscopy, we get set of values for intensity against concentration. In order to ascertain the composition of unknown sample its intensity is recorded, on *x*-axis against concentration on *y*-axis in uniform volume (e.g, μg/L) when we get a st. line graph. Only in case of standard addition method we get an intercept on *y*-axis of intensity. By knowing intensity of unknown sample its concentration can be evaluated from the calibration curve. Several problems are given in this chapter based upon the idea of calibration curve. Sometimes, log of (concentration) versus log of (intensity) is also plotted to evaluate the concentration of unknown sample.

The wavelength of resonance line for a metal is calculated from relation $\lambda = \dfrac{hc}{\Delta E}$, if $hc = 12{,}400$ and if ΔE the energy of excitation is known, then wavelength can be easily calculated.

At a times ratio of intensity at two wavelength is plotted against the concentration to get st. line plot for calibration.

It is also possible to identify the particular element by marking the maximum wavelength of emission. One or two lines are seen for each metal in the spectrum. The main advantages is that the trace metal determination can be done with accuracy by using solid samples by process of vaporisation. This is sort of non-destructive testing. However, the main limitations are that instruments are expensive, technique is time consuming, the method is generally semi-qualitative and the analysis cannot be done in absence of standards of known concentration for caliberation.

In flame photometry which is extensively used for the analysis of s-block metals, the solvent gets evaporated in flame leading to the vaporisation of salt and dissociation to atoms, which in turn are excited in flame to give radiation characteristic of their species. The graph of ($-\log P_O/P$) versus [C] gives st. line with negative slope. Very small concentration of metals can be analysed it e.g., Ca (0.00001), Li (0.000001), Na (0.00001), K (0.0002) and Cs (0.0003). The temperature of flame must be constant to get reproducible results. The rate of the introduction of the sample into the flame is also quite critical. In quantitative analysis internal standard is used; for instance fixed amount of Li in analysis of different samples of sodium. A method of standard addition involves, addition of same element which is to be analysed. In arc/spark spectroscopy sample to be analysed is made as a positive electrode *i.e.* cathode. Metals like copper (0.00008%) can be detected at lowest level while element like potassium (0.3%) have in comparison have low sensitivity.

A large number of elements are analysed by quantitative analysis of elements by AES and FES in analytical chemistry.

SOLVED PROBLEMS

28.1 The flame photometric analysis of lithium at 670 nm was carried out with sodium as the internal standard. The results obtained were as follows:

Con. of Li :	1.0	2.0	5.0	10.0	20.0	50.0	unknown (x)
Li-intensity :	10.0	15.3	22.2	35.4	33.4	77.5	38.5
Na-intensity:	10.0	10.5	9.5	10.0	11.0	10.0	10.5

Calculate the concentration of lithium in the unknown sample?

Ans: We should plot a graph of Log (Li/Na) intensity against Log (concentration).

The corresponding values for log were:

Log (Li/Na):	1.0	0.163	0.368	0.549	0.709	0.880	0.564
Log (Li):	0	0.3	0.69	1.0	1.30	1.69	Log(x)

The amount corresponding to Log (0.564) is 1.08

Antilog of 1.08 = 12-02 ppm lithium present.

28.2 What was the wavelength of resonance line of sodium atom if the excitation energy of the resonance level was 2.10 ev?

Ans: $\lambda = \dfrac{hc}{\Delta E}$ if h = planck, constant c = velocity of light, ΔE = diff. of potential in ev.

By substitution values since hc = 12,400; if ΔE = 2.10

$$\lambda = \frac{12{,}400}{2.10} = 587.2 \text{ nm.}$$

28.3 In analysis of potassium by flame photometry by standard addition method following readings were noted at 766.5 nm

Con. of [K] :	0	1.0	2.0	5.0	10.0	20.0	50.0	Unknown
Intensity of K :	18	19.5	21.0	25.5	33.0	48.0	93.0	18.0

Ans: Since blank gave intensity of 18.0 due to addition of the standard amount of K^+ for all values the readings were more by 18. If we plot the graph of (con. K) versus intensity, we get st. line graph with intercept of 18 on *y*-axis due to addition of standard.
The point corresponding to (18 + 18) = 36 units is 12 ppm of metal.

28.4 Calculate the concentration of calcium from sample of hard water if it was analysed by flame photometry at 422.7 nm. The reading were 0(4), 2(9.6), 4(15.0), 8(26.5) and 10(31.9) unknown (4). What was the concentration of calcium in unknown sample?

Ans: We plot graph of calcium concentration on *x*-axis against intensity on *y*-axis. The straight line graph had intercept of 4-0. So unknown sample will have intensity of (4+4) = 8 corresponding to this, the concentration of calcium was 26.5 ppm.

28.5 Strontium from milk sample was analysed by flame emission spectroscopy at 460.7 nm. The results obtained were as follows:
[Sr]/intensity 0(4.5), 4(21), 8(38.5), 12(56.3), 16(73) and 20(90) unknown (11.5). What is the concentration of strontium in milk in unknown sample?

Ans: As usual we plot a graph of [Sr.] against intensity. The st. line graph had intercept of 4.5 showing standard addition method was used. Now unknown concentration had an absorbance intensity of 11.5 so (11.5 – 4.5) = 7.00 active intensity corresponds to value of 9.1 ppm of strontium.

28.6 The excitation potential for the (4*p*) level in calcium was 2.95 ev.
Find out the wavelength of spectral line resulting from the transition from (4*p*) back to (4*s*) ground state?

Ans: $(E_1 - E_2) = \Delta E = 2.95$ eV electron returns from $4p \rightarrow 4s$ level.

Now $\Delta E = \dfrac{hc}{\lambda}$, if h = Planck's constant, c = Velocity of light

substituting the values $2.95 = \dfrac{12{,}400}{\lambda}$ or $\lambda = 420$ nm or 4200 Å.

28.7 What should be concentration of lead in the unknown sample of petrol when the following readings were obtained for analysis by AAS at 217 nm by atomic absorption spectroscopy.
1(0.07) 3(0.220) 5(0.390) 7(0.530) 9(0.660) Unknown (0.290)

Ans: We should first plot graph of concentration (*x*-axis) against absorbance on (*y*-axis) it will be from plot a point corresponding to 0.290 is 4 μg/ml which is present in the unknown sample of petroleum.

28.8 Calculate the concentration of strontium in unknown sample of grass if it was measured at 460.7 nm by flame photometry with calcium as the internal standard (2.50 μg/ml). The solutions when measured at 461 nm gave following readings?

Con. of Sr (μg/ml)	2	4	6	8	unknown
Int. at 423 nm for Ca	21.5	24.7	18.6	22.3	19.4
Int. at 461 nm for Sr	16.6	37.8	43.2	68.7	36.3
Int. ratio Sr/Ca	0.777	1.53	2.32	3.08	1.87

Ans: We calculate ratios of intensity of Sr/Ca. We have values as shown in Table. Now plot graph of ratio versus concentration of strontium which would give st. line graph. The concentration of the unknown sample was 4.85 μg/ml in grass.

28.9 What should be the amount of zinc in the unknown sample if it was analysed with copper as the internal standard (3 μg/ml of Cu) 10 ml sample + 10 ml of 7.5 μg Cu made up to 25 ml and measured at 213.9 nm (Zn) and 324.7 nm (Cu).

Zinc conc. (g/ml)	:	1.20	2.40	3.60	4.80	Unknown
Int. at 326.1 nm for Zn	:	31.5	32.7	29.8	30.4	30.2
Int. at 327.4 nm for Cu	:	18.7	38.6	52.7	71.7	45.3

Ans: First find ratio of intensity of Cu/Zn for all solution. Then plot a graph of ratio's of Cu/Zn versus con of zinc. The point is with (30.2) is for unknown. This actually corresponds to 4.80 μg/ml of Zinc.

28.10 Barium was analysed by AES at 553.6 nm with tungsten as standard solution (25 μg/ml) 50 ml of solution of barium was measured at 553.6 nm for barium and measured at 255 nm for tungsten.
Find out the concentration of tungsten?

Con. of barium	:	1.05	10.5	103.0	1030	10.500	Unknown
Int. 553.6 nm for Ba	:	0.67	3.40	18.0	115	739	2.5
Int. at 235.1 nm for W:		1.80	1.60	15.0	1.7	3.9	1.8

Ans: First find out ratio of intensity Ba/W at two wavelengths. Then plot a graph of ratio of intensity to concent of barium. The ratio of (2.5/1.8) corresponding to unknown Barium gives value as 6.2 μg/ml. Hence, 6.2 μg/ml of tungsten is present in unknown sample.

PROBLEMS FOR PRACTICE

28.11 Calculate the percentage of hafnium from the unknown sample if the emission intensity ratio of lines of hafnium/zirconium = 0.126 for 0.0045%. hafnium in zirconium metal and 1.29 for 0.070% hafnium in zirconium. Hafnium was measured at 307.3 nm and zirconium measured at 360.1 nm. The plot of log intensity against log [% concen] was st. line, what is the percentage of a hafnium in sample with ratio 0.598?
[**Ans:** 0.028%]

28.12 Microgram concentration of potassium was analysed at 766.5 nm from the sample of mineral. The reading for analysis were as follows: The mineral was added in each sample during preparation of the calibration curve. The values were

[K] mg/l	:	74.3	55.7	37.0	18.3	7.4	0	Unknown
Int. of 766.3 nm :		100	87	69.0	46	22	3	28

If one gram of sample in 100 ml water gave reading of 28, what was the amount of potassium in the unknown sample? [**Ans:** 0.104%.]

28.13 Calculate the two wavelength corresponding to two electronic transition as lithium shows as two transitions. viz., $3p \rightarrow 2s$ ($3.8s - 0$ev) and $4d \rightarrow 2p$ ($4.60s - 1.90$ ev)?
[**Ans:** 322 nm and 459 nm]

28.14 When high temperature is used in plasma spectroscopy it leads to ion formation which was not favourable for analysis. Can you rectify this problem?
[**Ans:** K has low ionisation potential when added the released electrons it would suppress ionisation of metal to be analysed.]

28.15 Can you guess which elements were present in alloy sample if it was analysed by atomic emission spectroscopy when following lines were observed in the spectrum: 242.7 nm, 265 nm, 280.2 nm, 328.7 nm, 327.3 nm, 460 nm, 780.0 nm, 852.1 nm and 670.8 nm?

[**Ans:** 242.7 (Au), 265.9 (Pb), 279.5 (Mn), 328.1 (Ag), 670 (Li), 780 (Rb) 852.1 (Cs) 327 (Ag), 460 (Sr)]

28.16 Can you predict what kind of metals were present in the unknown sample when analysed by atomic emission method when following lines were observed:
281.6, 326.1, 340.3, 346.6, 361, 372, 374, 375, 623, 624 and 643 nm?

[**Ans:** 281 (Pb), 326 (Cu), 340 (Ag), 346 (Cd), 361 (Co), 374 (Ni) and 623-643 nm (Li) were likely metals to be present in the sample.]

28.17 Calculate the concentration of chromium in the unknown sample if analysed in a sample of austentic stainless steel for chromium by AAS at 357.9 nm, (*i.e.*, 358 nm). The three standard samples contained.
(*a*) 4.7×10^{-2} (*b*) 1.0×10^{-1} (*c*) 2.54×10^{-1} intensity ratios of chromium to iron were (*i*) 1.1821 (*ii*) 0.531 (*iii*) 1.980 resp. unknown sample gave ratio of = 0.685?

[**Ans:** 0.121% or 1.21×10^{-1}]

28.18 Calculate the concentration of manganese in the unknown soil sample when it was determined with Fe as an internal standard by AES method at 279.5 nm. With log sector attachment height of various lines were compared:

% Manganese:	(*i*) 1×10^{-3}	(*ii*) 1×10^{-2}	(*iii*) 1×10^{-1}	(*iv*) 1×0,	(*v*) unknown
Height of Mn (mm):	(*i*) 7.6	(*ii*) 9.0	(*iii*) 9.9	(*iv*) 10.7	(*v*) 9.0
Height of Fe (mm):	(*i*) 4.2	(*ii*) 4.5	(*iii*) 4.4	(*iv*) 9.0	(*v*) 4.3

[**Ans:** 0.16% or 1.6×10^{-1} Mn.]

28.19 There was one line between 304.2 nm and 304-5 nm for iron. What was the wavelength if it showed 1.5 nm line with 2.3 nm away from the second line?

[**Ans:** λ = 304.37 nm.]

28.20 Find out the concentration of barium in unknown sample if chromium was used as internal standard for analysis of barium at 553.6 mm when following results were obtained:

Con of Ba	2	3.0	4.0	5.0	6.0	7.0	8.0	Unknown
Int. of Ba at 553.6 mm	16.9	22.0	29.9	41.2	54.7	64.7	74.7	36.1
Int. of Cr at 357.9 mm	35.7	34.4	33.1	35.6	38.5	38.9	39.3	35.4

[**Ans:** 9.94×10^{-5} M of Barium]

28.21 The ionisation energy for potassium was 4.32 eV. Is it possible to have an atomic emission lines below about 3000 Å? [**Ans:** No]

28.22 Calculate potassium concentration in unknown sample if 1.2456 gm of potassium was dissolved and diluted to 100 ml and analysed by flame photometry when following results were obtained at 766. nm 0.5 (24), 10 (49), 2 (103), 2.5 (120), 4 (190) unknown (121)?

[**Ans:** 0.02%)

28.23 What should be the concentration of manganese in the original alloy if an alloy (0.9421 weight containing microgram concentration of mineral was analysed by standard addition method by AES at 279.5 mm. The sample was dissolved and made up to 100 ml. To aliquat of 25 ml, 4.5 ppm of Mn was added, when absorbance was 0.220 while standard had absorbance 0.310?

[**Ans:** 263 ppm of Manganese]

28.24 Can you guess approximate components of the given sample if AES analysis of a certain sample following lines were obtained:

3247 Å, 2139 Å, 2288 Å, 2170 Å, 2483 Å, 2320 Å, 2407 Å, 2852 Å,

[**Ans:** Cu 3247 Å, Zn 2139 Å, Cd 2288 Å, Pb 2170 Å, Fe 2483 Å, Ni 2320 Å, Co 2407 Å and Mn 2852 Å.]

28.25 What should be the ratio of sodium atoms in the 3*p* excited state to 3*s* to the number of sodium atoms in the ground state at temperature of 2500-2510 K?

[**Ans:** 1.67×10^{-4} at 2500 K, 1.74×10^{-4} at 2510 K.]

REFERENCES

28.1 S.M. Khopkar, *Basic Concepts of Analytical Chemistry,* 3rd edn., New Age International Publishers (P) Ltd. (2007).

28.2 J.S. Fritz and G.H. Schenk, *Quantitative Analytical Chemistry,* Allyn & Bacon Inc. pp. 441-2; 475, 491 (1979) .

28.3 R.D. Braun, *Introduction to Instrumental Analysis,* McGraw-Hill & Co. pp. 231-2 (1987).

28.4 D.J. Pietzyke and C.W. Frank, *Analytical Chemistry: An Introduction,* Academic Press, p. 185 (1979).

28.5 J.A. Dean, *Flame Photometry*, McGraw-Hill & Co. (1960).

28.6 D.A. Skoog and D.M. West, *Principles of Instrumental Analysis*, 2nd edn., Holt Saunders, p. 310 (1980).

28.7 G.D. Christian, *Analytical Chemistry,* 5th edn., John Wiley & Co. (1998).

28.8 D.A. Skoog and J.J. Leoroy, *Principles of Instrumental of Analysis*, Harcourt Brace & Co., London (1997).

Chapter 29

Inductively Coupled Plasma—Atomic Emission Spectroscopy

INTRODUCTION

The techniques involving molecular absorption spectroscopy like UV visible, IR spectroscopy or atomic absorption spectroscopy (AAS) have become very popular with analyst. In comparison the techniques of atomic emission spectroscopy was not very popular. The only exception to this generalisation of acceptance was flame emission spectroscopy method. This was because low temperature for atomisation was necessary and further elements belonging to s-block exhibited emission in visible region of the spectra. The molecular emission techniques like fluorescence or phosphorescence is slowly gaining acceptability with analyst with the availability of proper fluoroscence agents promoting complexation and generation of fluorescence.

The main limitations of atomic emission spectroscopy technique involving arc, spark or flame as the source of excitation was that the cheap instruments were not easily available although the emission method as a rule had better sensitivity, e.g., flame photometry had become popular with analyst. However its principal limitations was the loss of 85% sample in drain and need to use low temp for atomisation. Most of the transition and main group elements needed very high temperature for atomisation. The process involved desolvation, vaporisation, atomisation, excitation and decay needing very high temperature. In order to circumvent these difficulties plasma spectroscopy was discovered to generate temperature as high as 9000 K facilitating atomisation of any kind of element from periodic classification of elements.

Amongst plasma spectroscopic, methods inductively coupled plasma was most popular. This technique when coupled with radio frequency generator facilitated the ionisation of inert gases like argon. An ion pair of argon with analytical element is formed, which in turn is atomised producing excitation. The micromilliliter samples could be handled with no loss of fluid during atomisation. The electrons after excitation would return back to the ground state with liberation of some energy in form of emission at a particular wavelength. The intensity of such emitted radiation was proportional to the concentration of the analyt sample. An element could have more than two wavelengths resonance line for energy of excitation but for practical purpose first line was preferred, e.g., Na $1s^2$, $2s^2$, $3s^1$, $2p^6$ showed transition of electrons from $3s \rightarrow 3p$ at 590 nm while second electron from $3p \rightarrow 4p$ occurs at 330 nm. We prefer all measurement at 590 nm as it furnish reproducible results. This was called resonance wavelength or frequency. All the sample is consumed with no loss whatsoever in plasma torch. The details of this technique will be found in book in reference as 26.1.

The main advantages of ICP-AES over AAS or FES is that very small samples (µg/l *i.e.*, ppb) can be analysed. The linear dynamic range (LDR) *i.e.*, upper limit of analysis is satisfactory and acceptable by a chemist. Large number of transition elements like Cu, Ni, Fe, Cd, Mo, V, Zn from various sources like environment, biological, soil samples were analysed very rapidly at trace concentration. The range of analysis of cobalt by ICP-AES was 0.02 µg/ml in comparison with AAS (0.002 µg/ml) or (0.1 µg/ml by FES). Only atomic fluorescence technique will suppress this limit to 0.001 µg/ml. Unfortunately atomic fluorescence models are not easily available and are expensive hence not much used.

In this chapter large number of multiple choice questions are furnished to test knowledge on basic principles, classification, comparison with other emission methods, kind of line selected for measurement, type of instrumentation used to generate plasma, role of radio frequency generator, LDR, flow of argon gas, limit of detection, type of electronic transition and role of nebuliser in generation of plasma.

PROBLEMS

Multiple Choice Questions

Tick mark correct answer from following multiple choices given:

29.1 Atomic emission spectroscopy is not popular in comparison with atomic absorption spectroscopy because:

(*a*) cheap instrumentals are not easily available
(*b*) it needs trained manpower to operate
(*c*) it is applicable only at milligram concentrations
(*d*) its sensitivity is not satisfactory.

29.2 Of all the emission methods one that is widely used is:

(*a*) AC arc spectroscopy (*b*) DC arc spectroscopy
(*c*) AC spark spectroscopy (*d*) flame emission spectroscopy.

29.3 Flame photometry had become popular because

(*a*) it was useful in visible region of spectra
(*b*) it used samples in liquid state
(*c*) it was used for analysis of voltile sample
(*d*) it was useful only at microgram concentration of sample.

29.4 The most important limitation of flame photometry was

(*a*) it needed air as oxidant
(*b*) sample injected was only 15% which was consumed by flame
(*c*) temperature of atomisation was very high
(*d*) volume of sample required was very large.

29.5 The element showing highest degree of ionisation at 3000°C is

(*a*) Mg (*b*) Ca
(*c*) Sr (*d*) Ba.

29.6 With air as oxidant and acetylene as the fuel the maximum attainable temperature is

(*a*) 2100°C (*b*) 1950°C
(*c*) 2200°C (*d*) 2700°C.

29.7 With oxygen as oxidant and acetylene as the fuel gas the temperature attained is
(*a*) 2700°C (*b*) 2800°C
(*c*) 3050°C (*d*) 2100°C.

29.8 During the process of atomisation, emission occurs at a stage when electrons in sample is
(*a*) desolvated (*b*) vaporised
(*c*) ionised (*d*) excited.

29.9 For an element the emission, occur at more than one wavelength but one preferred for the measurement the wavelength as resonance line is:
(*a*) ultimate line of emission (*b*) second line of emission
(*c*) first line of emission (*d*) any line of emission.

29.10 If by excitation and emission of electrons in sodium from 3*s* to 3*p* and 3*s* to 4*p* level showed emission at 589 nm and 330 nm we preferred first line because
(*a*) it gives reproducible results and is specific for sodium
(*b*) is easy for measurements
(*c*) is having long lifetime for measurements
(*d*) is easy to identify in spectrum of particular metal.

29.11 The flame as well as the graphite furnace were used in analysis with technique like:
(*a*) flame photometry
(*b*) atomic emission spectroscopy
(*c*) atomic absorption spectroscopy
(*d*) molecular absorption spectroscopy.

29.12 Temperature beyond 8000° K would be attained by use of
(*a*) flame with mixture of gases (*b*) furnace like graphite furnace
(*c*) electric discharge tube or (*d*) plasma torch.

29.13 The plasma in ICP-AES consisted of:
(*a*) electrons
(*b*) positive ions
(*c*) neutral molecules
(*d*) electrons, positive ions and neutral molecules.

29.14 The gas which was capable of producing plasma at high temperature is
(*a*) oxygen (*b*) nitrogen
(*c*) argon (*d*) carbon dioxide.

29.15 The significant advantage of using plasma as source of radiation is
(*a*) it allowed use of solid sample at low temperature
(*b*) it permitted use of liquid sample at high temperature
(*c*) it was useful for gases sample at critical temperature
(*d*) it was fast to dissociated sample at critical pressure.

29.16 The application of radio frequency generator in the plasma instrument is
(*a*) to produce charged ions to generate argon ions
(*b*) to raise the temperature to high value
(*c*) to ionise an element instantaneously
(*d*) to dispense sample in the plasma uniformly.

29.17 The transformation of the sample aerosol in plasma under goes
(*a*) desolvation and vaporisation
(*b*) atomisation and ionisation
(*c*) excitation and emission
(*d*) desolvation, atomisation, excitation, etc.

29.18 At the high temperature of excitation
(*a*) electrons are elevated to higher energy level with insignificant change in energy
(*b*) electrons remain stationary in the low orbital
(*c*) electrons are excited with further secondary reaction
(*d*) electrons are excited to emit radiation.

29.19 The linear dynamic range (LDR) represented, the
(*a*) smaller measurable concentration
(*b*) optimum measurable concentration
(*c*) upper limit of measurable concentration
(*d*) average level of concentration.

29.20 The radio frequency generator was used in ICP-AES model to
(*a*) deflect plasma source
(*b*) raise high temperature
(*c*) provide power for generation of plasma
(*d*) to measure intensity of radiation.

29.21 The flow of plasma argon in auxiliary nebuliser unit is in ratio of air to argon as:
(*a*) 1:1 (*b*) 1:2
(*c*) 1:50 (*d*) 1:15.

29.22 The elements like As, Sb, Se, Te, Hg could not be directly analysed by ICP-AES as
(*a*) they fouled the system
(*b*) they needed high temperature for atomisation
(*c*) they needed hydride generator for atomisation and ionisation
(*d*) they were non-volatile for measurement.

29.23 The alkali and alkaline earths were best analysed by
(*a*) atomic absorption spectroscopy (*b*) atomic emission spectroscopy
(*c*) flame photometry (*d*) ICP-AES-spectroscopy.

29.24 The analysis of uranium was unsatisfactory with low sensitivity as 0.138 μg/ml with
(*a*) atomic absorption spectroscopy (*b*) atomic emission spectroscopy
(*c*) plasma spectroscopy (*d*) flame emission spectroscopy.

29.25 During the analysis by plasma spectroscopy the sample should
(*a*) not be more than detection limit
(*b*) be five times detection limit
(*c*) be same as prescribed by the detection limit
(*d*) have no specific limit of analysis.

29.26 The analysis of metals at $10^{-1} - 10^{-8}$ gm could be done by
(*a*) flame photometry (*b*) atomic absorption spectroscopy
(*c*) atomic emission spectroscopy (*d*) plasma emission spectroscopy.

29.27 Plasma emission technique needed following gas:
(*a*) argon (*b*) oxygen
(*c*) acetylene (*d*) nitrogen.

29.28 The optimum attainable temperature in plasma spectroscopy was
(*a*) 200 K (*b*) 1000 K
(*c*) 15,000 K (*d*) 7500 K.

29.29 During the process of excitation at moderate temperature the electrons in sodium were pushed from 3s to
(*a*) 3*p* orbital (*b*) 4*p* orbital
(*c*) 3*d* orbital (*d*) 3*f* orbital.

29.30 Nebuliser was used in plasma spectroscopy to
(*a*) produce plasma
(*b*) convert liquid to aerosol and generate plasma
(*c*) to spray liquid in the chamber
(*d*) to detect the emitted radiation.

ANSWERS

29.1 a	29.2 d	29.3 a	29.4 b	29.5 d	29.6 c	29.7 c
29.8 d	29.9 c	29.10 a	29.11 c	29.12 d	29.13 d	29.14 c
29.15 b	29.16 a	29.17 d	29.18 d	29.19 c	29.20 c	29.21 d
29.22 c	29.23 c	29.24 a	29.25 b	29.26 d	29.27 a	29.28 c
29.29 a	29.30 b					

REFERENCES

29.1 S.M. Khopkar, *Basic Concepts of Analytical Chemistry,* 3rd edn., New Age International Publishers (P) Ltd. (2007).

29.2 A. Montaser and D.W. Golightly, *Inductively Coupled Plasma in Analytical Absorption Spectroscopy*, VCH Publisher (1992).

29.3 A. Varma, *CRC Handbook of Inductively Coupled Plasma Emission Spectroscopy*, Boca Raton Florida CRC Press (1991).

29.4 G.L. Moore, *Introduction to Inductively Coupled Plasma Atomic Emission Spectroscopy,* Elsevier Sci. Publisher (1989).

29.5 G.D. Christian, *Analytical Chemistry,* 5th edn., John Wiley and Co., New York (1994).

29.6 D.A. Skoog and J.J. Leory, *Principles of Instrumental Analysis,* Harcourt Brace Pub., London (1997).

29.7 M. Thompson and J.N. Walsh, *A Handbook of Inductively Coupled Plasma Spectroscopy,* Blace, Glasgow (1983).

29.8 G.F. Wallace and P. Barrett, *Analytical Methods Development for Inductively Coupled Plasma Spectroscopy,* Perkin Elimer (1981).

29.9 P.W.J.M. Boumans, *Inductively Coupled Plasma Emission Spectroscopy,* John Wiley (1987).

29.10 E. Metcalfe and F.E. Prichard, *Atomic Absorption and Emission Spectroscopy,* John Wiley (1987).

Chapter 30

Chemiluminscence, Atomic Fluoroscence and Ionisation Spectroscopy

INTRODUCTION

These are rated as the simplest methods for the quantitative analysis of component because the intensity of chemiluminscence is directly proportional to the concentration of species. The technique is coupled with chromatography, electrophoresis or enzyme reaction. It facilates detection of sulphur or nitrogen in their compounds, such as SO_2 and NO-NO_x from the atmosphere or in biochemical products like acids, amines, DNA polymers can be analysed by this method. Most of the measurements are carried out in liquid state; containing water or solvent. Amongst serious limitation is it is a destructive analysis technique and cannot be easily used for multicomponent determinations. The actual detection limit is very low (~10 ρg) or even (~10 fg) femtogram concentrations. The reaction always involves chemical luminescence reaction. Several of the organic reagents such as luminol, dioxetane, acridium ester; bis (tri-chlorophenyl oxalate or tris (bipyridine) ruthenium are used with gas phase or solution phase. The photodetector is used to produce chemiluminscence. Flowing stream methods are used to measure the intensity of emission.

(*a*) Also electrochemically one can generate chemiluminscence, but for this one needs electrochemical devices like potentiometer and good electrodes. The chemiluminscence reaction can undergo quenching like fluoroscence quenching. A quantitative relationship is used for anlaysis of these compounds as chemiluminscence intensity (CI) is

$$(\text{CI}) = \phi_{CL} \frac{dp}{dt} = \phi_{EX}\, \phi_{EM} \frac{dp}{dt} \qquad ...(1)$$

wherein (dp/dt) is rate of chemiluminscence, ϕ_{CL} = chemiluminscence quantum yield, ϕ_{Ex} = excitation/emission quantum yield.

This technique has numerous applications in biological as well as in environmental science. The atmospheric gases like SO_2, O_3, NO-NO_x can be monitored by chemiluminscence technique. It can be also used in conjunction with HPLC or capillary electrophoresis. The analysis of nucleic acid, immino enzyme and characterisation of biological materials can be done by chemiluminscence method. For qualitative analysis one can use Luminol in the presence of oxidants, transition metals and organometallic compounds. Instruments for static and flowing samples are available in the industry.

(*b*) The atomic fluorescence spectroscopy (AFS) is comparable with atomic absorption or emission methods however, it is not superior to these methods. A radiative excitation (or absorption) and deexcitation (or fluorescence) is observed in atomic fluorosence methods. Hg, Sb, As, Se, Te are best determined by AFS rather than by AAS or AES methods. Laser induced technique gives better sensitivity (~10^{-18} atogm). It is single element analysis technique. The process involves absorption of energy to get excited and latter emission of energy as fluorescence comparable to molecular fluorescence methods. The resonance transition is more intense. Two steps excitation in double resonance fluorescence uses two dye lasers. A cold vapour cell with hydride generator is used for the analysis of mercury. The chemical interferences reduces the sensitivity of the analysis. The spectral interference sometimes is encountered during analysis. The detection limits are lower than those obtained by AAS e.g. for Hg detection limit is 20 ηg/L while for AFS its higher viz. 1 μg/L. For selenium it is 50 ηg/L by AAS and lower 10 ηglL by AFS. Many elements like Ag, Al, Au, Cd, Co, Fe, Mn, Pb, Tl can be analysed by AFS, in concentration range of 0.5-5 μg/L Hg, Cd, As, Sc from atmosphere can be analysed easily by atomic fluourescence spectroscopy.

(*c*) Finally the resonance ionisation spectroscopy or laser induced ionisation spectroscopy is special class of ionisation spectroscopic technique. The use of ionisation suppressor such as the lanthanum or caesium is used in order to minimise the ion formation in AAS. While in Auger Spectroscopy the knowledge of ionisation cross-section as well as of back scattering facilitates accurate quantitative analysis. Such is the phenomena in spectroscopy of ionisation. In laser induced ionisation spectroscopy laser beam is invariably used to promote ionisation before spectral analysis. The non voltile or low voltile compounds are on top for the electron spray ionisation (ESI) and matrix assisted laser desorption/ionisation (MALDI) techniques. These are newly invented technique of recent origin and have as yet to find significant applications in Bioanalytical chemistry.

PROBLEMS

Multiple Choice Questions

Tick mark correct answer from the following multiple choices:

30.1 The detection limit for the element in attogram (10^{-18} gm) range of concentration is:

(*a*) Thallium (*b*) Gallium

(*c*) Indium (*d*) Aluminum

30.2 The atomic fluorescence spectroscopy is used for the chemical speciation in environmental science in combination with:

(*a*) Gas chromatography (*b*) HPLC

(*c*) Capillary Electrophoresis (*d*) Atomic absorption spectroscopy.

30.3 Mercury can be analysed by atomic fluorescence method at one of the following levels

(*a*) 1 μg/L (microgram) (*b*) 20 ηg/L (nanogram)

(*c*) 1 mg/L (milligram) (*d*) 1 ρ/L (picagram)

30.4 The element determined by hydride generation in atomic fluorescence spectroscopy is:

(*a*) Antimony (*b*) Bismuth

(*c*) Arsenic (*d*) Germanium

30.5 In combination with graphite furnace, laser excited atomic fluoroscence spectroscopy can analyse down to the concentration level of

(*a*) microgram level (*b*) nanogram level
(*c*) femagram level (*d*) milligram level

30.6 The preferable state of sample for analysis by atomic fluorescence spectroscopy is

(*a*) Solid state (*b*) Liquid state
(*c*) Gaseous state (*d*) Powdered form.

30.7 The volume of sample solution used for routine commercial analysis by atomic fluorescence spectroscopy is as small as:

(*a*) 1-5 litres (*b*) 10-100 ml
(*c*) 100-500 μl (*d*) 1-10 ηgl.

30.8 The gaseous sample cannot be analysed directly by atomic fluorescence spectroscopy as it is

(*a*) deposited on same metal followed by thermal release.
(*b*) dissolved in liquids
(*c*) condensed to vapor form
(*d*) cooled to liquid state before analysis.

30.9 Atomic fluorescence spectroscopy suffers from the limitation of the

(*a*) chemical and spectral interference
(*b*) need to get sample in the gaseous state,
(*c*) analysis at milligram concentration
(*d*) applications to exclusively for metalloids analysis.

30.10 The detection limit of analysis by atomic fluroscence spectroscopy in comparison to atomic absorption spectroscopy is

(*a*) at same level
(*b*) is better than atomic absorption spectroscopy
(*c*) permits analysis at ηg level
(*d*) good for analysis of transition metals.

30.11 The smallest concentration of that element which is measurable by atomic fluorescence method is

(*a*) Selenium (*b*) Tellurium
(*c*) Antimony (*d*) Mercury

30.12 The elements needing hydride generation in atomic fluorescence spectroscopy analysis are:

(*a*) Cadmium and zinc (*b*) Nickel and cobalt
(*c*) Arsenic and selenium (*d*) Gallum and indium

30.13 The application of laser has facilitated ultra trace analysis by the technique of:

(*a*) Inductively couple plasma atomic emission spectroscopy
(*b*) Molecular luminscence spectroscopy
(*c*) Laser enhanced ionisation spectroscopy
(*d*) Atomic emission spectroscopy

30.14 The most sensitive technique for trace level analysis of certain elements is

(*a*) Resonance ionisation spectroscopy
(*b*) Atomic absorption spectroscopy

(*c*) Atomic fluorescence spectroscopy
(*d*) Inductively coupled plasma atomic emission spectroscopy.

30.15 The laser which operates in the visible region of spectrum is
(*a*) Argon laser (*b*) Nitrogen laser
(*c*) Krypton laser (*d*) Ruby laser

30.16 Resonant fluoroscence spectroscopy facilitates the analysis of magnesium at the wavelength of
(*a*) 285.2 nm (*b*) 460.7 nm
(*c*) 193.7 nm (*d*) 369.2 nm

30.17 Laser enhanced resonance ionisation spectroscopy suffers from the limitation of
(*a*) chemical interference
(*b*) lack of specificity
(*c*) lack of sensitivity
(*d*) absences of selectivity in analysis.

30.18 Tunable dye laser consists of the use of following organic dyes as:
(*a*) rhodanine (*b*) alizarine
(*c*) quinanzarine (*d*) erythrosine

30.19 The energy required for chemical reaction to excite electrons in analyte is gained from phenomena of
(*a*) Electrochemiluminscence (*b*) Phosphorescence
(*c*) Fluoroscence (*d*) Chemiluminscence

30.20 The fragmentation of a molecule leads to exhibition of the
(*a*) Bioluminenscence (*b*) Electrochemiluminscence
(*c*) Chemiluminscence (*d*) Triplet-triplet annihilation

30.21 Luminol is used in quantitative analysis by chemiluminscence for
(*a*) HCl (*b*) HOCl
(*c*) $HClO_4$ (*d*) $HClO_3$

30.22 The triad of elements which were analysed with luminol fluorescence techniques are:
(*a*) Silver, gold, copper (*b*) Zinc, cadmium, mercury
(*c*) Iron, cobalt, nickel (*d*) Gallium, indium, thallium

30.23 One of the gaseous pollutant from atmosphere which is commonly analysed preferably by chemiluminscence technique is
(*a*) NO-NO_x (*b*) O_2
(*c*) CO_2 (*d*) Cl_2

30.24 The gas in chemiluminscence titration but not involving gas phase for analysis is
(*a*) Oxygen (*b*) Nitrogen
(*c*) Hydrogen (*d*) Ozone

30.25 In liquid phase titration the kind of indicator used for the analysis of arsenic is
(*a*) Luminol (*b*) Gallic acid
(*c*) Alkaline hydrogen peroxide (*d*) Lophine

ANSWERS

30.1 a	30.2 a	30.3 b	30.4 c	30.5 c	30.6 b	30.7 b
30.8 a	30.9 a	30.10 c	30.11 d	30.12 c	30.13 c	30.14 a
30.15 d	30.16 a	30.17 a	30.18 a	30.19 d	30.20 c	30.21 b
30.22 c	30.23 a	30.24 d	30.25 a			

REFERENCES

30.1 R.D. Braun, *Introduction to Instrumental Analysis*, McGraw-Hill and Co., p. 233 and 299. Indian edition (2001).

30.2 F. Settle (Ed.) *Handbook of Instrumental Techniques for Analytical Chemistry,* Pearson Education Publishers, p. 441, Indian edition (2004).

30.3 D.A. Skoog, F.J. Holler and T.A. Nieman, *Principles of Instrumental Analysis*, 5th Edition, Harcourt Brace College Publishers (1998).

30.4 S.M. Khopkar, *Basic Concepts of Analytical Chemistry*, 3rd Edition, New Age International Publishers (P) Ltd. (2007).

Chapter 31

Molecular Luminescence Spectroscopy

INTRODUCTION

The phenomena of chemiluminscence results when two species chemically interact to produce a compound which shows luminescence. The electronic luminescence results in the formation of luminscent compound when two species are interacting electrochemically. The phenomena of fluorescence or phosphorescence belongs to the category of molecular luminescence methods. The effect of fluorescence or phosphorescence can be ascribed as the phenomena of delayed emission by the molecules for instance the transition of electrons from $\pi^* \rightarrow \pi$ or $n^* \rightarrow \pi$ leads to fluorescence.

Let us imagine electrons are in the ground state (S_0) by excitation with Xenon lamp or Raman devices the electron gets excited and migrate to (S_1) state and if further energy is supplied electron migrates further to (S_2) state. In both the states electron is an unpaired and hence is called singlet state which is at higher energy level. Now if some time lapses say $10^{-7} - 10^{-11}$ seconds between excitation and emission; this is followed by the phenomena of relaxation. In this process the electron releases energy in form of delayed emission and returns back to steady state, *i.e.*, ($S_2 \rightarrow S_1 \rightarrow S_0$) state. The kind of emission displayed by electrons after $10^{-7} - 10^{-10}$ seconds is called as 'fluorescence' the process occures by simultaneous interconversion and relaxation.

However, some time the electron undergoes intersystem conversion (ISC) to convert from singulet state to triplet (paired) state. The triplet state is at lower energy level than singulet state. Now after relaxation and ISC the electron returns back to ground state. This occurs in time span of $10^{-3} - 10^{-2}$ seconds, this phenomena is designated as 'Phosphorescence'. Now if c = concentration, b = optical path, a = molar absorbivity. So $F = K\phi abc$.

This shows that the fluorescence intensity (F) is proportional to the concentration (C). This equation is also valid for phosphorescence effect. In above equation K = constant, ϕ is measure of efficency of production of fluorescene. It is defined as that fluoresence intensity which is proportional to the ϕ which is measure of efficency of production of fluorescence as:

$$p/p_0 = e^{-ebc} \text{ or } \left(\frac{p - p_0}{p_0}\right) = (p - p_0) = (1 - e^{-ebc})$$

$$F = (p_0 - p)\phi p_0 (1 - \varepsilon^{ebc})$$

$F = K \phi P_0 2 \varepsilon bc$ or $F \propto C$ if b is constant. This shows that fluorescence intensity is proportional to the concentration. This equation is also valid for phosphorescence. In this if K = constant and ϕ is measure of efficiency of production of fluorescence which is defined as:

$$\phi = \frac{\text{No. of photons emitted}}{\text{No. of photons absorbed}}$$

If $\phi = 1$, all emited photons emitted are equal to absorbed photons in a particular system. These molecules which exhibit high fluorescence intensity have high molar absorbivity (ε) because they have number of conjugated double bonds or have high resonance stability; or they possess an electron donating groups like NH_2 or OH or that they are rigid structures like metal complexes with fluorophoric ligands. The decrease in fluorescence intensity is called as quenching which is due to the collisional deactivation by solvent or component or species in the solutions.

The emission intensity wavelength is longer with red shift from the excitation wavelength. Such shift is called activation of energy of fluorescence. This is also true for phosphorescence phenomena with exception the time lag is longer. However in chemiluminescence energy is supplied to the molecule by chemical reaction.

Several compounds exhibit fluorescence, e.g., napthalene, anthracene and many flurophor ligands as are alizarine, benzoinoxime, flavonol, 8-hydroxylquinoline, morine, salicyleneor-thoaminophenol. The electronic transitions ($S_0 \rightarrow T_1$) is forbidden transition due to low probability of spin flip of the excited electron. The inner filter effect is due to disproportionate absorption of the excited radiation by concentrated solution, Lead poisoning in blood can be detected fluorimetrically by excitation at 424 nm and emission measurement at 625 nm specially at nanogram levels.

Number of problems are presented in this chapter to find out concentration of an unknown solution mostly from Beer' Law plot. Some questions pertain to identification of the flurophoric ligands used for quantitative analysis of metals.

SOLVED PROBLEMS

31.1 An unknown compound was measured by fluorimetry with standard (1 μg/ml) having fluorescence intensity of 73. If unknown solution showed intensity of 62 what was its concentration?

Ans: Since 73 corresponds to 1 μg/ml

62 will correspond to unknown concentration of $\frac{62}{73} \times 1.0 = 0.8493\ 1\mu\text{g/ml}$.

31.2 A beryllium-morine complex (Be = 200 mg/ml) had an intensity of 78.8 with two units for blank solution. What was the concentration of unknown if it had an intensity of 63.5 units?

Ans: Since 2 unit blank value was to be deducted from all readings;

(78.8 – 2.0) = 76.8 was for 200 μg/ml of beryllium complex

(63.5 – 2.0) = 61.5 was for Be/in unknown sample

$$\frac{61.5}{73.8} \times 200 = 0.016\ \ \mu\text{g/ml}.$$

31.3 What should be the concentration of cadmium in unknown sample if cadmium complex was analysed fluorimetrically. The observed values for known and unknown concentration of cadmium was as follows:

Con [Cd] mg/100	4.0	8.0	12.0	16.0	200	Unknown
% The fluorescence	20.1	39.8	61.0	85.7	100	3.7

Ans: We plot graph of % T intensity on (y-axis) with concentration of cadmium as on x-axis for 3.7 T corresponding point was 7.4 mg cadmium so we have 7.4 mg/100 ml or 74 mg/ml in unknown sample.

31.4 Calculate the concentration of vitamin riboflavin when analysed fluorimetrically which had reading of 52.3% T when 5 ml of standard (2 mg/ml vitamin) were added to 5 ml of sample corresponding reading was 67% T?

Ans: As x-mg/ml gave reading of 52.3% T for unknown
5 ml stand (2 μg/ml) + 5 ml of unknown = 67% T.
i.e., $x = 52.3$, $(x + 1) = 67.0$
We plot graph of concentration of vitamin against %T we get st. line graph for 10 ml $(x + 1)$ mg/ml = 67% × corresponds 52.3% T in original solution.
(67 – 52.3) = 4.7%T is due to unknown, *i.e.*,
0.13 mg/L or 0.0013 μg/ml.

31.5 State whether following elements can be analysed fluorimetrically with suitable ligand? (*a*) Iron (III), (*b*) Zinc (II), (*c*) Cadmium (II), (*d*) Strontium, (*e*) Copper (II), (*f*) Copper (I) and (*g*) silver (II)?

Ans: Let us consider their outermost electronic configuration
Fe → $3d^6\ 4s^2$, Fe (III) $3d^5$, Zn – $3d^{10}\ 4s^2$, Zn (II) $3d^{10}$, Cd $4d^{10}5s^2$, Cd (II) $4d^{10}$, Sr $4d^{10}5s^2$, Cu (II), $3d^9\ 4s^2$, Cu (I) → $3d^{10}\ 4s^1$, Ag (II) $4d^{10}\ 5s^1\ 4d^9$
The element which could exhibit no fluorescence due to electronic transitions are Cadmium or copper (II) Silver (III), as half filled or fully filled (d^{-10}) orbitals never encourage fluorescence.

31.6 Can you identify a heteroatom in compound which promotes flurescence if compound had molar absorpivity ε = 11,000 while for compound-B has molar absorptivity was 665?

Ans: Second compound has heteroatom which promotes flurescence.

31.7 Identify the compounds which can exhibit fluorescence?
(*a*) Benzene (*b*) Toluene (*c*) Anthracene (*d*) bromoanthracene (*e*) iodoanthracene

Ans: Out of the compounds shown above only iodoanthracene can exhibit the fluorescence due to conjugation and presence of heteroatom like iodine.

31.8 Write down number for compounds in order of increasing intensity of the fluorescent. (*i*) Nitronapthene (*ii*) hydroxynapthol (*iii*) methynalnapthlene (*iv*) carboxy napthalene (*v*) napthalene

Ans: The order of fluorescence intensity increases in order as:
(*ii*) > (*v*) > (*iii*) > (*iv*) > (*i*)

31.9 What should be the concentration of beryllium in unknown sample if beryllium was determined fluorometrically in chloroform with morine. The following result were obtained:

[Be] mg/100 ml :	4	8	12	16	20	Unknown
Fluorescence intensity :	10	19	28	37	45	30

Ans: We should plot graph of [Be] mg/100ml against fluorescence intensity. The concentration corresponding to 30 intensity was 13 mg/100 ml.

31.10 Find out the concentration of anion in the unknown sample if it results fluorescence quenching for anion fluorescence results were as follows:

[Anion] × 10^{-6} M:	0	0.22	0.44	1.99	unknown
Fluorescence Intensity:	100	94	89	42	64

Ans: For 64 fluorescence intensity was below between 0.22 × 10^{-6}M of anion as obtained from graph.

PROBLEMS FOR PRACTICE

31.11 Aluminium with oxine showed strong fluorescence. A standard solution of 5 mg/ml of aluminium showed an intensity of 0.450, what was the concentration of aluminium in unknown sample if its intensity was 0.990. **[Ans:** 11.0 μg/ml]

31.12 A 0.150 mg of selenium showed fluorescence of 0.250 with 2.2 diaminobenzidine. If 50 mg of gallium was added fluorescence rose to 0.750. Determine % of gallium in selenium if blank showed fluorescence of 0.05 units? **[Ans:** 1.3 × 10^{-2}% of Ga]

31.13 The fluorescence of uranium with dibenzylmethane was given by c = 2 × 10^{-5} as 5 × 10^{-4} = logK where c = concentration F = fluorescence intensity. Calculate the concentration of uranium (g/ml) if its fluorescence intensity was 50 units? **[Ans:** 8.7 × 10^{-4} g/ml. of uranium]

31.14 In polluted atmosphere the benzopyrene from air was collected by passing air through high volume sampler tube and it was washed with 100 ml of hexane and if 10^{-8} M solution was measured, what was the smallest concentration of benzopyene which could be collected (M.W. = 300)? **[Ans:** 0.3 mg]

31.15 Is it possible to analyse phenanthrene fluormetrically in presence of napthalene if phenanthrene had UV band at λ_{max} = 331 nm (log ε = 2.5) while napthalene had fluorescence peak at 310-370 nm? Calculate the amount in unknown sample from following data:

μ (mg/ml)	0.12	0.14	0.16	0.18	unknown
Intensity	61	72.3	85.7	95.3	37

[Ans: 0.074 mg/ml, *i.e.*, 74 μg/ml]

31.16 Calculate the amount of the antibiotic in blood (if one ml serum had reading of 25.75) An injection was given to an animal (7.5 mg/kg) the retention dose after fixed interval of time was as follows:

Conc μg/ml	10.3	7.0	4.3	3.5	2.7
Ret time mts:	30	90	180	300	480

[Ans: 89 mg]

31.17 Can anthracene show better phosphorescence in benzene in comparison with chloroform? Will wavelength of naphthalene be shifted to longer region as solvent is changed from benzene to diethyl ether? **[Ans:** No; Yes]

31.18 What was the quantum efficiency if 4.9×10^{17} photons were absorbed during excitation and 2.9×10^{17} photons were emitted in fluorescence? **[Ans:** 0.59]

31.19 If the absorbance of a solution was 0.35 find out the ratio of the second term in the bracket to first term in a given equation with usual notions:

$$Ia = \phi I_0 2.303\,abc - \left[\frac{2.303\varepsilon bc}{2} - \frac{(2.303)^2}{2} + \frac{(2.303\varepsilon bc)^3}{3}\right]$$

[Ans: = 0.040]

31.20 An absorbed radiation was reemitted with no change in the intensity such that the typical emission at the longer wavelength was named by definite name? What was name and what was kind of shift noted? **[Ans:** Resonance fluorescence and Stoke's shift]

31.21 (*a*) Identify the process which involved intermolecular process wherein molecule passed from singulet to lower energy electronic triplet state without emission?

(*b*) Name the process in which spin of excited electron was reversed from singulet state to a triple state? **[Ans:** (*a*) Internal conversion (*b*) Intersystem crossing]

31.22 Identify compounds in each set of which exhibit resonance fluorescene

(*a*) Propylbenzene, (*b*) fluorobenzene (*c*) chlorobenzene (*d*) nitrobenzene in toluene, xylene, hexane.

(*a*) Chlorobenzene (*b*) bromobenzene (*c*) iodobenzene (*d*) nitrobenzene and anisole in aniline, pyridine.

[Ans: Propylbenzene intoluenc and chlorobenzene in aniline shows resonance fluoroscence]

31.23 (*a*) Name the organic ligand used for fluorimetric analysis of (*i*) Aluminium (*ii*) Zirconium (*iii*) Zinc (*iv*) Beryllium.

[Ans: (*i*) Al-Alizarine/Oxine, (*ii*) Zr-pentahydroxy flavenol *i.e.* (morine), (*iii*) Zinc-benzoin oxime, (*iv*) Beryllium morine]

(*b*) Match metals with following fluorinating agents Al, Be, Ga, Ge, Se and Sn. Ligands are morine, alizarine, hydroxyquinoline, benzoin, flavnol. salicyaldchyde. Specify also the wavelength for such measurement.

[Ans: Al (Alizarine) 520 nm, Be (Morine) 620 nm, Ga (Oxine) 492 nm, Ge (benzoin) 410 nm Sc(salicyldehyde) 455 nm Sn-(flavenol) 470 nm]

REFERENCES

31.1 S.M. Khopkar, *Basic Concepts of Analytical Chemistry*, 3rd edn., New Age International Publishers (P) Ltd. (2007).

31.2 D.J. Pietrzyke and C.W. Frank, *Analytical Chemistry: An Introduction*, Academic Press, pp. 184, 197 (1979).

31.3 J.S. Fritz and G.H. Schenk, *Quantitative Analytical Chemistry,* Allyn and Baccon, pp. 459-60 (1979).

31.4 R.D. Braun, *Introduction to Instrumental Analysis*, McGraw-Hill and Co., pp. 344-5 (1987).

31.5 H.H. Willard, L.L. Merrit and J.A. Dean, *Instrumental Methods of Analysis*, D. Van Nostrand Co., p. 393 (1985).

31.6 D.M. Hercules, *Fluorescence and Phosphorescene Analysis,* Wiley-Interscience (1966).

31.7 G.D. Christian, *Analytical Chemistry*, 5th edn., John Wiley and Co. (1997).

31.8 D.A. Skoog and J.J. Leoray, *Principles of Instrumental Analysis*, Harcourt Brace Co. (1997).

31.9 G.W. Ewing, *Instrumental Methods in Chemical Analysis*, 5th edn., McGraw-Hill and Co. (1990).

31.10 R.U. Bumbloy, *A First Course in Quantitative Analysis*, Addison Wilsey (1990).

Chapter 32

Nephelometry and Turbidimetry

INTRODUCTION

The turbidimetry and nephelometry are used in quantitative chemical analysis.

If we have suspended particles in solution on passing beam of light in it 'Tyndall Scattering' is observed, when the particles are large. If the particles are small, then the scattering is called as the 'Rayleigh Scattering', where in wavelength of incident radiation remain same as scattered radiation. However, if the wavelength for incident and scattered radiation are different, then it is called as the 'Raman Scattering' as discussed in Chapter 26.

All scattered radiation can be measured quantitatively using equation as

$S = -\log \frac{P}{P_0} Kbc$ If P = intensity of transmitted light is P_o = intensity of incident radiation.

The turbidimetric methods are based upon absorption of radiation. We measure absorbance (or transmittance) of turbid solution. While in case of nephelometry we measure scattering of radiation. This scattering is proportional to the concentration of the solution. While in Raman spectroscopy there is change in frequency of scattering in relation to incident photons. We have $V_R = V_i + \Delta V$. If V_R = frequency of Raman Scattering, V_i = frequency of incident radiation and ΔV is change in frequency of radiation due to change in wave number.

When we consider Raman Scattering, we observe Stoke lines ($-\Delta E$) and anti-stoke lines ($+\Delta E$), we have $E = (hv - \Delta E)$ while in anti-stokes line $E = (hv + \Delta E)$. In Rayleigh scattering ΔE is negative.

For the purpose of scattering it is used in turbidity or nephlometry where in

$$S = \log \frac{P_0}{P} = \frac{Kbcd^3}{S4\alpha\lambda^4}.$$

If S = turbidence, I_o = intensity of incident light, I = intensity transmitted light, λ = wavelength, c = concentration, b = path length.

d = diameter and K is constant so also α constant.

The particles should be of uniform dimensions.

This is similar to Beers Lamberts law. Only in nephelometry just like in fluorimetry the source of radiation should be perpendicular to sources of light striking the detector in the measuring instruments.

For quantitative analysis of turbid solutions, we use turbidimeter e.g., ammonia by Nessler meter, Phosphate measured as canary yellow precipitate with (NH_4) MoO_4 and HNO_3. While scattering of radiation is measured to ascertain concentration of barium sulphate finally Raman Spectroscopy is used for the elucidation of structure of compound. Its special feature is it can be used for aqueous solution as water has no inferring radiations. This method is a good substitute for IR spectroscopy. If v_s is frequency of scattering, ΔE is difference in energies between ground and first vibration level we have an equation as $(hv_s - \Delta E)$ or $\Delta E = (hv_s - hv) = \Delta v$; $\Delta v = \Delta E/h$. In stokes line $(-\Delta E)$ is negative while in anti-stokes line is positive $(+\Delta E)$ as $E = hv \pm \Delta E$ for both stokes and anti-stokes line.

In this chapter problems in turbidimetry and nephelometry are based upon plot of intensity versus concentration with intensity of unknown sample given. Similar methodology is adopted for nephelometry. Raman spectroscopy is rarely used for the quantitative analysis as in most of the cases unlike Infra red technique it is used for the elucidation of the structure of the complex.

SOLVED PROBLEMS

32.1 Calculate the Raman shift for each Raman line and identify the normal mode of vibrations being Raman active?

(1) C—C bond in benzene ($K = 12.3 \times 10^5$ dynes/cm).

(2) C—C bond in ethane ($K = 4.5 \times 10^5$ dynes/cm).

Ans: Both normal modes of vibrations were Raman active the corresponding Raman shift could be calculated from Hook's law. It states $F = -Kx$ if distance displaced is x, F = force which exert to restore atom to original form K = force constant. If oscillation was simple harmonic then fundamental vibration frequency was given by equation.

$$v = \frac{1}{2\pi}\sqrt{K\left(\frac{m_1 + m_2}{m_1 m_2}\right)} = \frac{1}{2\pi}\sqrt{\frac{K}{\mu}}$$

, if $m_1 m_2$ are masses of atom, K = force constant while $(m_1 + m_2)/m_1 m_2$ is called reduced mass (μ) of diatomic oscillator.

Since, we have K = force constant and $\mu = \left(\frac{m_1 m_2}{m_1 + m_2}\right)$ represented by C-atom,

$$i.e., \quad \frac{24}{12 \times 12} = \left(\frac{24}{144}\right) = \mu$$

We substitute the values as

$$\gamma = \frac{1}{2 \times 3.14}\sqrt{\frac{12.3 \times 10^3}{1/6}} = 17.44 \text{ cm}^{-1}$$

for second case also 2143 and 1128 cm^{-2}.

$$\gamma = \frac{1}{2\pi}\sqrt{\frac{K}{\mu}} = \frac{1}{2 \times 3.14}\sqrt{\frac{4.5 \times 10^5}{1/6}} = 1465 \text{ cm}^{-1}.$$

32.2 Can you evaluate the turbidimetric coefficient like molar absorptivity (ε) if a sample was analysed by turbidimetry with transmittance as 0.368 cell path of 1 cm and concentration was 112 mg/L?

Ans: Since $A = \varepsilon bc$ or $-\log T = \varepsilon bc$ we substitute the values.
$b = 1$, $c = 0.112$ g/L, $A = -\log T = -\log T = -\text{Log } 0.368$ we have value for ε.
$\varepsilon = A/bc = -\log T/1 \times 0.112 = 0.00388$ mg/cm^2/l.

32.3 Calculate the concentration of sulphate in the unknown sample if it was analysed turbidimetrically as sulphate by addition of the barium chloride to varying concentration of sulphate. The reading were as follows:
[SO_4] mg/L(%/T): 15(83.8), 26(69.2); 38.7(49.0), 55.4(28.3), unknown (37.2).

Ans: We plot graph of sulphate concentration on *x*-axis and % transmittance on *y*-axis. We get st. line graph passing through the origin. Now the con. of sulphate corresponding to 37.2% was 49 mg/L on the graph.

32.4 Find out the concentration of silver from unknown sample by nephelometric analysis if precipitation was carried out with common salt. The reading observed were as follows:
[Ag] mg/L(% T): 1(16), 3(37.3), 5(51.5), 7(63.8), 9(79.7) unknown (56.8)?

Ans: We plot graph of [Ag] against [% T] to get st. line graph passing through origin. Now silver concentration corresponding to 56.8% T was 5.6 mg/L. Hence unknown solution contained 5.6 mg/L of silver.

32.5 Calculate the turbidence value if the intensity of incident radiation was 100 and the relative intensity of transmitted radiation was 45.2?

Ans: $P_0 = 100$, $P = 45.2$ as per equation Turbidence (*S*) value is

$$S = -\log\left(\frac{P}{P_0}\right) = -\log\frac{45.2}{100} = 0.345$$

Hence, the turbidence of unknown solution was 0.345.

32.6 Calculate the concentration of lead from unknown sample of lead chromate in nephelometry by calibration curve which was constructed in total volume of 50 ml. The readings were:
Vol. of soln. (absorbance): 2(0.150), 4(0.250), 6(0.320), 8(0.390) and 10(0.480) Unknown (0.350).
The standard solution contained 2500 μg/ml of Pb?

Ans: First we must convert ml to mg/ml Pb in all readings so 5(0.150), 10(0.250), 15(0.320), 20(0.390) and 25(0.480). Now we plot a graph of conc. of lead against absorbance which will be st. line graph. On plotting absorbance 0.350 on plot the corresponding concentration is 17.9 mg/ml.

32.7 A 100 ml of hydrochloric acid ($d = 1.5$) was made up to 250 ml and 25 ml of an aliquat was taken and silver nitrate was added and solution was made up to 50 ml. When unknown solution measured had an absorbance of 0.350. What was its concentration if calibration graph was:
Cl(A): 0.05(0.050), 0.10(0.100), 0.140(0.200), 0.19(0.300) and 0.24(0.400)?

Ans: We get st. line graph passing through zero now concentration corresponding to a 0.350 was 0.21 mg/ml. This was present in 25 ml of solution so in 250 ml of solution $0.21 \times 10 = 2.1$ mg/ml was the chloride concentration.

32.8 What should be the cross-sectional area of blackening of 1.0 gm of a solid with density of 1.50 g/cm^2 if the solid was distributed as a mono layer of spherical particle across a light beam and that the particles had radii of either 0.1 or 10 μm?

Ans: 1 Mm = 10^6m = 10^{-4} cm volume of particle with radius 10^{-5} cm is $4/3\pi\ (10^{-5})^3$ = 4.19×10^{-15} cm^2 was volume of particle

The total number of particle in one gram was

$$1.0 \times \frac{1}{1.50} \times \frac{1}{4.19 \times 10^{-5}} = 1.0 \times \frac{1}{\text{density}} \times \frac{1}{\text{Volume of particle}} = 1.59 \times 10^{14}$$

Now blackening area of 1.00 gm was

(area of particle) (nos. of particles) = $[\pi(10^{-5})^2]\ (1.59 \times 10^{14})$

= 4.99×10^4 cm^2 if radius was 10^{-3} cm blackening area was 4.99×10^2 cm^2.

32.9 Find out the concentration of sulphate if turbidimetric analysis of sulphate was carried out at 355 mm with optical path of 1 cm with the turbulence (*i.e.*, absorption) of 0.121, with turbidity coefficient was 1.73×10^3 1/mole/cm at 355 cm?

Ans: If $A = Kbc$ if A = Turbulence, b = optical path, K = Turbulence coefficient c = concentration. Since, $A = 0.121$, $K = 1.73 \times 10^3$, $b = 1$ cm, $c = x$ unknown sample, we substitute these values in above equation and have

$$0.121 = 1.73 \times 10^3 \times 1 \times c \text{ or } c = \frac{0.121}{1.73 \times 10^3} = 6.9 \times 10^{-3} \text{ moles/L.}$$

32.10 What should be the % T with optical path of 100 mm when measured at 500 nm, the suspension of silver chloride in water had uniform particles of 0.40 μm in diameter, with density 5.56 g/cm^3, and refractive index 207?

Ans: $S = \log \frac{P_0}{P} = Kbn$, if S-turbidence, K = turbiditly coefficient, b = optical path, n = number of scattering particles per mm.

This shows $K = 0.4343\left[\frac{2}{3}\pi^5 d^6 \lambda^{-4}\left(\frac{m^2-1}{m^2+2}\right)^2\right]$, if d = diameter of particle λ = wavelength; m = ratio.

PROBLEMS FOR PRACTICE

32.11 Calculate % of sulphur if a sample of mineral (0.556 g) acid was made to 250 ml and of this 5 ml was mixed to 50 ml of barium which gave an absorbance of 0.450. The reading for calibration plot were mg/ml (absorbance): 0.05(0.100) 0.100(0.200) 0.160(0.300) 0.22(0.400) 0.270(0.500)? **[Ans: 53.1%]**

32.12 Calculate amount of sulphate present in the unknown sample if a standard solution containing 7.59 mg/250 ml of had a reading of 5.0%. The unknown solution had reading of 4.26%? **[Ans: 35.3 ± 0.7%]**

32.13 What should be concentration of selenium if 0.750 gram of photosensitive material was dissolved in 10 ml of which one ml made up to 25 ml showed the fluoresence intensity of 35 units. The calibration readings were (μg/5 ml/sc) absorbance): 2(25) 4(50) 6(75) 8(100)? **[Ans: 11 μg/ml]**

32.14 How would you calculate the percentage purity of solution containing a suspension 5 ml of 1.34 M barium chloride added to unknown solution which gave turbidence at 355 nm as 0.295. If standard containing 1.200 mg of sulphate gave turbidence of 0.322? [**Ans:** 7.8 mg]

32.15 In Raman spectroscopic analysis, can you estimate relative sensitivity at 275 nm as against one at 776 nm? [**Ans:** 62.9]

32.16 Calculate the wavelength and wave number of the IR radiation in cm^{-1} with following frequencies. (*a*) 1.096×10^{14} H_2 (*b*) 9.431×10^{13} H_2.

[**Ans:** (*a*) 3656 cm^{-1} (*b*) 3146 cm^{-1} in nm (*a*) 2.735 μm (*b*) 3.19 μm]

32.17 What should be the concentration of barium in the unknown sample if in nephelometry following readings were obtained:
[Ba mg/ml] (absorbance): 10(0.050), 20(0.100), 40(0.200), 80(0.400) unknown (0.130)?

[**Ans:** 30 mg/ml of Barium]

32.18 What will be the concentration of chloride if it was determined by turbidence measurement with silver nitrate with the following readings [Cl^-) (absorbance) 10(0.05), 20(0.100), 40(0.200), 80(0.400) unknown (0.154)? [**Ans:** 35 μg/ml chloride]

32.19 What was the concentration of barium in unknown sample with a standard for analysing in solution of 0.20 mg/ml had matching with unknown if following reading were obtained
1st = 5 cm = 4.55, 4.53, 4.53, 4.56 cm
1st = 7.5 cm = 4.80, 6.85, 6.83 cm [**Ans:** C = 0.022 mg/ml]

32.20 Calculate the real absorbance of the coloured solution if in spectrophotometric measurement absorbance was 0.150 with 10 mm cell. Later solution was rendered cloudy and coloured, subsequent analysis had turbidity was 0.012? [**Ans:** 0.145]

32.21 Assuming turbidimetric coefficient to be 1.70×10^3 with optical path of 10 mm, and absorbance of 0.120 what should be the actual concentration of the solution?

[**Ans:** 69×10^{-2} moles/liter]

32.22 What should be the relative efficiency in a Rayleigh scattering experiment with violet light of 400 nm was scattered as a compared to green light of 550 nm? [**Ans:** 3.57]

32.23 What should be the period for visible radiation with wavelength of 600 nm if the forced oscillations which occurred in particles during the process of scattering existed for the period of the incident radiation? [**Ans:** 2×10^{-15} seconds]

32.24 Can the mutual exclusion be extended to the rotation and vibrational spectra if oxygen and water vapour furnish an IR or Raman rational spectra? [**Ans:** Yes]

REFERENCES

32.1 S.M. Khopkar, *Basic Concepts of Analytical Chemistry*, 3rd edn., New Age International Publishers (P) Ltd. (2007).

32.2 D.G. Peters, J.M. Hayes and G.M. Hieftze, *A brief Introduction to Modern Chemical Analysis*, Saunders and Co. (1980).

32.3 D.G. Peters, J.M. Hayes and D.M. Hieftze, *Chemical Separation and Measurements*, Saunder Golden Series (1974).

32.4 J.D. Braun, *Introduction to Instrumental Analysis*, McGraw-Hill & Co., pp. 443-4 (1987).

32.5 E.D. Olesen, *Modern Optical Methods of Analysis*, McGraw-Hill & Co., pp. 461-81, 479, 354 (1975).

32.6 Y. Lyalikov, *Physico Chemical Analysis*, Mir Publishing Co., p. 77 (1964).

Chapter 33

Refractometry

INTRODUCTION

In analytical chemistry information on refractive index provides useful information e.g. HPLC method. If 'n' represent refractive index while $\hat{i}$ represents angle of incidence of light and if $\hat{r}$ is angle of refraction of light we have a relationship

$$n = \frac{\sin i}{\sin r} \quad \text{(This term is called as the refractive index)} \qquad ...(1)$$

The numerical value of $n>1.0$ when a beam of light is entering from rarer to denser media, while $n<1.0$ if beam of light is entering from denser to rarer media. The value of 'n' depends upon temperature as well as on wavelength of light used if we use the sodium vapour lamp (D-line at 586 nm) at room temperature viz 20°C, then we write n_D^{20} is index of refraction at 20°C and 586 nm wavelength.

We use another relationship called 'specific refraction' (r_D). This is represented by an expression

$$r_D = \left(\frac{n^2 - 1}{n^2 + 2}\right) \cdot \frac{1}{\rho} \qquad ...(2)$$

If this term is multiplied by atomic weight we call it term as atomic refraction. In above equation 'n' is refractive index while 'ρ' is density of medium. So one can predict property of medium from the knowledge of (r_D) *i.e.*, specific refraction. Now if this value is multiplied by molecular weight we get what is termed as 'Molar refraction'. The specific refraction (r_D) can easily identify a compound as every substance has its own typical value for specific refraction. We also use a term called "Dispersion". This given by the following equation:

$$v = \frac{(n_D - 1)}{(n_f - n_c)} \qquad ...(3)$$

If in the expression n_f and n_c are refractive index for 'f' and 'c' lines of hydrogen at wavelength of 486 nm and 656 nm respectively. The term atomic refraction refers to specific refraction when it is multiplied with atomic weight e.g., $H = n_{20}^{D} = 1.0$, Double bond (n) = 1.73, Triple bond (n) = 2.4.

Now based upon the above expressions many numerical problems can be easily solved. Prior to that we would consider importance of refractive index in different kinds of analysis. In UV-visible spectrophotometer Kortum and Seilor discovered that coloured solutions with high concentration needs correction in absorptivity measurement due to variations in refractive index as $a = a\text{-}n/(n^2+2)^2$ where a = absorptivity and n = refractive index. At concentration of 0.001 M the value of refractive index is constant. This is only at high concentration refractive index may vary influencing value of 'a' also.

Secondly in atomic absorption spectrometry the phenomena of occlusion into refractory compounds occur. If matrix is refractory (e.g., Lanthanide elements) a small amount of element is trapped in the matrix. However, this can be minimised using a hotter flame during analysis.

In high performance liquid chromatography refractive index detectors are used. Any solution in column or eluant can be detected so long as refractive indexes of solute to be detected is different from that of mobile phase when refractive index is considered. Though good RI detectors have some limitations on account of small differences in refractive index of common substance the sensitivity of RI is not good in comparison to UV-detectors in HPLC methods. Further as refractive index depends upon the temperature it is necessary to use the thermostat both with cryogenic liquids to cool in order to maintain constant temperature. The sensitivity of detector to pressure variation is not favourable as pulses flow is retarded. A gradient elution technique cannot be used with RI detectors as they are affected by mobile phases of different concentration unless two columns and two RI detectors are used to nullify the effect.

The refraction is defined as change in its magnitude when a beam of light passes from one to another medium of different density. The refractive index decreases with rise of wavelength. This is called as 'dispersion'. From the knowledge of specific refraction $(r_D) = \left(\frac{n^2-1}{n+2}\right)\cdot\frac{1}{\rho}$ if ρ = density one can easily predict purity of substance which is being analysed. When r_D is multiplied by 'M' the molecular weight we get molecular refraction but if it is multiplied by atomic weight 'A' we get atomic refraction. The measurement of concentration by refractive index is more dependable in comparison to density determination. In physiological chemistry the composition of serum is analysed by refractive index information. The relationship in refractive index and circular dichromism as well as optical rotatory dispersion is given elsewhere.

SOLVED PROBLEMS

33.1 A typical solvent with refractive index n =1.37 with density of 1.04, has molecular weight (mw = 60), Calculate the value of specific refraction of the solvent and molar refraction?

Ans: The specific refraction is given by an expression as

$$r_D = \left(\frac{n^2-1}{n^2+2}\right)\cdot\frac{1}{\rho} \quad \text{if } n = 1.37,\ \rho = 1.04$$

we substitute the values

$$r_D = \left(\frac{(1.37)^2-1}{(1.37)^2+2}\right)\times\frac{1}{1.04} = 0.2155 \text{ cm}^2\text{ mol}^{-1} \text{ (Since Mol. wt. = 60)}$$

r_D × M.W. = Molar refraction so 0.2155 × 60 = 12.93 $\text{cm}^2\text{ mol}^{-1}$

33.2 Calculate the magnitude of the molar refraction of a compound like methyl formate? (Given atomic refraction of A = 2.418, H = 1.0, > C = 0 group 2.21 and ester oxygen 1.643?

Ans: In methyl formate we have 2C, 4H, 1 Ester group, 1 Carbonyl group so we substitute values of atomic refraction as (2 × 2.418) + (1 × 4) + (1 × 2.21) (1 × 1.64) = 13.090 $\text{cm}^3\text{ mol}^{-1}$.

33.3 Calculate a refractive index of compound which has angle of incidence at $\widehat{45°}$ with angle of refraction as $\widehat{90°}$?

Ans: We use above simple equation if $\hat{i} = \widehat{45}$, $r = \widehat{90°}$ we substitute in equation,

if $$n = \sin\hat{i}/\sin\hat{r} = \sin 45°/\sin 90° < 1.0.$$

As refraction occurs from denser to rarer media.

PROBLEMS FOR PRACTICE

33.4 What is the refractive index of solution containing 65% mole of 1-propanol and 35% mole propanol if $n_{20}^{D} = 1.36$? **[Ans:** 1.3]

33.5 A mixture of two fatty acids was analysed by refractometer with resulting $n = 1.41$. What is the composition of the mixture? ($n_1 = 1.42$, $n_2 = 1.40$ resp for two acids) **[Ans:** 68.1, 32.9%]

33.6 A typical compound has $n = 1.4573$ with $\rho = 1.59$, with M.W. = 152. What is its molar refraction? **[Ans:** 26.28]

33.7 What is the molar refraction for 1-5 hexadiene? (Given formula $CH_2 = CHCH_2CH_2$ if $>C = C< = 1.733$, H = 1.0, C = 2.48 are the values for the atomic refraction of groups) **[Ans:** 28.98]

33.8 Calculate the molar refraction of diethyl ether? (Given atomic refraction $\varepsilon = 2.48$, $H = 1.0$ and $O_2 \nless$ ether) = 6.4? **[Ans:** 22.32]

33.9 An unknown compound has refractive index (n) = 1.37 while its density is 0.7908g/cm². If the molecular weight of the compound is = 58. What will be its molar refraction?
[Hint: Use an expression as

$$R = \left(\frac{n^2 - 1}{n^2 + 2} \cdot \frac{M}{d}\right),$$ where M = mol wt, d = density, n = refractive index] **[Ans:** 16.09]

33.10 The refractive index of CCl_4 is 1.4573 at 20°C and density is 1.395 cm^{-1}. Calculate the molar refraction of the solvent? **[Ans:** 26.28]

REFERENCES

33.1 R.D. Braun, *Introduction to Instrumental Analysis*, McGraw-Hill & Co. p. 445 (1987).

33.2 H.H. Willard, L.L. Merritt, J.A. Dean and F.A. Settle, *Instrumental Method of Analysis*, CBS Publishers India, 6th Edition p. 403 (1986).

33.3 F. Settle (Ed.) *Handbook of Instrumental Techniques for Analytical Chemistry,* Pearson Education India p. 73 (2004).

33.4 L.F. Hamilton and S.G. Simpson, D.W. Ellis, *Calculations in Analytical Chemistry,* 7th Edition. McGraw-Hill & Co., p. 363 (1969).

33.5 S.M. Khopkar, *Basic Concepts in Analytical Chemistry*, 3rd edn., New Age International Publishers (P) Ltd. (2007).

Chapter 34

Polarimetry, Circular Dichromism and Optical Rotatory Dispersion

INTRODUCTION

The optically active compounds like carbohydrate can be measured by these technique. In such optically active compounds, we measure the specific rotation for glucose, it is +52.7 but for fructose it is –92.4. The other term normally encountered in such studies are CD and ORD. The change in optical activity with reference to wavelength is called optical rotatory dispersion (ORD). The refractive index is closely related to optical activity. If refractive indices (n_L) for left and right side (n_R) are different then angle of rotation $(\hat{\beta})\hat{\beta} = \frac{\pi}{\lambda}(n_L - n_R)$ while (n_L – n_R) is called as 'circular birefringence' while $\phi = \frac{1}{4}(\varepsilon_L - \varepsilon_R)$ where in (ε_L – ε_R) is called as 'circular dichromism'. One can experimentally measure $(\hat{\beta})$ angle of rotation by polarimeter while (ϕ) which actually represents elliptical (with eccentricity of ϕ) vector diagram. The ORD curves and CD curves are not independent, e.g., mid-point of ORD curve coincides with absorption peak of circular dichromism curve, while ORD trough matches with UV maxima of both ORD and CD.

The refractive index (n) is defined as the ratio of the velocity of light in electromagnetic radiation in vacuum (c) to the velocity (v) of the radiation in the medium:

$$n = \frac{c}{v} = \text{refractive index.}$$

As per Snell's law if $\hat{i}$ is angle of incidence, $\hat{\gamma}$ = angle of refraction

$$\frac{\sin i}{\sin v} = \frac{V_1}{V_2} = \frac{n_2}{n_1} = n_{21}. =$$

V_1 = Velocity in medium No. 1,

V_2 = Velocity in medium No. 2,

n_1 = Refractive index No. 1,

n_2 = Refractive index of medium No. 2.

Now, the molar refraction can be calculated from following relationship, viz.,

$$R = \frac{(n^2 - 1)M}{(n^2 + 2)d} \quad \text{if } M = \text{mole wt. of substance, } d = \text{density, } n = \text{refractive index.}$$

This formula was discovered by Lorent and Lorents. This is also called as 'specific rotation' if M is deleted then we have,

$$\text{Specific refraction} = \frac{(n^2 - 1)}{(n^2 + 2)}\frac{1}{d} \text{ while molar refraction is defined as}$$

$$\text{Molar refraction } (R) = \frac{(n^2 - 1)}{(n^2 + 2)}\frac{1}{d} \times M = \frac{(n^2 - 1)}{(n^2 + 2)}\frac{M}{d} = R.$$

These are specific properties of solutions in terms of refractive index. It is used to identify the substance and adjudge the purity. The refractive index of the water is (1.3325), chloroform (1.4429) and benzene (1.4970). These are taken as the standard values. For quantitative analysis we use an expression as follows:

$(n_s - n_o) = Kc$ if n_s = refractive index of solution, n_o = refractive index of the solvent $K^1 = \frac{(n_s - n_o)}{C}$

where C = concentration (g/100 ml) K = constant or specific refractive increment. A plot of $(n_s - n_o)$ α C will be straight line graph from which unknown concentration can be found. From the Table of components with increments of mole refraction one can also calculate it for a compound. This method considers, > C = C < H, C, etc., while counting increments of mole fraction so R = (contribution by double bond, triple bond, rings and contribution by atoms.

In this chapter number of problems are given, to calculate the concentration from value of specific rotation (α) by using relationship simplified as $\hat{\beta} = \frac{\alpha L c}{1000}$ if α = sp. rotation, L = length of tube, C = con. g/100 ml. $\hat{\beta}$ = angle of rotation. One can also calculate circular birefringence $(n_1 - n_2)$ if angle of rotation ($\hat{\beta}$) is known as $\hat{\beta} = \frac{180L}{\lambda}(n_1 - n_2)$ if L = path length, λ = wavelength, $\hat{\beta}$ = angle of rotation. One can also calculate % transmittance or can also find out unknown concentration of sample. In above equations C is concentration while small c = velocity of light.

SOLVED PROBLEMS

34.1 The specific rotation of nicotine was 162° at 590 nm. What was the concentration of solution (mole/L) if the angle of rotation was 0.52° with the length of tube was 10 cm *i.e.* 1 dcm? (Given M.W. of nicotine = 162)?

Ans: The equation used for specific rotation calculation is $\hat{\beta} = \frac{\alpha L C}{1000}$

where $\hat{\alpha}$ = specific rotation; $\hat{\beta}$ = angle of rotation, L = length of tube in cms C = concent. g/100 ml.

We substitute

$$0.52 = \frac{162 \times 10 \times C}{1000}$$

$$i.e., C = \frac{0.52 \times 1000}{162 \times 10} = 32.09 \text{ gms/100}; \frac{32.9 \times 10}{162} = 0.020 \text{ moles/L}.$$

34.2 The specific rotation of glucose containing C = 5 g/L was to be determined when angle of rotation was 1.3° and length of tube was 25 cm. Calculate the specific rotation of glucose solution?

Ans: Since β = angle of rotation

$$\beta = \frac{\alpha LC}{1000} = \frac{\alpha \times 25 \times 5}{1000} = 1.3$$

$$\alpha = \frac{1.3 \times 1000}{25 \times 5} = \frac{1300}{25} = 140 \text{ specific rotation} (\alpha).$$

34.3 The sample of fructose had a specific rotation (α) 45.2°. A solution showed angle of rotation (β) as 13.20° with tube of (L) 10 cm length. What was the concentration of compound?

Ans: $\beta = \frac{\alpha Lc}{1000}$ if $\alpha = 45.2; \beta = 13.20; L = 10\,\text{cm}; C =$ unknown concentration on substitution of these values in given equation, we have

$$13.20 = \frac{45.2 \times 10 \times C}{1000} = \text{or } C = \frac{13{,}20 \times 1000}{45.2 \times 10} = 29.299 \text{ gm/100 ml or } 0.29 \text{ g/ml}.$$

34.4 A sample of raffinose with specific rotation (α) of 45.2 showed the angle of rotation (β) as 12.12 with 10 cm tube. Calculate the concentration of raffinose?

Ans: α = 45.2, β = 12.12, L = 10, c = Unknown.

$$\beta = \frac{\alpha Lc}{1000} = \frac{45.2 \times 10 \times C}{1000} = 12.12$$

$$c = \frac{1000 \times 12.12}{45.2 \times 10} = 26.8 \text{ gm/100 ml}.$$

34.5 Calculate the difference in refractive indices of medium for dextro and laveo rotatory form in circular polarised light with sodium D line showing angle rotation β of 100° for 10 cm cell?

Ans: We have λ = 589 nm, *i.e.*, 5.89×10^{-5} cm. We have equation

$(n_1 - n_2)$ is circular birefringence, d = rotation angle, l = length of path, x = wavelength;

$$\beta = \text{angle of rotation as } \beta = \frac{180l}{\lambda}(n_1 - n_2)$$

$$100 = \frac{180 \times 10}{5.89 \times 10^{-5}}(n_1 - n_2)$$

$$(n_1 - n_2) = \frac{5.89 \times 10^{-5} \times 100}{180 \times 10}$$

i.e., $(n_1 - n_2) = 3.3 \times 10^{-6}$ was difference in refractive index or circular birefrigence.

34.6 How will you find out the circular birefringence if aqueous solution measured in 1-dm cell at $\lambda = 500$ nm when it showed angle of rotation of plane polarised light as $\beta = 10$?

Ans: Since $\alpha = \dfrac{180l}{\lambda}(n_1 - n_2)$ we calculate $(n_1 - n_2)$ circular birefrigence

as $\lambda = 500$, *i.e.*, 5.0×10^{-5} cm, $L = 10$ and $\alpha = 10$. We have

$$\beta = \frac{180l}{\lambda}(n_1 - n_2),\ 10 = \frac{180 \times 10}{5 \times 10^{-5}}(n_1 - n_2) = 50 \times 10^{-5} = 1800\,(n_1 - n_2)$$

So $(n_1 - n_2) = \dfrac{1800}{5 \times 10^{-4}} = \dfrac{1800}{0.0005} = 0.000036$ or 3.6×10^{-5} circular birefrigence.

34.7 What is the magnitude of specific rotation (α) if a solution of sucrose filled in (l) 20 cm tube had angle of rotation (β) as 66.5 with concentration of 25 g/100 ml?

Ans: The relation used is

$$\beta = \frac{180 \times l \times c}{100} = \frac{180 \times 20 \times 25}{100} = 66.5$$

$[\alpha]^{L}_{\lambda} = \dfrac{\beta}{lc}$ if α = specific rotation β = angle of rotation, l = optical path in *deci* meter and C = concentration as gm/100 ml = 0.025 g/ml as $\alpha = 66.5$; $d = 20$; $C = \dfrac{259}{100} = 0.025$ g/ml.

$$\alpha = \frac{66.5}{20 \times 0.02} = \frac{66.5}{0.40} = 166.1.$$

34.8 Calculate the specific rotation of the morphine of 15 mg/ml (C) in alcohol if its angle of rotation (β) is 4.92 and tube length (l) is 2.5 decimeter.

Ans: Since $\hat{\beta} = 4.92,\ l = 2.5$ dcm, $C = \dfrac{0.45}{30} = \dfrac{4.5}{300} = \dfrac{4.54 \times 1000}{300} = \dfrac{45}{3} = 15$ gm/lt. concentration.

Now $\beta = \dfrac{\alpha LC}{1000}$, We substitutes these values C = 15 g/L or C = 0.0151 g/ml.,

l = 2.5 dcm = 25 cm

$$4.92 = \frac{\alpha \times 25 \times 15}{1000} \text{ or } \frac{4.92 \times 1000}{25 \times 15} = 131.2.$$

PROBLEMS FOR PRACTICE

34.9 Calculate the specific rotation (α) with angle of rotation $\hat{\beta} = -3.30$ when 9.88 grams of solute was dissolved in 94.6 g of solvent with tube of length of 5 cm? **[Ans: −85°]**

34.10 Calculate the concentration of the solution if the specific rotation of strychine hydrochloride (α) was −104° with angle of rotation was – 1.56° and the length of tube was 2.5 decimeter? **[Ans: 0.6 gm/100 ml]**

34.11 Find out the concentration of malic acid if malic acid rotates the plane of polarisation of light to the right while malic acid (–) rotates plane of polarisation to the left when the specific rotation was 2.3°C for both acids with angle of rotation as +0.8° and if the total concentration of the acid was unity? **[Ans: 0.679/100 ml]**

34.12 Calculate the specific rotation at 590 nm; 486 nm and 434 nm if specific rotation of glucose at various wavelength was given as follows:

λnm:	447 nm	479 nm	508 nm	535 nm	656 nm
α:	+96.62	+83.88	+73.61	+65.35	+41.89?

[**Ans:** 52 at 589 nm, 80° at 486 nm; 104 at 434 nm]

34.13 Calculate graphically the magnitude of specific rotation at room temperature if it was caused by a 40% solution of malic acid at different temp which had as α = + 1.31 at 10°C, α = 0.54 at 20°C and α = –0.12°C at 30°C? [**Ans:** α_{23} = +0.25°]

34.14 Find out the amount of tartaric acid graphically with the following results:

Tartaric acid (con. %)	10	20	30	40	Unknown
Scale reading, nm:	9.9	19.1	27.3	3.49	10.5

[**Ans:** *C* = 10.73%]

34.15 What should be the specific rotation of the substance when 2 gm of it was dissolved in 150 ml of water. The solution in a 20 cm long tube read + 2.787 in a polarimeter with sodium D line at 589 nm. The distilled water read + 0.018°? [**Ans:** ±0.30°]

34.16 What should be % transmittance in 100 nm cell at 500 nm if a suspension of silver chloride aqueous solution contained uniform particles of 0.040 μm in diameter, density 5.56 g/cm^3 and refractive index of 2.07?

34.17 Will it be possible to calculate the molar refraction of acetone if refractive index of a sample at room temp. is given when measured with sodium line was 1.3721. The density was 0.7908 gm/cm^3.

34.18 Identify the organic acid if the refractive index was 1.3721 for a sample with its density as 1.0491? [**Ans:** Acetic acid]

34.19 What is the amount of acetone % (w/w) in the unknown sample if a mole fraction of acetone with refractive index at 20°C temp. with sodium D line were given as mole fraction (n) 0(1.3562), 0.2 (1.3750), 0.3 (1.3840) and 0.4 (1.3940) unknown (1.3760)?

[**Ans:** 11.6% with mole fraction of 0.2025]

34.20 Calculate the % of nitrogen in unknown sample of oil with refractive index 1.4673 if % nitrogen (against refractive index) in brackets value were given as:

%N (n) 0.2(0.4598), 0.4(1.4629), 0.60(1.4658), 0.8(1.4689)? [**Ans:** 0.69% nitrogen]

34.21 What should be molar refraction at 20°C if the refractive index of carbon tetrachloride was 1.4573 and its density was 1.595 g/cm^2? [Ans: 26.51]

34.22 What should be the molar refraction and specific rotation if at 20°C refractive index of benzene was 1.4979 and the density was 0.879 g/cm? [**Ans:** 28.5, ±0.20°]

34.23 If acetic acid had refractive index of 1.3698 at 20°C. What should be its specific refraction and molar refraction (*R*) if its density was 1.049? [**Ans:** α = 0.2154 cm^3/g, *R* = 12.93 cm^3/mol]

34.24 What should be the specific refraction for methyl formate if it has formula ($HCOOCH_3$) if α = 13.090? [**Ans:** *R* = 12.97]

34.25 What should be molar refraction and specific refraction for nitrobenzene if its refractive index was 1.210 g/cm^2 and density of 1.554 with molecular weight = 124? [**Ans:** 31.0; 0.25°]

REFERENCES

34.1 S.M. Khopkar, *Basic Concepts of Analytical Chemistry*, 3rd edn., New Age International Publishers (P). Ltd. (2007).

34.2 Y. Lyolikov, *Physico-Chemical Analysis,* Mir Publishing House (1964).

34.3 E.D. Olsen, *Modern Optical Methods of Analysis,* McGraw-Hill and Co., p. 449 (1975).

34.4 G.W. Ewing, *Instrumental Methods of Chemical Analysis,* 5th edn., McGraw-Hill and Co., pp. 159-78 (1985).

34.5 C. Djerass, *Optical Rotatory Dispersion – Applications to Organic Chemistry,* McGraw-Hill and Co. (1960).

34.6 L.M. Velluz, M. Legrand and M. Grosjean, *Optical Circular Dichromism,* Academic Press (1965).

34.7 D.C. Walker, *Origin of Optical Activity in Nature,* Elsvier Publisher (1979).

34.8 G.D. Christain, *Analytical Chemistry*, 5th edn., Harcourt Butler and Co. (1997).

34.9 D.A. Skoog and D.M. West, *Principles of Instrumental Analysis,* 2nd edn., Saunders Holt (1980).

34.10 R.D. Braun, *Introduction to Instrumental Analysis,* McGraw-Hill and Co., p. 445 (1987).

Chapter 35

Nuclear Magnetic Resonance Spectroscopy

INTRODUCTION

The atom is composed of electrons, protons and neutrons. The atomic number of an element represents total number of either electrons or protons inside the nucleus, while the number of neutrons present in nuclei are give by difference of mass number (Z) and atomic number (N) we may thus have same number of neutron as protons e.g., C_6^{12} has neutrons, protons and electrons, are same in number. $n_0^1 = (Z - N)$ while number of electrons or protons are same as atomic number (N).

Supposing if we keep this atom in two magnetic fields, out of these fields if first magnetic field is fixed while second magnetic field varies, we observe typical phenomena, viz., neutrons absorbs energy and orient themselves in particular order. The varying magnetic field may be easily applied from radiofrequency generator or solenoid. The extent of orientation also depends upon the strength of applied magnetic field.

The atom has four quantum numbers of which magnetic quantum number has value of ±1/2. It may be +1/2 or −1/2 depending upon the clockwise or anticlockwise movement of nuclei around its own axis. This movement is called spin and this spin gives rise to magnetic field. The orientation occurs such that spin axis is parallel to the applied field. They have an energy given as $E = \mu_x H_o$. If H_o applied magnetic field and μ_x is magnetic moment, the quantum of energy level (E) shall therefore be expressed by a equations: $E = \dfrac{-m\mu\beta H_0}{I}$ if β is constant called "nuclear magneton", I = quantum number since proton has magnetic quantum number of +1/2 or −1/2 we have energy for magnetic quantum number (m + 1/2) as −1/2 so $\dfrac{\mu\beta H_0}{1/2} = -\mu\beta H_0$ and for other m = 1/2 is + μβ H_o. These are quantum energies. So we have energy difference between two levels as $\Delta E = 2\mu\beta H_o$.

For all molecules the energy difference is given by $h\nu = \mu\beta \dfrac{H_0}{I}$. The lower energy level has slightly more energy than higher energy level when nucleii rotates the pressional velocity which is given by an expression:

$$w_0 = \frac{2\pi\mu\beta'}{I\lambda} H_0$$

$$\gamma = \frac{\mu\beta}{1(\lambda h/2\pi)} \text{ where } \gamma \text{ is magnetic ratio}$$

The 'Larmov frequency'; is given by a relationship as

$$\gamma_0 = \frac{w_0}{2\pi} = \frac{\gamma H_0}{2\pi} \; i.e., \; h\nu_0 = \frac{\mu\beta H_0}{I}$$

The correlation between frequency of absorbed radiation and magnetic field is given by $\gamma = \frac{rH_0}{2\pi}$ by considering precession of particle in the field.

Now we consider absorption process. The potential energy of precessing particle is $E = -\mu_x H_o = -\mu H_o \cos\theta$ an absorption of energy, when angle of precession changes. This precession is reversible so excited particle can return back to ground state by reemission. This leads to process of relaxation the rate of relaxation influences quality of NMR. The rapid relaxation is preferable. These are two, viz., spin lattice and transverse (spin-spin) relaxation. The spin lattice relaxation is designated by time (T_1) this represents average life time of nuclei. The proximity of other element (with unpaired electrons) reduces this time. In spin relaxation with time (T_2) various factors reduce this time, leading to line broading phenomena. The absorption of energy is although small can be easily measured. Such study of absorption of radio frequency radiation by nucleii is called as NMR or nuclear magnetic resonance. In 1946 Bloch and Purcell first observed and won Nobel prize for their discovery. The two nuclei have to same precession rate but have different magnetic field represented by lower and upper states. One can consider this in terms of environmental effect on proton NMR spectra.

The NMR can be measured by low resolution (60,100 MHz) or wide line high resolution (270, 400 MHz) instruments. The latter models give fine peaks in the spectra. If we take a spectra of CH_3CH_2OH, we have three peaks due to CH_3, CH_2 and HO but if we use deutrium compound, viz., CD_3CD_2OD only two peaks appear with the disappearing OH peak. Such effect is called 'chemical shift' observed due to what is termed as spin-spin splitting. The chemical shift originates due to circulation of electrons round nucleus under influence of magnetic field. Such moment of electrons generate small magnetic field. This generated magnetic field or precession is given by $H_o = (H_{app} - \sigma H_{app}) = H_{app}(1 - \sigma)$. If H_{app} is applied field, H_o = resultant field, σ = shielding parameter, it varies for proton in different function of groups.

The spin-spin splitting is due to effective increase in field around nucleus by a hydrogen nuclei bonded to an adjacent atom. Internal standard like TMS (*i.e.*, tetra methyl silanc). $(CH_3)_4$ Si is used to compare relative field changes in nucleii. Now the field strength H_{ref} needed to produce TMS resonance frequency '*r*' is given by an equation.

$$H_o = H_{ref}(1 - \sigma_{ref}) \text{ or } \sigma_{ref} = \left(\frac{H_{ref} - H}{H_{ref}}\right) \text{ or } \sigma_{sn} = \left(\frac{H_{sn} - H_0}{H_{sm}}\right) \text{ where } (sm)\text{-represents sample.}$$

So we define chemical shift as

$$\delta = \left(\sigma_{ref} - \sigma_{sm}\right) \times 10^6 \text{ or } \delta = \frac{H_0\left(H_{ref} - H_{sm}\right)}{H_{sm}\, H_{ref}} \times 10^6. \text{ But } H_{sm} H_0.$$

$$\text{So } \delta = \left[\frac{H_{ref} - H_{samp.}}{H_{ref}}\right] \times 10^6$$

This (δ) is dimensionless but explains the shift in ppm for a given peak. The value lies in between (1-12) but if we use tau (τ) then values are $J = (10 - \delta)$. It will always posses positive magnitude. σ for $J = (10 - \delta)$ TMS is very large. Thus chemical shift arises from secondary magnetic fields produced by circulation of the electrons in the molecule. The chemical shift is used to identify functional groups in the compounds. It depends upon solvent used and concentration of a solute employed.

The chemical shifts reagents have the effect of dispersing the absorption peaks over a large frequency range. It helps to separate overlapping peaks, e.g., dipivalomethane complex of Pr written as $Pr(DPM)_3$ is used in solvent CCl_4, $CDCl_3$, C_6H_6. Eu $(DPM)_3$ is also used. Apart from proton one can use F_9^{19}, P_{15}^{31} in NMR spectroscopy C_6^{13} is also useful to get reliable information.

In this chapter several problems are given to classify isotopes according to their spin. The calculation of frequency for resonance; magnetic field for resonance, evaluation of magnetogyric ratio, prediction of chemical shift, ascertain the composition of the mixture, calculation of frequency of protons, energy needed for excitation, are presented. Many problems are based upon equations given in the text. NMR is rarely used for quantitative chemical analysis. The main role is to elucidate chemical structure of organic or inorganic compounds containing *H*. However this method must be supplemented alongwith NMR with UV, IR, Mass spectrometry and elemental analysis in order to get reliable information.

SOLVED PROBLEMS

35.1 Asign spin quantum number to isotopes depending upon their spins:

$^1_1H,\ ^3_1H,\ ^6_3Li,\ ^7_3Li,\ ^{10}_5B,\ ^{11}_5B,\ ^{12}_6C,\ ^{13}_6C,\ ^{14}_6C,\ ^{14}_7C,\ ^{14}_7N,\ ^{13}_7N,\ ^{16}_8O,\ ^{17}_8O,\ ^{19}_9F,\ ^{31}_{15}P$

Ans: The isotopes are classified as to have integral, half or zero spin as

Integral spin: $^6_3Li,\ ^{10}_5B,\ ^{14}_7N,\ ^{32}_{15}P$, (Even mass odd number but even atomic number, 1, 3 spin integral nos)

Half Spin: $^1_1H,\ ^3_1H,\ ^7_3Li,\ ^{13}_6C,\ ^{15}_7N,\ F_9^{19}$ (odd mass number 1/2, 3/2, 5/2 spins)

Zero Spin: $^{12}_6C,\ ^{14}_6C,\ ^{16}_8O$ (Even atomic and mass number, zero spin).

35.2 Calculate the frequency of radiation necessary for the resonance absorption by H_1^1 in magnetic field of (H_o) 14,092 Gauss. If the magnetic movement of $H_1^1 = 2.793 \times 5.05 \times 10^{-24}$ ergs/G, $h = 6.63 \times 10^{-27}$ and $I = 1/2$ is magnetic quantum number?

Ans: The magnetic moment of proton is calculated as follows:

$\mu = (2.793) \times \left(5.05 \times 10^{-24}\right) = 1.41 \times 10^{-2}$ ergs/G. $= H_1^1/G$.

Now frequency was calculated by knowing μ, *H*, *h* and *I* so we have

$$v = \frac{\mu H}{hI} = \frac{(1.41\times10^{-2})(14{,}092)}{(6.63\times10^{-27})(1/2)} = 60\times10^{6}\,S^{-1},\ \textit{i.e.},\ 60\text{ MHz}.$$

35.3 Calculate the magnetic field if resonance absorption of proton at 100 MHz was studied? (Given $I = 1/2$, $h = 6.63 \times 10^{-27}$, $\mu = 1.41 \times 10^{-2}$, $\gamma = 100 \times 10^{6}\ S^{-1}$)

Ans:
$$H = \frac{Ihr}{\mu} = \frac{(1/2)(6.63\times10^{-27})(100\times10^{6}\,S^{-1})}{(1.41\times10^{-2})} = 23.5\times10^{3}\,\text{G}.$$

35.4 What is the magnetogyric ratio for ${}^{3}_{1}H$ (Given $\pi = 3.14$, $\mu = 1.41 \times 10^{-27}$, $h = 6.63 \times 10^{-27}$ and $I = 1/2$) with usual notations?

Ans: $\gamma = \dfrac{2\pi\mu}{hI}$, if γ = magnetogyric ratio on substituting the values

$$\gamma = \frac{2(3.14)(1.41\times10^{-27})}{(6.63\times10^{-27})(1/2)} = 2.67\times10^{4}\ \text{GS}^{-1}.$$

35.5 Calculate the ratio of (n_2/n_1) in low and high energy states in a magnetic field of 10,000 G at room temperature?
(Given $\mu = (2.79 \times 5.05 \times 10^{-24}$ ergs); $H = 10{,}000$, $T = (27 + 273°)$, $k = (1.3805 \times 10^{-16}$ er).

Ans: $\Delta E = h\nu = \mu H/I$; $I = 1/2$ so $\Delta E = 2\mu H$.

$$\frac{\Delta E}{kT} = \frac{2\mu H}{kT} = \frac{2\times(2.79\times5.05\times10^{-24})\times(10{,}000)}{(1.3805\times10^{-16})(300)} = 6.80\times10^{-6}$$

35.6 In the organic compounds you can predict the chemical shift with low and high magnetic field on CH_3, $CH_3 - NH$, CH_3O, CH_3F, predict the shift?

Ans: In the above series $CH_3 - N' -$ will have high shift while $CH_3 - F$ will have low chemical shift, as this is related to the electronegativity of nitrogen and fluoride atoms.

35.7 What should be the average residence time (t) for OH proton in CH_3CH_2OH where OH– triplet just coalesces in broad absorption peak?

Ans: Since average time (t) $= \sqrt{2}/\pi J$ if J = sepn of peak with no exchange = 5

So
$$t = \frac{\sqrt{2}}{3.14\times5} = 0.090$$

So average residence time = 0.090.

35.8 The derivatives of butanol exhibited shifts as shown below:

(*a*) 5% 2.2 dimethyl - 2 butanol : 47 : 4.7 ppm (65)
(*b*) 8% 2.2 dimethyl - 3 butanol : 75 : 4.3 ppm (55)
(*c*) 6% 2.3 dimethyl - 2 butanol : 57 : 4.0 ppm (20)

A mixture exhibited integrated intensity as 65 at 47 ppm 55 at 4.3 ppm and 20 at 4 ppm. Can you calculate the composition of the mixture?

Ans: (*a*) $= \frac{5 \times 65}{47} = 6.91\%$ is 2-butanol.

(*b*) $= \frac{8 \times 55}{75} = 5.86\%$ is 3-butanol.

(*c*) $= \frac{6 \times 20}{57} = 2.10\%$ is 2-butanol.

35.9 β-alanine and normal alanine can be analysed in heavy water by NMR. A 3% solution of alanine and β-alanine in heavy water showed reading of 980 at 1.48 ppm and 650 at 2.54 ppm respectively. The first peak is due to α-methyl and second peak is due to α-methylene group. An unknown sample mixture shows a reading of 377 at 1.48 ppm and 400 at 2.54 ppm. What is the % of alanine?

Ans: The standard and unknown are measured under similar conditions.

$$\text{amt. of alanine} = \frac{377 \times 100}{980} = 38.46\% \text{ at } 1.48 \text{ ppm}$$

$$\text{amt. of } \beta\text{-alanine} = \frac{400 \times 100}{650} = 61.53\% \text{ at } 2.54 \text{ ppm.}$$

35.10 Calculate energy and frequency of radiation required to excite proton from lower energy level to higher energy level if magnetic flux is 14,092 G?
(Given magnetic moment of H = 2.7927; I = 1/2, B_o = magnetic flux = 14,092 G)

Ans:

$$\Delta E = \frac{\mu_\beta B_0 (m_1 - m_2)}{I} \quad \text{since } \mu = 2.7927(5.05 \times 10^{-3}),\ B_0 = 14{,}092 \text{ G.}$$

$$\beta = \pm 1/2$$

$$\Delta E = \frac{(2.7927 \times 5.05 \times 10^{-3})(14{,}092)}{1/2} = 3.98 \times 10^{-26} \text{ J}$$

The frequency was calculated from relationship, ($\Delta E = hv$) $h = 6.626 \times 10^{-31}$ (Planck constant).
So $3.98 \times 10^{-26} = 6.626 \times 10^{-31} \times v$

$$\nu = \frac{3.98 \times 10^{-26}}{6.626 \times 10^{-31}} = 6.00 \times 10^{-7} \text{ S}^{-1}, \textit{ i.e.}, 60 \text{ MHz.}$$

35.11 Find out the chemical shift in 60 MHz instrument if reference nucleus showed absorption at magnetic flux density of 0.063 G presuming the sample also absorbs?

Ans: For 14×10^3 magnetic flux density need 60 MHz chemical shift is converted from gauss to Hertz.

$$(B_{\text{ref}} - B_{\text{sample}}) = 0.063 G \times \frac{60 \times 10^6 \text{ Hz}}{14 \times 10^3 G} = 2.7 \times 10^2 \text{ Hz.}$$

$$\text{The shift}(\delta) = \frac{(B_{\text{ref}} - B_{\text{sam}})}{\nu} = \times 10^{-6} = \frac{2.7 \times 10^2}{60 \times 10^6} \times 10^{-6} = 4.5 \text{ ppm is chemical shift.}$$

35.12 What is Larmov frequency in NMR spectroscopy?

Ans: The frequency with which a nucleus precess around the applied magnetic field was called Larmov frequency. It depends upon the strength of applied magnetic field and on magnetic ratio of nucleus. It is $V_0 = \frac{\omega_0}{2\pi} = \frac{rH_0}{2\pi}$ as $hv_0 = \frac{\mu\beta H_0}{I}$.

35.13 Select one of the following conditions which would be preferable:

(*a*) more sample - on lower NMR (60 MHz)

(*b*) less sample - on higher NMR (300 MHz)

Ans: High MHz was preferred due to high frequency solid particles if present cause increase T_1 and their would be line broadening of NMR proton - NMR works on lower conc. fraction and not on C_6^{13} – NMR.

35.14 Which should be the ideal probe in the following; NMR studies

(*a*) Biochemical work and natural products

(*b*) Organometallic and Phosphorus compounds

(*c*) Organic synthesis on large scale

(*d*) Nitrogenous bases.

Ans: (*a*) small diameter probe (5 mm size) e.g., 2.5 mm would prove be best.

(*b*) broadband multinuclear probe as different kind of nuclie were encountered

(*c*) large probe (10-15 mm) size in organic synthesis

(*d*) a broadband probe specific for nitrogen should be used.

35.15 If the signals used in NMR are H_1^1, Li_3^7 and F_9^{19} state whether same nucleus for same element be used?

Ans: No. The signal as lock signal from the lock transmitter which would interfere the signal we desire to record.

35.16 Identify the hard and soft pulses in NMR spectroscopy on basis of width?

Ans: They depend upon pulse width. The shorter width of pulse will give hard pulse which would be powerful. While longer width pulse gives soft pulse, which were weak.

35.17 Why FT-NMR was preferable to pulse NMR technique?

Ans: In pulse NMR we measure time domain which in turn is subjected to FT to convert them into frequency domain.

35.18 Explain signal to noise ratio and show how it is improved by getting large noise of FIDs?

Ans: This is S/N ratio, can be improved by signal averaging, *i.e.*, record many spectras and combine them. For 'n' scans signal will increase 'n' times while noise amplitude will increase by $n^{1/2}$.

35.19 What is the meaning of terms sensitivity and resolution in NMR spectroscopy?

Ans: The sensitivity represent S/N ratio, alternatively height of NMR peak gives idea of sensitivity. Resolution is extent to which the line shape deviated from ideal line. It can be determined by measuring width of signal at half height in hertz.

35.20 Can you state various factors which could improve NMR sensitivity?

Ans: The sample concentration temperature, magnetic field strength, R_f power of B, number of scans, relaxation rate and polarisation offer sensitivity. Sample amount should be large. Low temp, was preferable, increasing magnetic field (60 M → 200 MHz) gave better resolution.

R_f increased signal strength. More scans were better. Nuclear over house effect (nOe) increases sensitivity. Large time between pulse gave relaxation, or relaxing agents could be added to the sample.

35.21 Define the nuclear over house effect (nOe) in NMR spectroscopy?

Ans: If magnetic nucleus was held at short distance from irradiated nucleus, an additional relaxation is seen where in irradiated nucleus transfer its excess energy to neighbouring nucleus through dipole-dipole interaction. Through space and not through bond nucleus energy is transferred. This gives rise in peak height hence it is called nOe. It is used for determining structure of compound and also to locate hidden spectral lines. This happens only if two nuclei were in proximity of one each another.

PROBLEMS FOR PRACTISE

35.22 What was the frequency of ^{1}H if it had magnetic moments of 1.45×10^{-27} with $h = 6.625 \times 10^{-27}$ in field of 14,092 G?

[**Ans:** Use $\Delta E = h\nu = \mu H = 60 \times 10^6$ sec^{-1}]

35.23 Find out the % by weight of fluorine in unknown sample if with fluorobenzene as an internal standard with CCl_4 as the solvent following results were obtained amt. of fluorine = 1.94 gm. wt of unknown = 0.57 gm, volume of CCl_4 = 5, reading due to fluorobenzene = 784 reading of unknown sample 387? [**Ans:** 24.6% F]

35.24 Can you ascertain the concentration of cryptopine in the baine if the baine showed three peaks in NMR spectra due to protons at 4.8-5.75 ppm, 5.75-7.15 ppm but it had no peak in the first region namely 4.8-5.75 ppm and it had an absorption in second region due to the six protons?

35.25 Find out the spatial configuration for β diketone if there were five isomers as NMR proton showed two signals of equal intensity, assuming the empirical formula was $C_4H_4O_2$?

[**Ans:** $CH_2{=}C{-}CH_2O{-}C{=}0$]]

35.26 Show the probable structure for the NMR spectrum for the compound C_3H_6O had a triplet centered at 84.73 (J = 5.27) and quartet at S = 2.72 (J = 7.28)?

35.27 Calculate the amount of hydrogen in the unknown sample if sample containing hydrogen was prepared. 0.320 g/ml of *n*-decane in CCl_4 gave peak at 922. An unknown sample containing 0.285 gm/ml had peak of 325? [**Ans:** 6.14% H]

35.28 Predict the relative shift in protons of ethylene and acetylene in the unsaturated hydrocarbons?

[**Ans:** Ethylene and acetylene; resp 7:65 > 4.67]

35.29 Is it possible to enhance field strength rise in compounds such as CH_3F, CH_3Cl, CH_3Br, and CH_3l? [**Ans:** CH_3F, CH_3Cl, CH_3Br, CH_3l is the order of increase in field strength]

35.30 Arrange in the order of increasing strength the derivatives of CH_4 and mono, di and tri-substituted phenyl derivatives? [**Ans:** $(C_6H_5)_3CH < (C_6H_5)_2CH_2 < (C_6H_5)\ CH_3 < CH_4$]

35.31 What was the order strength of saturated hydro carbon for the following:
CH_4 RCH_3, R_2CH_2 and R_3CH? [**Ans:** $(CH_3)_3CH > (CH_3)_2\ CH_2 > (CH_3CH_3) > CH_4$]

35.32 Predict the PMR spectrum for:
2-chloro 3-nitrothiophene (–) and 2, 5 dichlorothiophen?

[**Ans:** Proton in 3, 4 are not same in first 4 is lower than
3 Proton in 3, 4 position equivalent in single peak]

35.33 Calculate the ratio of number of H-nuclei in higher energy level to those of nuclei in lower energy level at 25°C in magnetic field of 14092 G? [**Ans:** 0.99999]

35.34 Calculate the energy of radiation that was required to excite $^{13}_{6}C$ nuclei from lower to higher energy level if magnetic flux density = 1.0×10^4 G? **[Ans:** 7.093×10^{-27} J]

35.35 Predict the ratio of $^{13}_{6}C$ nucleus in upper energy level to lower energy level at 303 K in magnetic field with flux density of 25,100 G? **[Ans:** 0.99999575]

35.36 Find out the ratio of $^{19}_{9}F$ nucleus in higher energy to lower energy at 298° K in a magnetic field with flux density of 1.3×10^4? **[Ans:** 0.99999033]

35.37 Find out the frequency of radiation required to interact with $^{19}_{9}F$ nuclei in the field of 14,092G? **[Ans:** 56.446 MHz]

35.38 What should be the magnetogyric ratio for $^{11}_{3}B$ nuclei and show how you would calculate the frequency of radiation that would be needed for resonance of $^{11}_{3}B$ in a magnetic field of 10 KG? **[Ans:** 13.66 MHz]

35.39 If in a wide line NMR spectroscopy a typical line width of IG, can you find out the degree of homogeneity a magnetic field should have for wide line proton resonance studies at 60 MHz? **[Ans:** About 30 ppm or better]

35.40 Ascertain the molar concentration range for which NMR is sensitive if from the weight/volume % sample composition typically used in NMR technique? **[Ans:** About 0.1 to 2M]

35.41 Find out the relative population of high and low energy rotational states at room temperature, if rotational energy level spacing of 1.24×10^{-4} ev by using Boltzman distribution law? **[Ans:** 0.9952]

REFERENCES

35.1 E.D. Oleson, *Modern Optical Methods of Analysis,* McGraw Hill and Co., pp. 492-7; 501-5 (1975).

35.2 S.M. Khopkar, *Basic Concepts of Analytical Chemistry*, 3rd edn., New Age International Publishers (P) Ltd. (2007).

35.3 R.D. Braun, *Introduction to Instrumental Analysis,* McGraw Hill and Co., pp. 469, 475, 505 (1987).

35.4 H.H. Willard, L.L. Merritt and J.A. Dean, *Instrumental Methods of Analysis*, 5th edn., D. Van Nostrand and Co. (1996).

35.5 A.U. Rahman and M.I. Chaudhary, *Solving Problems with NMR Spectroscopy*, Academic Press (1996).

35.6 G.D. Christen, *Analytical Chemistry*, 5th edn., Harcourt Butler and Co. (1997).

35.7 R.J. Bell, *Introductory Fourier Transform NMR Spectroscopy*, Academic (1972).

35.8 J.A. Pople, W.G. Schnieder and H.J. Beinstein, *High Resolution NMR,* McGraw-Hill and Co. (1950).

35.9 L.M. Jackman, *Applications of NMR Spectroscopy in Organic Chemistry*, Pergamon and Co. (1959).

35.10 D.A. Skoog and D.M. West, *Principles of Instrumental Analysis,* 2nd edn., Saunder College, USA, p. 376 (1980).

Chapter 36

Electron Spin Resonance Spectroscopy

INTRODUCTION

Electron Spin Resonance (ESR) is also called as Electron paramagnetic resonance spectroscopy (EPR). The principle is similar to NMR. In NMR technique we had considered all the changes in energy and magnetic properties of protons, while in ESR we consider such changes with reference to electrons. The revolving electron is subjected to variable magnetic field (9500 MHz) with perpendicular electromagnetic field with fixed frequency of 3400 G. Since the electron also have magnetic moment (m_s) of +1/2 and −1/2 depending upon their movement whether it is clockwise or anticlockwise direction, these electrons on exposure to magnetic field absorb energy in 'microwave' radiation region. Those of the substances which contain *unpaired* electrons, or atoms, molecules having *odd number* of electrons, free radicals, transition metal complexes and triplet state molecule show ESR spectra. Similarly lanthanide complex, exhibit fine ESR spectra. The ESR spectral lines depends upon four parameters, viz., their intensity, width, position and multiple structure. The intensity varies with concentration of free radical. The magnetic moment of electron (μ) is 1836 in comparison with that of proton which is only 2.8. This causes large scale splitting of energy level as $\Delta E = h\gamma = \frac{\mu H}{I}$ if I = magnetic quantum number: H = applied magnetic field.

$\Delta E\ h\gamma = (E_i - E_j) = 2\ \mu\beta_0$ as h = Planck constant, μ = magnetic moment of electron β_0 = constant

$\Delta E = h\gamma = \mu\beta N\frac{H_0}{I}$. Here we have g = splitting factor and β_N = Bohr magneton = 9.27×10^{21} ergs/G.

$$\Delta E = g\beta_N H_o \text{ if } g = 2.0023\ \beta_N.$$

Secondly the position of spectral line is proportional to the magnetic field (H or H_o) so $\Delta E = g\beta H$. Since electrons are spinning around their axis $\gamma = 2\beta_o\mu$ and also circulate around the nucleus of atom, we have two spin orientation parallel and opposite in the direction of external field (+1/2 and −1/2 resp). The third factor, viz., multiple structure of ESR spectra is of two kind namely fine structure and hyperfine structure. If paramagnetic material has two or more unpaired electrons level in same molecule with parallel spins (S = 1, 0) they show fine structure while the hyperfine structure is due to coupling of unpaired electron in the neighbouring nuclear spin. This hyperfine structure gives ESR spectra with line separations of 50MHz (1-20 G).

The effect of magnetic moment of the electron on choice of applied frequency and magnetic field strength is obtained from equation as: $\Delta E = hv = 2\mu H$ (since μ = 1836 mag The moment of electron (H = 5.05 × 10^{-24} erg) we can evaluate magnitude of resonating frequency (v) as by equation: $\gamma = \left(\frac{2\mu H}{h}\right) = 2.8 \times 10^6 H$. This works out to be 9400 MHz if H = 39.5 G Hz. Since magnetic moment of electron is large (in comparison with proton) we can have high sensitivity for ESR given by equation as n_i = upper level, n_j = lower level orbital $\frac{n_i}{n_j} = e^{-\Delta E/KT}$ from Boltzman distribution equation)

$\frac{n_i}{n_j} = e^{-\Delta\mu\beta E/KT}$ $\frac{n_i}{n_j} = e^{-\Delta E/KT}$ (This equation can be used for kinetic studies).

We can also express frequency as

$$Hz = \frac{2g\mu\beta}{I} = \Delta E = g\mu Ho$$

$$v = \frac{2(g\mu/2)Ho}{h} = \frac{g\mu Ho}{h} = 2.9 = g$$

The energy level corresponding to each value of m_s for unpaired electron is equal to $E = 9\ g\mu\beta_o m_s + A_{m1m2}$. In analysis we consider the concentrations of sample and internal standard like DPPH conc. of sample/C_{ref} = H_{sample}/H_{ref} = A Sample/A_{ref} where A = Peak area $A = h\,(\Delta B)^2$.

$$v = \frac{C_{samp}}{C_{ref}} = H_{samp}\left(\Delta B_{samp}\right)^2 / H_{ref}\left(\Delta B_{ref}\right)^2$$

The molecules containing even number of electrons do not show ESR (because the spins are paired) such substances are usually diamagnetic due to orbital precession. In hyperfine structure spin of electrons combines with spin of the neighbouring proton. ESR is used to study chemical, photochemical and electrochemical reactions, proceeding via free radical mechanism. ESR is also used for detection and for the semiquantative analysis of paramagnetic ions at low concentration in biological system, relaxation is essential to present saturation of upper energy level electron during absorption. Such relaxation occurs either by spin lattice relaxation or spin relaxation. Spin lattice relaxation originates from the interaction between an electrons in excited state and the surroundings. While spin-spin relaxation occurs when a spinning electron in upper level transfers its energy to an electron in lower energy level. In spin lattice relaxation lowering of temperature favours the exhibition of narrow lines spin-spin relaxation shortens relaxation time. The spectral lines also narrows when DPPH is used as internal standard where g = 2.0036. It contains 1.53 × 10^{21} unpaired electrons per gram of compound. In quantitative analysis peak height or peak area of sample and reference material is used. Reference has known number of unpaired electron. ELDOR and ENDOR stand for electron double resonance's and electron nuclear double resonance resp. which are used in analysis ENDOR is used to find out value of g factor while ENDOR is used to identify overlapping resonance ESR peaks. Transition metals at trace concentrations can be analysed in metal complexes, e.g., Cr^{3+}, Cu^{2+}, Gd^{3+}, Fe^{3+}, Mn^{2+}, Ti^{3+}. ESR can be also used to measure surface area. Chemical shift is not observed in ESR. Such task of identification can however be done by examining value of g-factor.

This chapter contains numerous problems covering calculation of frequency of radiation, magnetic field strength, evaluation, of line width, frequency of an electron, calculation of resonance frequency, magnetic flux density measurements of total peaks of ions, or compounds, concentration of free radicals, etc. All the formulas and equations given in beginning of this chapter are useful for solving such problems.

SOLVED PROBLEMS

36.1 Calculate the frequency required for the resonance absorption by H in the magnetic field of 14.092 G?

Ans: The magnetic moment of proton is = $2.79 \times 5.05 \times 10^{-24}$ ergs/G
Hence magnetic moment (μ_p) of proton is
$\mu_p = (2.793)(5.03 \times 10^{-24}$ ergs/G$) = 1.41 \times 10^{-23}$ ergs/G

Since $\Delta E = h\nu = \dfrac{\mu H}{I}$ with usual notations, we have

$\mu = 1.41 \times 10^{-23}$ erg/G, $H = 14092$ G, $h = 6.63 \times 10^{-27}$ erg and $I = 1/2 = 0.5$

$$\nu = \frac{\mu H}{hI} = \frac{(1.41\times10^{-23})(14.092)}{(6.63\times10^{-27})(0.5)} = 60\times10^{+6}, \text{ i.e., } 60\,\text{MHz}.$$

36.2 Evaluate the magnetic field (H) essential to have an absorption by proton at 100 Mz?

Ans: Since $H = \dfrac{Ihr}{\mu}$ if $I = 0.5$, $h = 6.63\times10^{-27}$, $r = 100\times10^{6}$

And $\mu = 1.41 \times 10^{-23}$ as given in above example

$$H = \frac{(0.5)(6.63\times10^{-27})(100\times10^{6})}{(1.41\times10^{-23})} = 23.5\times10^{3}\,\text{G}.$$

36.3 Calculate the geromagnetic ratio (γ) for a proton assuming
$\mu_p = 1.41 \times 10^{-23}$, $h = 6.63 \times 10^{-27}$, $I = 1/2 = 0.5$?

Ans: Since geromagnetic ratio $(\gamma) = \dfrac{2\pi\mu_p}{hI}$ we substitute above values so

$$\gamma = \frac{2(3.14)(1.41\times10^{-23})}{(6.63\times10^{-27})(0.5)} = 2.67\times10^{4}\ \text{G/s}.$$

36.4 Calculate the ratio of n_i/n_j for H-nucleus in low and high energy states with the magnetic field of 10,000 G at 300° K?
(Assume $\mu_p = 2.793 \times 5.05 \times 10^{-24}$, $I = 1/2$, $k = 1.3805 \times 10^{-10}$)

Ans: Since $n_i/n_j = e^{-\Delta E/KT}$ if n_i = upper state, n_j = lower state
$\Delta E = h\nu = \mu\nu = \mu H/I$ as $I = 0.5$, $\Delta E = 2\mu H$. So $T = 300$°K

$$\frac{\Delta E}{kT} = \frac{2\mu H}{kT} = \frac{2\times2.793\times5.05\times10^{-24}}{1.3805\times10^{-10}\times300} = 6.80\times10^{-9}$$

$$\left(n_i/n_j = e^{-\Delta E/kT} = e^{-6.80\times10^{-6}} \text{ as } kT = 6.80\times10^{-6}, \text{ so } n_i/n_j = 0.9993\right).$$

36.5 Calculate the line width (frequency unit) of ESR spectra?

Ans: Since, $\Delta E = hv$ or $\Delta E = h\Delta v$ with both equations

$$\frac{\Delta E}{kT} = \frac{1}{2\pi\Delta t} = \Delta p \text{ so } \Delta r = \frac{0.1}{\Delta t}$$

As per Hiesenberg uncertainly principle $\Delta E\ \Delta T = h/2\pi$

If $\Delta t = 10^{-8}$ $\Delta r = 10^{7}$ Hz.

36.6 Predict the possible nature of ESR spectra of compound such as 2-3 dichlorobenzoquinine?

Ans: The only nucleus of non-zero spin were 2-protons which in turn are identical, ESR would by triplet of relative intensity ratios of 1:2:1. This was experimentally confirmed by chemists.

36.7 Ascertain the frequency of an electron?

(Assume $\mu = 1836$, $I = 1/2$) and $H = 50.05 \times 10^{-24}$ and $h = 6.63 \times 10^{-23}$)

Ans: Since $\Delta E = hv = 2\mu H$

$$\mu = 1836,\ H = 5.051 \times 10^{-24}\ h = 6.63 \times 10^{-27}$$

we substitute the values

$$v = \frac{2\mu H}{h} = \frac{2 \times 1836 \times 5.051 \times 10^{-24}}{6.62 \times 10^{-27}} = 2.808 \times 10^{-6} \text{ is frequency.}$$

36.8 What is the gyromagnetic factor in ESR spectroscopy? (assume ΔE is the energy of splitting, μ = magnetic moment, H = magnetic field, $I = 1/2$ and $\Delta E = \mu H/I = g\beta H$. Where $\beta = 9.2732 \times 10^{-24}$)?

Ans: Since $\Delta E = \mu H/I = gpH = \text{or } \dfrac{\Delta E}{H} = g \text{ or } v = \dfrac{z\mu\beta}{h}$

The gyromagnetic ratio for an electron was 1760 compared to the proton of 2.68, if

$$\beta = 9.2732 \times 10^{-24}.$$

So
$$g = \frac{\Delta E}{\beta H} = \frac{\Delta E}{(9.2732 \times 10^{-24}) \times H} = 2.0023$$

$$\Delta E = \frac{2g\mu}{2} B_0 \text{ and } v = \frac{2(g\mu/2)B_0}{h} = \frac{g\mu_B B_0}{h} \qquad \text{if } B_0 = 4$$

$$M \text{ for electrons was } 9.285 \times 10^{-24} \text{ or } \gamma = \frac{2\mu H}{h} = 2.8026 \times 10^{6}\,\text{H}.$$

36.9 What should be the frequency for unpaired electron? (Given splitting factor is 2; β for B_o magnetron has value of 9.27×10^{-21} when the magnetic field employed = 3400 G with Planck constant ($h = 6.63 \times 10^{-27}$)?

Ans: The resonance frequency was calculated from equation $v = \dfrac{g\beta H_0}{h}$ if g = 2.0,

$\beta = 9.27 \times 10^{-21}$, $H_o = 3400$ G and $h = 6.63 \times 10^{-27}$.

$$\text{The frequency was } \gamma = \frac{2.0 \times (9.27 \times 10^{-21}) \times (34.50)}{(6.63 \times 10^{-27})} = 9.51 \times 10^{9} \text{ Hz.}$$

The field was 9500 MHz in usual ESR.

36.10 Find out the resonance frequency of free electron?
(Given nuclear magnetic unit = 5.051×10^{-24} ergs/s).

Ans: We know energy of splitting ΔE is given by equation

$$\Delta E = h\nu = \frac{\mu H}{I} \text{ if spin of electron } m = 1/2 \text{ we have } (I = 0.5)$$

$$\Delta E = h\nu = \frac{\mu H}{0.5} = 2\mu H \text{ or the resonance frequency would be}$$

$r = \dfrac{2\mu H}{h}$. Now we know $H = 5.05 \times 10^{-24}$, $h = 6.6 \times 10^{-27}$ and $\mu = 1836$ on substituting the values in above equation we would have

$$r = \frac{2 \times (1836)(5.05 \times 10^{-24})}{(6.62 \times 10^{-27})} = 2.803 \times 10^{6} \text{ h.}$$

Since, $h = 14{,}092$ we would have resonance frequency as 39.35 GHz.

PROBLEMS FOR PRACTICE

36.11 Find out the ratio of free electrons in the upper region to the lower region at –35°C? (Given magnetic field with the flux density = 0.339, 0.821 and 1.2497?
[**Ans:** 0.998 at 0.339 T; 0.995 at 0.821 T; 0.9930 at 0.2497 T]

36.12 Ascertain the magnetic flux density when free electron $\mu_e = 9.285 \times 10^{-24g}/T$) resonates with ESR instrument operated on eight bands like *X, K* and *Q*? [**Ans:** 0.34, 0.82 and 1.2T]

36.13 Calculate the value of the magnetic factor if there was resonance for an unpaired electron at magnetic resonance flux density of 0.3158 T and resonance frequency of 9500 Ghz?
[**Ans:** 2.150]

36.14 Find out the energetic changes for an electron during free resonance as with ESR instrument operated on eight bands when *X, K* and *Q* bands were used? [**Ans:** 6.3×10^{-24}]

36.15 Calculate the magnetic flux density for resonance for electron at 9500 GHz when an unpaired electron had g-factor of 2.015? [**Ans:** 0.3368T]

36.16 Predict the kind of ESR spectrum in terms of number of lines and relative intensities for a compound para-benzosemiquinone anion?

36.17 Is it possible to denote line width in applied field gauss in ESR technique when line widths was of the order of 1-10 MHz assuming $E = hv$ or $\Delta E/h = 1/2\ \pi\Delta t = \Delta v$? [**Ans:** No]

36.18 Calculate the total numbers of peaks for the ions assuming the unpaired electron was coupled to only one nucleus for elements as: Ti_{47} (III), Gd_{157} (III), Cu_{63} (II) and Mn_{56} (II).
[**Ans:** (1) 6 (2) 4, (3) 4, (4) 6 resp]

36.19 How many peaks and net area in each peak in spectra of $(CH_3)_3$ C is possible?
[**Ans:** 10 peaks area 1, 9, 36, 84, 126, 125, 84, 36, 9 and 1]

36.20 Calculate the number of peaks existing in the spectra of $(CH_3)_3CH$? [**Ans:** 12 peaks]

36.21 Find out strength of free radical (if 0.25 ml solution) had if one unpaired electron with peak width is 0.15 G and height is 24.3. The spectrum of the 0.25 ml of 2.50×10^{-5} M the usual internal standard was 0.12 G with peak height of 31.6? [**Ans:** 3.0×10^{-5} M]

36.22 Calculate the total concentration of the unknown samples when the area for various concentration (1×10^{-5} M) were obtained for both the standard the analytical sample in following order. The peak areas were as follows:
1(19.5), 2(39.1), 3(58.7), 4(78.2), 5(97.8) unknown (42.7) [**Ans:** 2.19×10^{-5} M]

36.23 Predict the nature of the line pattern expected from an aqueous solution of 10^{-6} M of (*i*) manganese chloride (*ii*) vandylacetylacetonate.
[**Ans:** (*i*) Six line pattern with 5-unpaired electron
(*ii*) Since $I = 7/2$ for V^{51}, Eight line pattern.]

36.24 Can you predict the oxidation state of copper and silver which would show very strong ESR pattern? [**Ans:** Cu (II), Ag (I)]

36.25 Can you find out the correct structure, if the ESR spectrum of 8-quinoline complex of Copper is given?
[**Ans:** The structure is $Cu^{11}(C_9H_6NS)_2$]

REFERENCES

36.1 S.M. Khopkar, *Basic Concepts of Analytical Chemistry*, 3rd edn., New Age International Publishers, (P) Ltd. (2007).

36.2 E.D. Olesen, *Modern Optical Methods of Analysis*, McGraw-Hill and Co., pp. 568, 576 (1975).

36.3 G.W. Ewing, *Instrumental Methods of Chemical Analysis*, 2nd edn., McGraw-Hill and Co., p. 257 (1985).

36.4 R.D. Braun, *Introduction to Instrumental Analysis,* McGraw-Hill and Co., pp. 532-33 (1987).

36.5 H.H. Willard, L.L. Merritt, J.A. Dean, *Instrumental Methods of Analysis*, D. Van Nostrand & Co. 2nd edn., pp. 196 (1990).

36.6 L.J. Berliner, *Spin Labeling: Theory and Application*, Academic, N.Y. (1976).

36.7 P.B. Ayscough, *Electron Spin Resonance in Chemistry,* Methuen (1967).

36.8 J.E. Wertz, V.R. Bolton, *Electron Spin Resonance – Elementary Theory and Practical Applications,* McGraw-Hill (1972).

36.9 R.S. Alger, *Electron Para Magnetic Resonance – Technique and Application*, Inter Science (1968).

36.10 G.D. Christian, *Analytical Chemistry,* 5th edn., Harcourt Butler and Co. (1997).

36.11 D.A. Skoog and D.M. West, *Principles of Instrumental Analysis,* 2nd edn., Saunder, USA, pp. 412-25 (1980).

Chapter 37

X-ray Spectroscopy

INTRODUCTION

This technique broadly comprises of X-ray diffraction, X-ray Fluorescence and X-ray Photoelectron spectroscopy. *i.e.*, XRD, XRF and XPS respectively. The X-ray diffraction (XRD) is principally used for elucidation of the structure and quantitative phase analysis or identification of the species. It is very useful for the quality control in industries involving qualitative and quantitative analysis. It is used for the characterisations of polycrystalline films, polymers and metals. The thermal expansion stability and phase diagrams can be also obtained by verifying temperature. We require sample in powder form or single crystal or fiber form for analysis. Glass capillary is used for mounting in quantitative analysis for which little longer time is needed. It is useful for the analysis of crystalline phases many a times it overlaps peak and hinder peak recognition. The strong diffraction over shadows is weak diffraction. The accuracy is ±1% of analysis. The single crystal and powder are two modes of analysis. The anode material used is Cr, Co, Cu or Mo with filters of V, Fe, Ni, and Zr resp.

In X-ray fluorescence (XRF) of solids and liquids can be analysed both for quality and quantitative determination. The technique is non-destructive in the nature. It is used for analysis of metals, minerals, lubricating oil, Sulphur in diesel and fuel oils. The microgram to gram level analysis is possible. No special sample preparation is needed. XRF recording can be completed in just 5-minutes. Elements up to Boron can be analysed *i.e.*, atomic number 11. The sensitivity is in ppm specially if $N < 19$ or less. This is atomic spectrometric method. The emission is confined to X-ray region. The sample preparation needs minimum time. Energy and wave dispersive instrument are available in market. XRF gives spectrum from range of 0-40 Kev. In quantitative analysis we use the expression as

$$GI = mc + b$$

If I = intensity, C = concentration, m = slope, b = background.

X-ray Photoelectron spectroscopy is mainly surface analysis technique and is useful for depth profiling and in site analysis. The atomic concentration and chemical state can be easily identified. The diffraction of conductors and insulators can be studied. The phenomena of absorption and chemisorption can be also studied. Solid surfaces are analysed. Time of analysis is 1–5 minutes for conducting samples. Sampling depth used is monolayers with resolution of 0.2mm. The sensitivity is about ± 1.3%. This is called (XPS) method. One can encounter chemical shift like NMR spectroscopy. For quantitative analysis we use an expression as:

$$I = nf\sigma\phi\eta\, AT\,\lambda$$

if n = no. of atoms, f = flux, σ = photoelectric cross section, ϕ = angular efficiency, η = efficiency of production, λ = mean free path. We would consider certain expressions employed in the numerical analysis problems. Just as Beer's Law was valid for absorption spectroscopy. The very same law is also used in X-ray spectroscopy for the quantitative analysis. Thus in X-ray in absorption work we have:

$$P = P_0\, e^{\mu l}$$

If P = intensity of transmitted beam, P_0 = intensity of incident beam, μ = linear absorption coefficient in cm^{-1} and l = thickness of absorbing media in cms. This can be also expressed in form of common logarithms. We can also express it also as

$$\mu m = \mu/P \text{ (cm}^2\text{ per gm)}$$

where μm = mass absorption coefficient, ρ = density of absorber, μ = linear absorption coefficient. In diffraction work we use Bragg's equation, which states that

$$n\lambda = 2d{\cdot}\sin\theta$$

where in n = order of diffraction, (e.g., n = 1,2,3...) λ = wavelength of X-rays, $\hat{\theta}$ = incident angle, d = distance between atomic planes in the crystal. As stated earlier in X-ray absorption we use an equation as

$$\log(P_0/P) = \mu m \cdot r0/2.303 \text{ a}$$

where P_0, P are intensity of the incident and emited radiation, μm = mass absorption coefficient. a = area of the sample in cm^2.

Further the mass absorption coefficient is also expressed as

$$\frac{\mu}{\rho} = \frac{Cz^4\lambda^{5/2}N}{A}$$

where N = Avogadro's number,

A = atomic weight of element, z = atomic number, x = constant over narrow region of wavelengths. The μm of sample is also related to each element composing such sample so we have

$$\mu m = \mu m_1 + \mu m_2 + \mu m_3 \ldots \mu m_n$$

or in other words we write

$$\left(\frac{\mu}{\rho}\right)_n = \left(\frac{\mu}{\rho}\right)_1 w_1 + \left(\frac{\mu}{\rho}\right)_2 w_2 \ldots\ldots \left(\frac{\mu}{\rho}\right)_n w_n$$

if w = repression wt. of sample component = $w_1 m$ no. 1, 2, 3..... n and ρ = density of individual sample. The energy of incident electron is 'Ye' where 'e' is charge of electron while energy of emitted photon in short wavelength cut off is $hc/\lambda c$ we get by equation two equations as follows:

$$Ve = hc/\lambda_0$$

or

$$\lambda_0 = hc/V_e$$

The Duane-Hunt equation is

$$\lambda_0 = 1239.8/V$$

If λ = wavelength in nm and V_0 is volts. The absolute intensity (P) of emitted X-rays, at any Voltage is then

$$P = a \cdot Z(E_D - E) + bz^2$$

If *ab* are constant, E_D = Energy in volts, E = energy corresponding to particular emitted X-ray. The area '*A*' under intensity wavelength curve is expressed as:

$$A = kizv^2$$

if k = *proportionately constant*.

Finally in XRF in fluorescence analysis is the (P) intensity of the fluorescence radiation it is expressed as:

$$P = kW(\mu_i + \mu_c)$$

If k = proportionality constant, W = weight of element μ_i, μ_f are linear absorption coefficient of entire sample at wavelength of incident and fluorescent radiation resp.

We summarise important expression for quantitative analysis as

(*a*) XRD we use

$$P = \frac{K_i X_a}{\rho_a (\mu / \rho)_n}$$

if K_{ia} = Constant, ρ = density of phase, $(\mu/\rho)_n$ mass absorption coefficient; of the sample while in fluorescence measurement.

(*b*) In XRF we use,

$$P = mc + b$$

wherein I = Intensity of emitted X-rays, m = slope of the line, b = background and C = concentration.

(*c*) Finally in XPS in X-ray photoelectron spectroscopy we use an equation.

$$P = nf\sigma\phi\eta AT\lambda$$

where n = nos of atoms, f = flux of photons; σ= cross section of transition, φ = angular efficiency, η = efficiency of production, A = area of the sample, T= efficiency of detection, λ = mean free path.

By making use of several equations cited in this introductory portion, it will be possible to tackle problem in quantitative analysis involving X-ray diffraction (XRD), X-ray fluorescence (XRF) or X-ray photoelectron spectroscopy(XPS).

SOLVED PROBLEMS

37.1 What is the magnitude of the molecular absorption coefficient of an organic acid if the mass absorption coefficient for atoms of carbon, hydrogen and oxygen are 4.52, 0.48 and 11.1 resp at this wavelength?

Ans: From molecular weight we calculate percentage composition of each element. The percentage is 68.9% for carbon, 4.9% for hydrogen, 26.2% for oxygen. Since the (μm) mass absorption coefficient are given for each compounds in one gram sample. We have

$$\left(\frac{\mu m}{\rho}\right)_r = \underbrace{\frac{(0.689 \times 4.52)}{1}}_{\text{(For carbon)}} + \underbrace{(0.262 \times 11.1)}_{\text{(For hydrogen)}} + \underbrace{(0.262 \times 0.48)}_{\text{(For oxygen)}} = 6.03 \text{ cm}^2/\text{gm}.$$

37.2 What is short wavelength limit of X-ray tube at 50kv?

Ans: Since voltage is 50kv=50000 V. Let us assume that the shorter wavelength is in region of 12400 so $\lambda_a = 12400/50000 = 0.248$. This energy is adequate to excite electrons for element of atomic number 63 *i.e.*, Europium (63).

37.3 The amount of lead in aviation fuel was analysed by X-ray method. The readings for % of lead (absorption) are as follows with absorption in brackets 0.1% (27.5), 0.2% (32.8), 0.3% (38.6), 0.5% (49.4). If absorption reading of unknown sample is 41, What is the corresponding amount of lead in the sample?

Ans: Here, we plot a calibration graph of % lead versus the absorption value, we get a straight line graph but it is not passing through the origin. For the absorption of 41 the corresponding lead element is 0.34%.

37.4 A sample of steel contains nickel. It was analysed by the technique of XRF when the values for fluorescent intensity with % of nickel in bracket was as follows:
1.0(5.6); 2(10.6); 3(14.7); 4.0(18.2); 6(24.4); 7(26.6) and 8(28.3). If the unknown sample of steel shows fluorescence intensity of 20.4, what is % of nickel in the unknown sample?

Ans: We plot a graph of % of nickel against fluorescent intensity. We get a typical calibration curve. Now on plotting the values of 20.4 intensity on the graph, we find corresponding value of nickel as 4.7%.

Therefore, the amount of nickel in unknown sample is 4.7%

37.5 What is the mass absorption coefficient of an alloy cupronickel containing 60% copper and 40% nickel at wavelength of radiation for copper if mass absorption coefficient of copper and nickel resp. are 52.7 cm²/g and 49.3 cm²/gm?

Ans: We had seen that mass absorption coefficient viz. $\mu_m = \mu_1 w_1 + \mu_2 w_2 + \mu_3 w_3 \cdots \mu_n w_n$ so $\mu_m = \underset{\text{nickel}}{(0.40 \times 49.3)} + \underset{\text{copper}}{(0.60 \times 52.7)} = 51$ cm²/gm. So value of mass absorption coefficient of cupronickel alloy is 51 cm²/gm as $m_1 = 49.3$ nickel, $m_2 = 60$ copper, $\mu_1 = 49.3$, $\mu_2 = 52.7$.

37.6 With gonimeter the radius of 57.296 mm at incident radiation is for copper line and if the distance between two opposite plate is 58.88 mm, what is the distance between spacing (*d*) which was responsible for showing the X-ray diffraction?

Ans: If radius $r = 57.296$, and if angle is $4\hat{\theta}$ with $d = 58.88$ m. Then we find out angle $\theta = 58.88/4 = 17.72°$ as $(4\theta_{rad} = s/r)$. By using the Bragg's equation viz. $n\lambda = 2d \sin\theta$ if $n = 1$, $\lambda = 0.15$, $\hat{\theta} = 14.72$ on substituting these values in the equation we get value of *d* between spacing is (1)(0.154 nm) = 2*d* sin (14.72°). So, *d* = 0.303 nm. It is distance between spacing.

PROBLEMS FOR PRACTICE

37.7 A sample of an alloy has mass absorption coefficient of 4.75 cm²/g for line of copper at (1.5°A). What will be mass absorption coefficient for zinc line at (1.43°A)?

(**Ans.** 3.96)

37.8 An amount of cobalt in particular sample in X-ray spectroscopy gave following readings for the calibration curve as: Concn. (reading) 0.5(17), 1.0(23), 2.0 (34), 2.5(42). If unknown sample has reading of 22, what is the percentage of cobalt in the sample?

(**Ans.** 0.93%)

37.9 Evaluate the mass absorption coefficient at 0.436 nm of sample 85% iron, 3% nickel, 9% copper and 1% zinc. If the corresponding μm for pure metals is 0.436 nm are resp. 610 (Fe), 715 (Ni), 760 (Cu), 910 (Zn) cm²/km.

(**Ans.** 631.8 cm²/gm)

37.10 A mixture contains chlorides of copper and calcium. At 0.193 nm, what will be mass absorption coefficient if % of calcium is 60% and copper is 40%. The mass absorption coefficients for calcium is = 306, chloride = 198 and copper = 92.6 cm^2/gm at that wavelength?

[**Ans:** 202.2 cm^2/gm]

37.11 What should be the setting for the X-ray gonimeter for calcium in LiF using usual equation for angle $\hat{\theta}$ and for K_α? (**Ans.** 112.6)

REFERENCES

37.1 H.H. Willard, L.L. Meeritt, J.A. Dean, F.A. Settle, *Instrumental Method of Analysis,* 6th Edition, CBS Publishers India, p. 239 (1986).

37.2 L.F. Hamilton, S.G. Simpson, D.W. Ellis, *Calculations of Analytical Chemistry,* 7th Edition, McGraw-Hill & Co., p. 339 (1969).

37.3 R.D. Braun, *Introduction to Instrumental Analysis*, McGraw-Hill & Co., p. 535 (1987).

37.4 J.H. Harely, S.E. Wiberley, *Instrumental Analysis*, John Wiley & Co., p. 302 (1954).

37.5 D.A. Skoog, D.M. West, F.J. Holler, *Fundamentals of Analytical Chemistry*. Harcourt Brace & Co. (1992).

37.6 F. Settle, *Handbook of Instrumental Techniques for Analytical Chemistry,* Pearson Educational Publication (1997).

37.7 S.M. Khopkar, *Basic Concept in Analytical Chemistry*, 3rd Edition, New Age International Publishers (P) Ltd. (2007).

Chapter 38

Mössbauer Spectroscopy

INTRODUCTION

It is the method in which resonance absorption of γ-rays are measured. On absorption of γ-radiation, the nucleus in consideration is excited from ground state to higher energy state. Such energetic transitions of electrons match with nuclear rather than with electronic transitions within an element.

The source of γ-rays which are used for Mössbauer spectroscopy, are radionuclides which decay emission of particular nucleus by the process of electron capture. Such excited nucleus emits γ-rays when it relaxes to ground state during return journey of electrons. The γ-rays are characterised by kind of radionuclides used in the disintegration.

The γ-rays are absorbed by sample containing nucleus. A radionuclei in turn during relaxation process emits energy (ΔE) which is equal to the difference between the two energy levels viz. the ground and excited levels viz. $(E_c - E_g) = \Delta E = h\nu + E_r$, wherein E_c is excited state, E_g = ground state and E_r = recoil energy of the radionuclide.

As the part is emitted as γ-rays and other part is recoil energy (E_r) of the radio element. During the process of emission, momentum must be controlled. Such momentum is equivalent to recoiling nucleus. The recoil energy must be very small so as to facilitate resonance absorption in the sample. The emitted radiation is always large in comparison to recoiling (E_r) energy. The source and sample used are in solid state. The tightly bound nucleus like Fe^{56} or Co^{59} behaves as small part of heavy body. There should be any kind of vibrational excitations of lattice. As mass of solid is high the rate imparted to the body during recoil is minimum. This process is called as recoil emission (E_r). In order to have recoil less emission to below γ-ray must represents the difference of energy viz $(E_c - E_g) = \Delta E$ as per above equation. The emitted photon is absorbable by identical nucleus in the sample.

The detectors used for X-ray and Mössbauer are identical e.g., they are scintillation, proportional or semiconductors detectors used in radiochemistry. The absorption of energy depends upon the chemical environment. The situation is analogous to NMR. The small shifts or splitting process are measured and are correlated to chemical structure of the nucleus.

As the energetic shifts are small, instrument must have significant resolution. The Doppler shifts result in when energy is changed for incident radiation (E_c). Doppler shift involves movement of sample and source relative to one another and so the velocity of movement determines the extent of Doppler Shift and hence the frequency of the radiation impinging on sample. A positive velocity represents the movement of source towards sample. The plot of γ-ray intensity versus Doppler velocity provides a Mössbauer spectrum which can be used for qualitative and quantitative analysis of sample.

A drive coil in loudspeaker controls the motion of source in relation to the detector. Velocity is changed periodically when spectra is drawn. The drawing of spectrum takes hours or even few days. Liquid helium or nitrogen is used to reduce recoil energy by process of cooling. The splitting of spectra can be accompanied by magnetic field. Such splitting leads to nucleus hyperfine structure. These interaction may be due electrical quadrupole or magnetic dipole interactions or electric monopole interaction. The splitting occurs due to hyperfine interactions. A change in electronegative group; or change in electron density or variation in kind of ligand used for the complexations leads to shift in position of peak.

The important applications of the Mössbauer spectroscopy is to identify oxidation state of Mössbauer atom. Qualitative analysis is possible by investigating in details the chemical shifts. A specific substance can be recognised by studying Mössbauer spectrum. One can use calcium, barium or potassium to get Mössbauer spectra. Nickel, germanium or antimony can be also used as the target materials. The analysis of deep layer of surface is done by this spectroscopy. A gravitation of a red shift can be similarly studied. Synchroton can be used as source of radiation to provide γ-rays for impingement.

PROBLEMS

Multiple Choice Questions

Tick mark the correct answer from the following multiple choices.

38.1 Mössbauer spectroscopy was discovered in 1957 by R.L. Mössbauer who was in university as a

(*a*) Graduate student (*b*) Postgraduate student
(*c*) Research scholar (*d*) Professor of Physics

38.2 Mössbauer spectroscopy deals with Recoil free emission or absorption phenomena involving the emission of

(*a*) α-rays (*b*) β-rays
(*c*) γ-ray (*d*) electrons

38.3 The most commonly used nuclide material for generation the γ-rays is Mössbauer spectroscopy is

(*a*) U^{238} (*b*) C^{12}
(*c*) Fe^{57} (*d*) O^{16}

38.4 The following recoil free fraction determines the magnitude of Mössbauer spectra. It is

(*a*) Size (*b*) Shape
(*c*) Charge (*d*) Total fraction

38.5 The chemical isomeric shift can be obtained analogous to NMR spectroscopy by the following spectra

(*a*) ESR (*b*) Mössbauer
(*c*) UV-visible (*d*) Infrared

38.6 The interaction of the magnetic field with Mössbauer spectra leads to the formation of

(*a*) Symmetrical spectra (*b*) Unsymmetrical spectra
(*c*) Spectra with narrow peaks (*d*) Hyperfine structured spectra showing splitting

38.7 Excited state nucleus for ground state by emission of the γ-rays, in turn lead to conversion of electron Mössbauer spectra if

(*a*) emission of intershell atomic electron occurs

(*b*) K-electron is captured
(*c*) emission of valency electron takes place
(*d*) spalliation of electrons is noted.

38.8 Within s-block elements those exhibiting Mössbauer spectra are one of the following sets of elements
(*a*) Li, Na, Rb, Cs (*b*) Ca, Sr, Ra
(*c*) Be, Mg, Ca, Sr (*d*) K, Ca and Ba

38.9 Within transition elements those nuclei showing Mössbauer spectra are
(*a*) Fe, Co, Cr (*b*) Cu, Ag, Au
(*c*) Fe, Zn, Ni, Hg (*d*) None of these

38.10 In main group elements those elements which can show the Mössbauer spectra are
(*a*) Ge, Sn, Sb, Te (*b*) Al, Ga, Zn, Ti
(*c*) O, S, Po (*d*) N, P, As, Bi

38.11 With transition series of element Mössbauer spectra is exhibited by
(*a*) Transuranic elements
(*b*) Lanthanides (except La and Ce)
(*c*) Actinides (excluding Th, Pa, Np, etc.)
(*d*) None of them show sub spectra.

38.12 Mössbauer spectra shows one of the following phenomena when there is nuclear resonant absorption in the sample it is the
(*a*) dips (*b*) peaks
(*c*) trough (*d*) hyperfine

39.13 On account of typical nature of iron, Mössbauer spectroscopy is used in the analysis of
(*a*) pigments and clays in work of art (*b*) polymers and plastics
(*c*) carbohydrates and amino acids (*d*) oils, fat and waxes.

38.14 Apart from Fe^{57} one can also use following isotope for the generation of Mössbauer effect such as
(*a*) Nickel (*b*) Cobalt
(*c*) Zinc (*d*) Cadmium

38.15 One can get hyperfine spectra by one of the following interaction of the nuclei such as
(*a*) Isomeric effect interaction
(*b*) Quadrupole interaction
(*c*) Magnetic interaction
(*d*) Combined isomeric quadrupole interaction.

38.16 The analysis of the deep layer of the surfaces can be easily done by one of the following spectral methods viz.
(*a*) Mössbauer spectroscopy (*b*) NMR spectroscopy
(*c*) ESCA spectroscopy (*d*) Aüger spectroscopy

38.17 A study of ferric ion as function of temperature shows
(*a*) slowing of relaxation rate with rise of temperature
(*b*) slowing of relaxation rate with decrease in the temperature.
(*c*) no change is noted in temperature on relaxation.
(*d*) its enhances relaxation with increase in temperature.

38.18 Most exotic application of Mössbauer spectroscopy is the study of gravitional
(*a*) blue shift (*b*) red shift
(*c*) field change (*d*) prediction shift

38.19 In future serious efforts will be made to observe the Mössbauer effect of the following nuclei
(*a*) Ag^{107} Ag^{108} (*b*) Au
(*c*) Cd (*d*) Pb

38.20 The use of synchrotron radiation is Mössbauer effect will provide
(*a*) α-optics (*b*) β-rays
(*c*) γ-rays optics (*d*) γ-rays lasers

ANSWERS

38.1 a	38.2 c	38.3 c	38.4 a	38.5 b	38.6 d	38.7 a
38.8 d	38.9 c	38.10 a	38.11 b	38.12 b	38.13 a	38.14 b
38.15 c	38.16 a	38.17 a	38.18 b	38.19 a	38.20 c.	

REFERENCES

38.1 R.A. Meyer, Ed. *Encyclopedia of Physical Sciences and Technology* Vol. 10, G.R. Leoy Academic Press p. 465 (1992).

38.2 B.V. Thosar, P.K. Iyengar, S.K. Srivastava, S.C. Bhargawa, *Advances in Mössbauer Spectroscopy*, Elsevier (1983).

38.3 J.G. Stevens, G.K. Shenoy, *Mössbauer Spectroscopy and Its Chemical Applications*, American Chem. Soc. (1981).

38.4 G.J. Long Ed. *Mössbauer Spectroscopy as Applied to Inorganic Chemistry* Vol. 1-3 Plennum Press (1989).

38.5 D.P.E. Dickson, F. J. Berry, *Mössbauer Spectroscopy*, Cambridge Univ. Press (1986).

38.6 R.D. Braun, *Introduction to Instrumental Analysis,* McGraw Hill & Co. p. 641 (1987).

38.7 S.M. Khopkar, *Basic Concepts of Analytical Chemistry*, 3rd edn., New Age International Publishers (P) Ltd. (2007).

Chapter 39

Electron Spectroscopy

INTRODUCTION

Analytical chemistry was confined to the qualitative and quantitative analysis. In both the cases, it was necessary to have the sample in solution state. It was possible to characterise the analyt in solution by organic spotting agents. Further the determination of the constituent was done by wet analysis involving gravimetry volumetry or instrumental methods of analysis.

The main limitations of these analytical techniques were that none of them permitted non-destructive analysis. The methods did not permit analysis of surfaces. The reflection spectroscopy was the only one exception where analysis was carried out for the intensity of reflectance light from coloured solution. Further the composition of the surface and the internal bulk solid material was not always identical. This led to the brisk activity for the discovery of newer methods for surface analysis.

The surface analysis was carried out using probe to irradiate the surface of the sample. Such irradiation was carried out by different probes like electron probes, X-rays, UV-radiation, photon, ions, monoenergetic electrons, bombardment with either electrons or with ions with low energy or collusion of the gas phase excited atom. However, technique involving electron probes was mostly preferred as it permitted penetration in the outer surface of the object. The method using electrons from electron gun was called as Aüger spectroscopy while one involving electron probes from X-rays was called as ESCA, *i.e.*, electron spectroscopy for chemical analysis. The technique involving ions or photons as projectile material was less frequently used. Such technique was called SIMS, *i.e.*, secondary ion mass spectroscopy. Now in the Aüger or ESCA technique electron bombardment resulted in ejection of electrons with change in their orbital energy as well as in their kinetic energy.

In both cases, electron from *K* shell was promoted to *L*-shell and vacancy was created in that *K*-shell which was filled by another electron from *L*-shell. This was called KLL transition. In MV transition *V* is valancy shell was also possible to vacate and fill in. Now this transition of electrons lead to change in kinetic and orbital energy of an element. Since this orbital energy was specific with a particular element, this facilitated identification of element. Further if the intensity of energy was proportional to the concentration it provided a means for quantitative analysis of an element.

Aüger spectroscopy was commonly preferred to ESCA. Both mode indicate chemical shift in diatomic molecules like N_2, Cl_2, etc. This mode provided atom for further identification. However elements like H_2 or He were not analysed by electron spectroscopy methods.

In this chapter large number of multiple choice questions are provided to test the knowledge about kind of probe or projectile material, used various transformation of the electrons in *K*, *L* and *M* shell, the relaxation of electrons by returning from higher to lower energy orbital, changes in orbital and kinetic energy and its interrelation to the concentration of an element in the sample, consideration of vacant orbitals, absorption of an energy, electron ejection, formation of charged ion, penetration depth of electron, etc., are considered.

PROBLEMS

Multiple Choice Questions

Tick mark correct answer from following multiple choices:

39.1 The composition of surface of solid and inside of bulk solid is
(*a*) identical in nature (*b*) different in nature
(*c*) independent of surface composition (*d*) related to used source of electron probe.

39.2 In Aüger spectroscopy the probe beam consists of
(*a*) electrons (*b*) protons
(*c*) neutron (*d*) positrons.

39.3 Electron beam is preferred as probe in surface analysis as it
(*a*) has small mass (*b*) it penetrates to outermost layer
(*c*) is easy to handle (*d*) is having negative charge.

39.4 When photons or ions are used as probe beam such technique is called as
(*a*) electron spectroscopy for chemical analysis (ESCA)
(*b*) Aüger spectroscopy (AES)
(*c*) secondary ion mass spectroscopy (SIMS)
(*d*) electron impact spectroscopy (EIS).

39.5 It is not possible to exchange energy with electromagnetic radiation in
(*a*) Aüger spectroscopy (*b*) ESCA
(*c*) Penning ionisation spectroscopy (*d*) SIMS.

39.6 Electron beam is preferred as probe as it is
(*a*) giving better spatial resolution with good sensitivity
(*b*) very cheap
(*c*) having small charge
(*d*) easy to handle.

39.7 Excitation of surface by irradiation is best done with
(*a*) positron (*b*) electron
(*c*) ions (*d*) photons.

39.8 In X-ray photoelectron spectroscopy irradiation is done with
(*a*) ions (*b*) photons
(*c*) non chromatic X-rays (*d*) gamma radiation.

39.9 Electron spectroscopy fails to identify and analyse gases like
(*a*) argon (*b*) neon
(*c*) oxygen (*d*) helium.

39.10 X-rays in ESCA technique produces on bombardment
(*a*) vacancy in inner shell (*b*) occupies vacant orbital
(*c*) produces charge ion (*d*) reduces kinetic energy.

39.11 On excitation electron in lower shell in Aüger or ESCA technique:
(*a*) jump to higher shell (*b*) fall into vacant inner shell
(*c*) remain stationary (*d*) exhibits emissions.

39.12 Aüger effect involving electron transitions lead to
(*a*) ejecting electrons and resulting triply charged ions
(*b*) transition of electrons from inner to outer shell
(*c*) ejection of electron from sample producing doubly charged ion
(*d*) increase in energy in outer shell.

39.13 In ESCA energy of incident photons leads to
(*a*) loss of energy (*b*) gain of kinetic energy
(*c*) decrease of orbital energy (*d*) absorption of energy.

39.14 Since kinetic and orbital energy are interrelated, qualitative identification is possible as:
(*a*) such energy is peculiar of atom or molecules
(*b*) have no relation to kind of atoms
(*c*) has no relation to type of molecular
(*d*) are different for similar molecules.

39.15 The KLL Aüger process involves
(*a*) formation charged ion with no electronic transitions
(*b*) ejection of *s*-electron and relaxation of 2*s*-electron in vacancy of 1*s* shell
(*c*) ejection of *s*-electron and relaxation in M shell
(*d*) ejection followed by filling electron in *K* shell from same shell.

39.16 The MMV transition involves
(*a*) emission from valancy shell
(*b*) no vacancy created in *M*-shell
(*c*) electron ejection from *M*-shell, falls to vacancy and eject *M*-electron to *V*-shell
(*d*) no shell transitions at all.

39.17 The kinetic energy of ejected Aüger electron
(*a*) does not change net energy
(*b*) leads to loss of energy
(*c*) emission of energy matches with each another
(*d*) matches with difference between energy loss due it inner shell transfer.

39.18 The Aüger electron energy is represented by equation
(*a*) $E_A = (E_a - E_b) - E_C$ (*b*) $E_A = E_a + E_b + E_C$
(*c*) $E_A = (E_a - E_b) + E_C$ (*d*) $E_A = E_a + E_b - E_C$

39.19 The emission of second Aüger electron leads to
(*a*) formation of doubly charged ion (*b*) formation of triply charged ion
(*c*) formation of neutral molecule (*d*) formation of uncharged species.

39.20 In ESCA electron source for probing surface is
(*a*) electron gun (*b*) gas discharge tube
(*c*) X-ray tube (*d*) microprobe.

39.21 It is necessary to use vacuum in electron spectroscopy for chemical analysis as
(*a*) it weakens electron beam (*b*) prevents scattering of beam
(*c*) sample is not damaged (*d*) avoids attenuation of electron beam.

39.22 Dispersion model over retardation field is preferred as
(*a*) radius of deflected field is proportional to intensity of field
(*b*) circular path predicts field strength
(*c*) beam cannot be deflected in circular path
(*d*) focussing of electron on detectors is complicated.

39.23 ESCA is used more for characterisation than for determination as it
(*a*) finds out oxidation state of element
(*b*) is concerned with core electron binding energy
(*c*) permits depth profiling of surface
(*d*) allows analysis of helium or hydrogen.

39.24 Aüger spectroscopy fails in
(*a*) analysis of hydrogen
(*b*) ascertaing oxidation state
(*c*) elemental analysis with low atomic weight
(*d*) identify valancy of element.

39.25 Chemical shift is not possible in
(*a*) NMR (*b*) ESR
(*c*) ESCA (*d*) Aüger.

39.26 Electron generation in Aüger spectroscopy involves
(*a*) use vaccum UV radiation (*b*) collision of gas phase
(*c*) bombardment with electrons (*d*) low energy ions being used.

39.27 Mode of generation of electron in ESCA is
(*a*) electron gun (*b*) UV radiation
(*c*) IR radiation (*d*) monochromatic X-ray.

39.28 X-ray photoelectron spectroscopy is another name for
(*a*) ESCA (*b*) AES
(*c*) electron impact spectroscopy (*d*) penning ionisation spectroscopy.

39.29 The limitation of penetration of electrons in ESCA is restricted to
(*a*) 1 mm (*b*) 5 mm
(*c*) 10 mm (*d*) 100 mm.

39.30 ESCA provides information on relative number of each kind of atom, by chemical shift of
(*a*) S from 0 to +6 state (*b*) S from 0 to +4 state
(*c*) N from +1 to +5 state (*d*) N from +1 to 2 state.

39.31 The technique using ionising radiation in UV region for surface analysis is called
(*a*) ESCA (*b*) Aüger
(*c*) X-ray photoelectron (*d*) UV photoelectron spectroscopy.

39.32 In Aüger spectroscopy ionisation is best carried out by
(*a*) electron probe (*b*) positron probe
(*c*) neutron probe (*d*) UV radiation.

39.33 The technique dealing with inner shell in surface analysis is called as:
(*a*) UPS (*b*) PESS
(*c*) XPS (*d*) PESOS.

39.34 A satellite peak in ESCA is due to
(*a*) vacancy created in inner shell
(*b*) change in nuclear charge
(*c*) ejection of inner shell electron to outer shell which is exposed without shielding
(*d*) *f*-electron shake up leading to electron ejection in valancy-shell.

39.35 The probe used in UPS for irradiation of surface analysis is
(*a*) electron (*b*) ions
(*c*) photons (*d*) UV radiation.

39.36 The technique involving monoenergetic electron is:
(*a*) electron impact specroscopy (*b*) ESCA
(*c*) AES (*d*) SIMS.

39.37 In penning ionisation spectroscopy probe for electron generation is
(*a*) X-rays (*b*) collision of gas phase of excited atom
(*c*) UV radiation (*d*) electron gun.

39.38 Ion neutralisation spectroscopy has electron source as
(*a*) monoenergetic (*b*) photons
(*c*) low energy ions (*d*) UV-radiation.

39.39 The vacuum UV radiation is used as source of electron in
(*a*) SIMS (*b*) ESCA
(*c*) AES (*d*) ultraviolet photoelectron spectroscopy.

39.40 The chemical shift in nitrogen by ESCA analysis covers:
(*a*) 0 to 5 state (*b*) 0 to +3 state
(*c*) +3 to +5 state (*d*) +2 to +4 state.

ANSWERS

39.1 b	39.2 a	39.3 b	39.4 c	39.5 d
39.6 a	39.7 b	39.8 c	39.9 d	39.10 a
39.11 b	39.12 c	39.13 d	39.14 a	39.15 b
39.16 c	39.17 d	39.18 a	39.19 b	39.20 c
39.21 d	39.22 a	39.23 c	39.24 a	39.25 b
39.26 c	39.27 d	39.28 a	39.29 b	39.30 c
39.31 d	39.32 a	39.33 b	39.34 c	39.35 d
39.36 a	39.37 b	39.38 c	39.39 d	39.40 a

REFERENCES

39.1 S.M. Khopkar, *Basic Concepts of Analytical Chemistry,* 3rd edn., New Age International Publishers (P) Ltd. (2007).

39.2 R.D. Braun, *Introduction to Instrumental Analysis,* McGraw-Hill and Co. (1987).

39.3 G.W. Ewing, *Instrumental Methods of Anlaysis,* 5th edn., McGraw-Hill Book Co. (1985).

39.4 T.A. Carlson, *Photoelectron and Aüger spectroscopy,* Plenum NY (1975).

39.5 C.R. Brundle and A.D. Baker, *Electron Spectroscopy: Theory Technique and Application,* Academic Press (1977).

39.6 J.W. Rabalasis, *Principles of Ultraviolet Photoelectron Spectroscopy,* Wiley-Interscience (1977).

39.7 T.L. Barr, *Applications of ESCA in Industrial Research,* American Lab, p. 65 (1978).

39.8 D.A. Skoog and J.J. Lerroy, *Principle of Instrumental Analysis Harcourt,* Brace Publication, London (1992).

39.9 D. Briggs M.P. Sech, *Aüger and X-ray Photoelectron Spectroscopy,* John Wiley (1990).

39.10 R.H. Clark R.E. Hesler, *Spectroscopy of Surfaces,* John Wiley (1988).

Chapter 40

Electron Microscopy

INTRODUCTION

This technique is also designated as scanning probe microscopy (SPM). It is presumed that it covers several techniques of scanning such as scanning tunelling microscopy (STM), Atomic force microscopy (AFM), ultrahigh vacuum scanning tunelling microscopy (UHV-STM). Of these, the atomic force microscopy is most surface sensitive method with excellent resolution. These methods are used for surface to photographic profiling study of atomic arrangement of surface atoms; while STM which is preferably carried out in ultra high vacuum, provides real space imaging an electrically conducting surfaces mostly at fermi level. AFM is also used to assess roughness on semiconductor surfaces or to study ordering of polymer films. STM on the contrary does mapping of surface atomic structure, crystalline ordering on nanometer sized samples. The electrically conducting surfaces of metals and semiconductors are best probed by STM, while insulating and conducting surfaces are best probed by AFM. Thus the process of annealing, sputtering, cleaving is best examined by UHV-STM, but ATM can examine object without surface preparation including absorbed water layer. Time required is small, while STM needs more time for scanning and interpretation.

As regards the limitations of these two techniques, STM needs careful treatment of tunneling tip, because the spatial resolution pattern very much depends upon the nature of the tip. In UHV-STM the extent of pressure determines observation of specific sample surface. On an average the pressure employed is < 1E - 10 mbar. Elemental analysis is a problem as STM can get information from first or second top layer of the surface only.

With the advances in nanoscience and nanotechnology, scanning probe microscopy has become very useful for analysis. A quantitative work is possible with UHV-STM, while AFM is done under the existing laboratory conditions not requiring electrically conducting surfaces. One can get nanometer resolution by STM at atomic levels. In comparison with Aüger electron spectroscopy the depth access is limited in STM or AFM. It is restricted to first two layers. The ambient air condition is most efficient for AFM technique. It can be operated in two modes, viz. contacting and non-contacting conditions with surface with low load, high resolution surface profile is possible by former mode as STM while in latter in AFM the attractive force near contact plays very important role in analysis. STM facilitates atomic scale observation of metallic and semiconductor surfaces e.g., noble metals, silicon, etc.

A high resolution transmission electron microscopy (HRTEM) is more sensitive than normal STM which restricts the observation only to first or second layer of surfaces, HRTEM on the other hand can penetrate to many layer of surfaces viz. 50-100 layers of the sample thickness of 20-40 nm – AFM is ideal for flatness analysis, imaging of small magnetic or replica structures and polymers in pure surface science. While STM compliments electron diffraction technique.

As regards cost of these instruments UHV-STM is much more expensive than AFM as former cost more on account of UHV encloser and cost of vacuum pumps. Further the basic probe head is also expensive in STM A piezoelectric transducer is used for maintaining the tips A sample separation is effected via feed back from current sensing circuit. The other complimentary techniques used in combination with SPM are secondary ion mass spectrometry (SIMS). Aüger electron spectroscopy (AES), X-ray photo-electron spectroscopy (XPS); photoelectron spectroscopy (PES) and low energy electron diffraction (LEED) technique. AFM & STM are thus imaging techniques providing topography on atomic scale. The spatial resolution is 1 Å to 5 Å in AFM and only till 1 Å in STM method. In STM atomic pressure of 10^8 pascal provides excellent results. While AFM uses usual atmospheric pressure. STM is used for the analysis of DNA, proton enzyme interaction, membrane tissues, while AFM however permits underwater imaging of biological samples. Scanning electron microscopy furnishes morphologic topography in the formation of surfaces. STM and AFM are called SPM *i.e.*, scanning probe microscopy and useful for surface characterisation.

PROBLEMS

Multiple Choice Questions

Tick mark the correct answer from the following multiple choices provided:

40.1 The optical microscopy had its own limitations such as possessing the
(*a*) diffraction effects for the wavelength of light
(*b*) being very expensive technique
(*c*) permitting only analyse composition of sample
(*d*) very high resolution.

40.2 Modern methods of qualitative characterisation of physical nature of the surface of non-conducting solids is by
(*a*) Atomic force microscopy
(*b*) Aüger spectroscopy
(*c*) X-ray photoelectron spectroscopy
(*d*) ESCA-electronic spectroscopy for chemical analysis.

40.3 Scanning probe microscopy (SPM) includes the following method(s),
(*a*) Scanning tunnel microscopy (STM) and atomic force microscopy (AFM),
(*b*) Scanning electron microscopy (SEM)
(*c*) Secondary ion mass spectroscopy (SIMS)
(*d*) X-ray photoelectron spectroscopy.

40.4 The surface of solid sample is swept in 'Raster pattern' with focussed beam of electrons wherein
(*a*) It swifts across a surface upwards in x-direction and it is shifted downwards to y-axis direction by standard increments,
(*b*) Scanning pattern is not similar to cathode ray tube (CRT)
(*c*) Scanning is not analogous to TV screen with electron beam,
(*d*) Scanning signal is not received in *z*-direction

40.5 Scanning probe microscopes (SPM) has limitation as it
(*a*) is used for resolving details of surfaces at atomic level.
(*b*) is unsuitable to measure surface topography
(*c*) shows no details in *z*-direction but can show only in the *x* and *y* direction.
(*d*) the resolution cannot go less than 20 Å.

40.6 Scanning tunneling microscope (STM) is used to
(*a*) scan surfaces in raster pattern (in *x*/4 direction)
(*b*) scan in *z*-direction.
(*c*) scan in *x*-direction.
(*d*) scanning without details to atomic size scale

40.7 Scanning tunneling microscope
(*a*) has limitations of scanning conducting solid surfaces
(*b*) cannot provide morphological information.
(*c*) is incapable of telling about topographic plan
(*d*) tunneling current goes on changing in each step of scanning.

40.8 Atomic force microscope
(*a*) facilitates examination of conducting surfaces
(*b*) useful for examination of mainly non-conducting surfaces
(*c*) needs no piezoelectric tube for scanning
(*d*) has optical control system which is different from STM.

40.9 In AFM the scanning tip contacting surface of sample is
(*a*) easily damaged leading to image distortion
(*b*) good for examining biological samples
(*c*) is best for polymer analysis
(*d*) a silicon wafer which is minutely scanned.

40.10 The chief advantage of AFM is it
(*a*) examines sample under water in vacuum
(*b*) avoids distortion of images
(*c*) cannot detect defects in surface of silicon
(*d*) has no facility for imaging of DNA in biotechnology.

40.11 STM retrieves information from the
(*a*) topmost surface of sample (*b*) internal surface of sample
(*c*) bottom surface of sample (*d*) last layers of surface of sample.

40.12 Scanning tunnel microscopy (STM)
(*a*) is used for identification of electronic surface states near fermi level.
(*b*) facilitates quantitative analysis of sample
(*c*) allows rapid analysis of sample
(*d*) permits elucidation of structure of the sample.

40.13 Atomic arrangement of surface atoms at nanometer resolution is possible with
(*a*) Scanning tunneling microscope
(*b*) Atomic force microscope
(*c*) Optical microscope
(*d*) High spatial resolution tunnel electron microscope (HRTEM).

40.14 The surface topographic profiling is carried out in
(*a*) non-contact mode
(*b*) contact mode
(*c*) both contact and roaster contact mode
(*d*) remote sensing mode

40.15 The ordering polymeric film can be identified by
(*a*) Scanning tunneling microscopy (*b*) Atomic force microscope
(*c*) Auger electron spectroscopy (*d*) Photoelectron spectroscopy

40.16 The sub micro nucleation process in scanning tunneling microscope involves
(*a*) absorption of atoms (*b*) adsorption of atoms
(*c*) transmission of atoms (*d*) scattering of particles

40.17 The critical surface conditioning is not necessary in examination by
(*a*) scanning tunneling microscopy (*b*) Atomic force microscopy
(*c*) X-ray photoelectron spectroscopy (*d*) Photoelectron spectroscopy

40.18 The information from the top most layer of surface is procured by
(*a*) Scanning electron microscope (*b*) Secondary ion mass spectrometer
(*c*) Scanning Tunneling microscope (*d*) Low energy electron diffraction microscope

40.19 The scanning probe microscopy cover two techniques as are
(*a*) SEM and STM (*b*) AFM and ESCA
(*c*) XPS microprobe and AFM (*d*) STM and AFM

40.20 X-ray photons are used as primary beam in
(*a*) SIMS (*b*) UPS
(*c*) AES (*d*) XPS

40.21 Secondary beam of X-ray photons is observed in
(*a*) Electron microprobe
(*b*) Ultraviolet photoelectron spectroscope
(*c*) Electron spectroscope for chemical analysis
(*d*) Secondary ion mass spectroscope

40.22 The detailed knowledge of physical nature of surface of solids is vital in study of
(*a*) Heterogenous catalyst (*b*) Polymers
(*c*) Minerals (*d*) Lunar rocks

40.23 In the following method both primary and secondary beams contain the electrons. The technique is
(*a*) Electron microprobe (*b*) X-ray photoelectron spectroscopy
(*c*) Aüger electron spectroscopy (*d*) Laser microprobe mass spectrometry.

40.24 Which of the following techniques permits sub monolayer examination in nucleation?
(*a*) AFM (*b*) STM
(*c*) UHV-STM (*d*) UTV-AFM

40.25 Which technique permits crystalline ordering within ordered nanometer sized samples?
(*a*) STM (*b*) AFM
(*c*) SIMS (*d*) Fast atom bombardment.

ANSWERS

40.1 a	40.2 a	40.3 a	40.4 a	40.5 a	40.6 a	40.7 a
40.8 b	40.9 a	40.10 a	40.11 a	40.12 a	40.13 a	40.14 c
40.15 b	40.16 b	40.17 b	40.18 c	40.19 d	40.20 d	40.21 a
40.22 a	40.23 c	40.24 c	40.25 a.			

REFERENCES

40.1 D.A. Skoog, F.J. Holler and T.A. Nieman, *Principles of the Instrumental Analysis* 5th edn., Harcourt Brace Court Pub. p. 535 (1998).

40.2 F.A. Settle, *Handbook of Instrumental Methods In Analytical Chemistry,* Pearson Educational Publication p. 775 (2004).

40.3 R.D. Braun, *Introduction to Instrumental Analysis*, McGraw-Hill & Co. Indian Edition p. 585 (2003).

40.4 H.H. Willard, L.L. Merritt, J.A. Dean and F.A. Settle, *Instrumental Methods of Analysis*, CBS Publication Indian edition p. 379 (1986).

40.5 S.M. Khopkar, *Basic Concepts of Analytical Chemistry*, 3rd edn., New Age International Publishers (P) Ltd. (2007).

Chapter 41

Photoacoustic Spectroscopy

INTRODUCTION

The radiation of matter, originates on account of migration of electrons from ground state to higher energy state. Such excitation of electrons is possible with powerful source of radiation. When electron goes to higher orbital, they cannot remain there for ever. They have to return back to ground state with release of energy. Such return is accompanied by absorption of energy as seen in absorption spectroscopy or return is possible along with emission of radiation as in emission spectroscopy. There may be the time lag followed by delayed emission as it is noted in luminescence spectroscopy. However, there is one more possibility when there is absorption or emission. This is called non-radiative or radiation loss phenomena. This non-radiative process leads to generation of heat. This heat in turn expands gases leading to origin of pressure waves. Now if these pressure waves are in the region of range (10-100 Hz), we get sound waves of which intensity can be measured. The non-radiative process is proportional to the amount of sample excited and the heat generated is interrelated to pressure wave formation. It is also proportional to the concentration of solution. Much success of this method depends upon thermal diffusion. With longer wavelength strong signals are emitted. Alternatively to get strong signals it is necessary to increase the intensity of source of radiation with xenon or krypton lamp.

The role of detector is very important. Like UV-visible spectrometry, we do not use PMT, *i.e.*, Photomultiplier tubes. Since the signals relate to information of sound waves, *i.e.*, acoustic we either measure them by usual microphone in fluids or piezoelectric transducer in solids. In latter case change in charge which in turns changes potential is measured. If we can measure intensity of non-radiative source in form of pressure waves, it is very easy to relate it with concentration of the sample.

All solids, liquid or gases can be measured by PAS. Only in the presence of fluorescence or phosphoresces, PAS cannot be measured as PAS signals are too weak that they cannot be easily measured. We witness transition of electrons or fluids or lattice vibration in solids leading to perceptible changes in pressure waves which can be measured by microphone or piezoelectronic (PZI) indicator. The generation of heat is most essential step in analysis. For weak signal long thermal diffusion tube is employed. This however depends upon modulation frequency.

In PAS instruments we have xenon source, with chopper to modulate the frequency of radiation and detector like microphone. The sample holder can be made out of silver, gold, aluminium, to prevent saturation (inability to heat to diffuse to surrounding). It can be prevented by using KBr or Al_2O_3 with sample. For surface depth of 5-6 mm spin flip Raman laser is used, as it provides measurement over wide range (250-2500 nm) of wavelength.

This chapter therefore contains several multiple choice questions. They test intricate knowledge about source of non-radiative energy, type of cell used, length of thermal diffusion tube, kind of signal provided, type of detector used; the state of matter which is most suitable for analysis; intensity of sound pressure, vibration expansion fluids accompanied by heat generation leading to pressure wave for the measurement. For more detailed discussion of this topic you are advised to refer to number of authoritative monograph as cited in references at end of this chapter and in ref. 41.1.

PROBLEMS

Multiple Choice Questions

Tick mark correct answer from following multiple choices:

41.1 Photoacoustic spectroscopy is also known by other name as
(*a*) optoacoustic spectroscopy (*b*) luminescent spectroscopy
(*c*) scattering spectroscopy (*d*) resonance spectroscopy.

41.2 In photoacoustic spectroscopy we measure change in intensity of the
(*a*) radiation absorbed (*b*) sound pressure
(*c*) radiation emitted (*d*) velocity of light.

41.3 The photoacoustic spectrometry permits analysis of
(*a*) solids (*b*) liquid
(*c*) fluids and solids (*d*) gases

41.4 Opaque substance can be also analysed by following technique:
(*a*) Aüger spectroscopy (*b*) reflectance spectroscopy
(*c*) mass spectroscopy (*d*) photoacoustic spectroscopy.

41.5 The non-radiative process in a molecule leads to
(*a*) transition from higher vibrational level to ground level.
(*b*) absorption
(*c*) emission
(*d*) excitation of molecule.

41.6 Electronic transitions from higher to ground level in solids leads to
(*a*) drop in thermal measurement index (*b*) conversion to lattice vibrations
(*c*) drop in temperature (*d*) decrease in thermal movement.

41.7 In PAS radiation falls on sample and latter returns to ground state as the non-radiative process leads to change in
(*a*) intensity of incident radiation
(*b*) temperature of surrounding
(*c*) in expansion due to change in pressure
(*d*) change in intensity of pressure.

41.8 The source of radiation in PAS technique consist of
(*a*) hallow cathode (*b*) hydrogen lamp
(*c*) tungsten lamp (*d*) xenon lamp.

41.9 The kind of radiation used in PAS technique consists of
(*a*) UV/IR radiation (*b*) scattered radiation
(*c*) delayed radiation (*d*) X-radiation.

41.10 In PAS after excitation of molecule electrons return to the ground state by
(*a*) absorption of energy (*b*) non-radiative process
(*c*) emission phenomena (*d*) reflection of radiation.

41.11 The non-radiactive process in PAS leads to
(*a*) absorption of heat (*b*) change in temperature
(*c*) generation of heat (*d*) release of heat.

41.12 Heat generated in PAS technique is proportional to
(*a*) radiation source
(*b*) absorptivity of sample
(*c*) neither to source and absorpivity of sample
(*d*) absorpivity of sample and intensity of source in PAS.

41.13 One prefers high intensity source in PAS as it
(*a*) generates more heat to give strong PAS signal
(*b*) can produce weak signal
(*c*) would not produce heat
(*d*) would not promote relaxation of electrons.

41.14 In PAS, excitation of radiations as non-radiative source is modulated to
(*a*) reduce temperature of system (*b*) cause pressure wave to give signal
(*c*) promote absorption of radiation (*d*) create variation in intensity of radiation.

41.15 The conversion of heat to pressure wave in fluids leads to
(*a*) drop in temperature (*b*) no change in temperature
(*c*) rise in temperature (*d*) both rise and fall in temperature

41.16 PAS signal is directly proportional to
(*a*) temperature
(*b*) heat absorbed
(*c*) absorpivity with no relation to cell length
(*d*) absorpivity and diffusion length of sample.

41.17 If diffusion length of cell is 10-100 μm, then modulation frequency required is
(*a*) 10-100 kHz (*b*) 1-10 kHz
(*c*) 0.1-1 kHz (*d*) 100-1000 kHz.

41.18 Greater thermal diffusion length in PAS leads to
(*a*) no change in radiation process (*b*) absorption and warming of gases
(*c*) release of radiation to cool gas (*d*) no apparent change in gas temperature

41.19 PAS and UV-visible instruments are comparable as both of them have same type of
(*a*) photoelectric tube
(*b*) radiation source
(*c*) chopper to modulate frequency of radiation
(*d*) monochromatic source.

41.20 One of the following source of radiation is useless for PAS.
(*a*) xenon lamp (*b*) argon lamp
(*c*) tunable dye laser (*d*) deuterium lamp.

41.21 The sample holder in PAS technique is made up of
(*a*) gold, silver or aluminium (*b*) glass
(*c*) quartz (*d*) copper.

41.22 One has to use piezoelectric detector if sample in PAS is
(*a*) gas (*b*) solid
(*c*) liquid (*d*) inert material.

41.23 Piezoelectric detector readily causes change in
(*a*) temperature (*b*) volume
(*c*) pressure (*d*) physical state of sample.

41.24 The piezoelectric effect is exhibited by
(*a*) $PbCl_2$ (*b*) $PbCrO_4$
(*c*) PbO_2 (*d*) $PbZrO_2$.

41.25 Saturation effect in PAS technique can be eliminated by
(*a*) grinding sample with KBr or Al_2O_3 (*b*) taking aqueous sample
(*c*) using quartz cell (*d*) using suitable filter.

41.26 PAS signal is considerably reduced by phenomena of
(*a*) scattering (*b*) luminescence
(*c*) reflection (*d*) absorption.

41.27 In PAS, saturation zone can be visualised with jamming effect if the coefficient of thermal diffusion (α) is
(*a*) $> 10^2$ (*b*) $> 10^3$
(*c*) $> 10^4$ (*d*) $> 10^{-1}$.

41.28 Microphone laser detector cannot be used if sample is
(*a*) solid (*b*) gas
(*c*) opaque material (*d*) liquid.

41.29 In PAS if H = heat generated, Z = intensity, a = absorpivity, then
(*a*) $\text{H} \alpha\ (a \times I)$ (*b*) $H\alpha \dfrac{1}{(a)}$
(*c*) $H\alpha \dfrac{1}{(1)}$ (*d*) $H\alpha \dfrac{1}{(aI)}$.

41.30 In PAS, wavelength of acoustically resonating signal is given by
(*a*) $\lambda = (1 - n)$ (*b*) $\lambda = 2L/n$
(*c*) $\lambda = 2\ Ln$ (*d*) $\lambda = 2\ n/L$ (if L = diffusion length, n = integer.)

41.31 In PAS, radiation less relaxation lead to
(*a*) delayed emission similar to fluorescence
(*b*) scattering as in nephelometry
(*c*) liberation of heat leading to pressure wave
(*d*) reflection of radiation from sample.

41.32 The modulation of incident radiation in PAS leads to change in
(*a*) temperature
(*b*) intensity of radiation
(*c*) emission process
(*d*) expansion or contraction of pressure waves.

41.33 For a surface of depth 5-6 μm in PAS one uses
(*a*) spin chip Raman laser (*b*) tunable dye laser
(*c*) semiconductor diode (*d*) solid state emitter laser.

41.34 PAS method can be made sensitive by
(*a*) decreasing diffusion length of cell (*b*) increasing intensity source
(*c*) decreasing temp. of surrounding (*d*) use of better monochromatic source.

41.35 In PAS if modulation frequency is 10-20 Hz the thermal diffusion length is
(*a*) 1-110 μm (*b*) 100-1000 μm
(*c*) 10-100 μm (*d*) 0.1-1 μm.

41.36 For operating from 250-2500 nm the source of radiation in PAS is
(*a*) tungsten lamp (*b*) tunable dye laser
(*c*) krypton lamp (*d*) xenon arc lamp.

41.37 In piezoelectric transducer there is change in pressure resulting in change of:
(*a*) charge and potential (*b*) weight
(*c*) volume (*d*) state of sample

41.38 The phenomena of saturation is noticed in PAS instruments because
(*a*) change in temperature prevents diffusion of heat
(*b*) of limited rate of diffusion of heat from sample to detector
(*c*) change in pressure when heat cannot be absorbed to surface
(*d*) bad conduction of heat to surrounding gas.

41.39 The thermal diffusion co-efficient (αs) is related to frequency (v), actual thermal diffusion (2s) and 1-path of absorbing body the expression as
(*a*) $a_s = \pi v/\alpha s$ (*b*) $a_s = \pi v \alpha s$
(*c*) $a_s \left(\frac{\pi v}{\alpha_s}\right)^{1/2}$ (*d*) $a_s = \alpha s v$.

41.40 The sample showing one of the phenomena decreases signal:
(*a*) emission (*b*) absorbance
(*c*) scattering (*d*) fluorescence.

ANSWER

41.1 a	41.2 b	41.3 c	41.4 d	41.5 a
41.6 b	41.7 c	41.8 d	41.9 a	41.10 b
41.11 c	41.12 d	41.13 a	41.14 b	41.15 c
41.16 d	41.17 a	41.18 b	41.19 c	41.20 d
41.21 a	41.22 b	41.23 c	41.24 d	41.25 a
41.26 b	41.27 c	41.28 d	41.29 a	41.30 b

41.31 c	41.32 d	41.33 a	41.34 b	41.35 c
41.36 d	41.37 a	41.38 b	41.39 c	41.40 d

REFERENCES

41.1 S.M. Khopkar, *Basic Concepts of Analytical Chemistry,* 3rd edn., New Age International Publishers (P) Ltd. (2007).

41.2 R.D. Braun, *Introduction to Instrumental Analysis,* McGraw-Hill and Co. (1987).

41.3 G.W. Ewing, *Instrumental Methods of Chemical Analysis,* 5th edn., McGraw-Hill Book Co. (1985).

41.4 F.P. Schaper, *Dye Lasers,* Springer Verlag Berlin (1973).

41.5 Y. Youhon Pao, *Opto Acoustic Spectroscopy & Detection,* Academic Press (1977).

41.6 A. Rosenebarg, *Photo Acoustics & Photo Acoustic Spectroscopy*, Wiley N.Y. (1980).

41.7 G.H. Ayers, *Quantitative Chemical Analysis,* Harper and Brother N.Y. (1980).

41.8 H.H. Willard, L.L. Merritte and J.A. Dean, *Instrumental Methods of Analysis,* Van Nostrand and Co. 5th edn., (1994).

41.9 D.A. Skoog and J.J. Leroy, *Principles of Instrumental Analysis,* Harcourt Brace & Co. (1994).

41.10 R.M. Silverstin, G.C. Bassler and T.C. Moril, *Spectrophotometric Identification of Organic Compounds,* 5th edn., Wiley (1991).

Chapter 42

Mass Spectrometry

INTRODUCTION

This is unique technique largely used for the determination of molecular weight or mass number of a compound. The technique also facilitates identification of various isotopes of the elements. Mass spectrometer is coupled with gas chromatograph (GC-MS) or with liquid (LC-MS) chromatograph to permit simultaneous separation and determination. Such techniques are designated as hyphenated technique. In recent years even two mass spectrometers were coupled up together with varying ion source, and is called 'Tandem mass spectrometry' (MS-MS).

The basic principle employed in mass spectrometry is to segregate different ions based upon their differences in charge to mass (*e/m*) ratio. We plot a graph of relative concentration of the gaseous component as the function of (*m/e*) ratio of the component. One can analyse both cations as well as anions by mass spectrometry but latter is very commonly analysed.

The instrument mainly consists of inlet system, ion source, mass analyser and detectors with recording device. If a sample is in gaseous state it is directly injected into system however if it is solid or liquid both being volatile they are first converted into gases state in inlet system. This is followed by the process of ionisation. Several devices like electron bombardment ionisers arc/spark ioniser; thermal ioniser, photo ioniser, chemical ioniser with methane negative chemical ioniser, laser induced ioniser, ionic bombardment ioniser or atomic bombardment ioniser are available but chemical ioniser and electron bombardment ioniser are most frequently used. In second stage compound is ionised when it gets fragmented into different species with varying *m/e* ratio. These species are then passed through analyser. There are four single, double focussing, time of flight and quadrupole analyser as

$$m/e = \frac{H^2 r^2}{V}$$

if H = magnetic field, r = radius of path, V is accelerating voltage or potential of the system (also given as $m/e = \frac{H^2 r^2 V}{2v}$), if $\frac{1}{V} = \frac{r}{2v}$.

The velocity of moving ion can be calculated from the relationship as

$$V = \frac{Her}{m} \text{ or } \frac{m}{c} = \frac{KH^2}{E},$$

K = constant, H = magnetic field, E = applied potential or voltage. In double focussing analyser, r = radius of curvature, it is given by equation $r_e = 2\ E/V$, if V = potential difference between two plates and E = accelerating potential applied to ions. In a time of flight analyser, we find relationship as $m/e = 2\ Et^2/d^2$ wherein d = distance travelled by ion, t = time of travel of ion or $m/e = 2E/V^2$. If $V = (d/m/e\ t)$. In quadrauple analyser four concentric tubes are diagonally arranged and ion beam is allowed to pass through the interstitial space between these tubes. Finally in ion cyclone resonance analyser we have $V_d = E/B$ if $V_d = E/H$. In all cases, we have $\frac{mv^2}{r} = Her$ leading to $W_c = \frac{V}{r} = \frac{He}{m}$ if W_0 is frequency of the ion which complete the circle (V/r).

After passing through the analyser, the ions proceed to detector. Various kinds of detectors are used but photographic detector is most useful for qualitative analysis. The detector is attached to recorder to furnish read out. One can ascertain mass number by examining fragmentation pattern. The abundance of ions can also be ascertained by mass spectrometer. In fragmentation pattern the highest peak is called, base peak. All other peaks, are subsequently analysed. The other peaks are analysed in relation to this base peak'. It is thus possible to identify the compound from fragmentation pattern; because the ion current is proportional to the concentration of the ion in question. Good vacuum system (0.1 - 0.001 tor) is essential to get reproducible results. The vaccum inlet is 10^{-5} tor while in analyser it is 10^{-7} tor. The time of flight analyser $(t) = K\sqrt{\frac{m}{e}}$, if K = const. The magnitude of the beam current is of the order of 10^{-15} to 10^{-5} amps. If it is small it can be easily amplified. The resolution defines the ability of an instrument to differentiate between two ions of equal mass (e.g., $C_{12}H_3O$ and $C_{13}H_4$). It is expressed as $M/\Delta M$ if M = normal mass of well spaced peaks of same height at M and other as ($M + \Delta M$) Mass spectra contains several peaks of varying heights. Thus $M + e^- = M^+ + 2e^-$ if M is molecular and M^+ is ion with (EI) of 70 ev we get good splitting. The ion abundance is given in terms of % of base peak. The first ion appearing in fragmentation pattern is called 'Parent ion'. A small peak is called 'satellite peak', Mass spectrometer is calibrated with perfluoro kerosene (PFR) with peaks at CF_3 (69), C_3F_3 (93), CC_4F_3 (105), and C_3F_5 (135). The spark source mass spectrometer is useful for analysis of trace elements. The minimum requirement in each component must show at least one peak, the contribution of each component to peak is additive their sensitivity must be good and known standard for comparison are easily available.

In this chapter large number of problems are given of which half of them are solved and other half are given for practice along with answers. The problems pertain to calculating mole % of compounds in mixture; the resolution, or measure magnetic flux density required to focus a beam of anion, time necessary for fragmentation; kinetic energy of charged ion, or accelerating potential, voltage for scanning time for ions to pass through analyser, etc. The relevant mathematical relations used in such problems are provided in this chapter. A large number of books cited at end of chapter in the references are very useful to get clear in sight of the subject.

SOLVED PROBLEMS

42.1 The mass spectrum of the nitrogen produced from air was characterised by the presence of the peaks of isotope of nitrogen $N^{14}N^{14}$ and $N^{14}N^{15}$ equal to 680 and 5 mm respectively. What was the % of nitrogen (N^{15}) in the sample?

Ans: The amount of nitrogen was obtained from following equation if x = amount of nitrogen

$$x = \frac{100}{2R+1} \text{ if}$$

$$R = \text{ratio of peak height of two isotopes which was } = \frac{680}{5} = 136$$

$$R = \frac{hN^{14}}{hN^{15}} = \frac{680}{5} = 136$$

$$x = \frac{100}{(2 \times 136) + 17} = \frac{100}{273} = 0.3666\%.$$

42.2 A mixture of compound A and its saturated analogue compound B were anlaysed by mass spectrometer. Calculate the mole % of each compound in a mixture from the following data as well as mole ratio of compounds?

m/e	Compound *A*	Compound *B*	Unknown Sample
81	0.3	49 µg	48.9
95	0.2	43	43
137	0.1	100	100
141	100	0.1	9.3
338	48	0	4.4
348	0	8.4	8.4
Sensitivity	0.797	0.910	–

Ans: The compound A had peak of 338 for e/m while for the compound B had peak at 137 for e/m ratio.
Now Uncorrected partial pressure was $100 = 0.1\ (xA) + 100(xB)$;
$4.4 = 48\ (xA) + 0(xB)$

So $xA = 0.08$ and $xB = 0.92$ Now ratio was

$xA/S = 0.08/0.797 = 0.10$ press are for A if S = Sensitivity

$xB/S = 0.92/0.910 = 0.01$ press for B. P = pressure

The partial pressure was converted into % as

$$\%A = \frac{0.1 \times 100}{1.11} = 9 \text{ moles.}$$

$$\%B = \frac{0.01 \times 100}{1.11} = 91 \text{ moles.}$$

The ratio of two % A/%B = 9/91.

42.3 Find out the resolution in a mass spectra if a peak at centre was 245 amu with its width at 5% of the height was 0.52 amu?

Ans: $R = \frac{M}{\Delta M}$, if M = average mass and ΔM change in mass.

$\frac{\Delta H}{H} = 0.1$ is just resolved so $R = \left(\frac{M}{W_{0.52}}\right)$ is peak at 5% height.

In this problem, we have therefore,

$$R = \frac{M}{W_{0.05}} = \frac{245}{0.52} = 4.7 \times 10^2 \text{ if } W_{0/05} \text{ mass peak 5\% height.}$$

The resolution (R) was 4.7×10^2.

42.4 Calculate the magnetic flux density required to focus a $C_4H_9^{2+}$ ion on the detector, with potential of 2kV and radius of the curvature as 30 cm?

Ans: Since $m/e = 57$ This in terms of S.I. units was expressed as $57 \times 1.66 \times 10^{-27} \times 1/1.6 \times 10^{-19} = 5.9 \times 10^{-7}$ kg/C = m/e. In magnetic analyser

$$m/e = \frac{B^2 r^2}{2E}$$

if B = magnetic flux, r = radius of curvature, E = potential so

$$5.9 \times 10^{-7} = \frac{B^2 \times (0.300)}{2 \times (2 \times 10^3 \text{ V})}$$

$B^2 = 2.6 \times 10^{-2}$ kg^2/S^4A^2 from the values of magnetic fluxes.
$B = 0.16$ kg/S^2 if $A = 0.16T$ (as a tesla T is kg/S^2 A) we can also use

$$\frac{amu}{e} = 4.824 \times 10^7 \times \frac{B^2 r^2}{E} \text{ as } r, E. \text{ are given so}$$

$$m/e = \frac{4.824 \times 10^7 \times B^2 \times (0.3)^2}{(2 \times 10^3)} \text{ so } B^2 = 2.6 \times 10^{-2} \text{ and } B = 0.16 \text{ kg/S}^2.$$

42.5 Calculate the time needed for fragmentation pattern with $m/e = 100$, and 101 with an accelerating potential of 2.5×10^5 V and flight path of 1000 mm is recorded with time of flight analyser in mass specrometer?

Ans: m/e values should be converted to S.I. units.

$$m/e(100) = 100 \times 1.66 \times 10^{-27} \times \frac{1}{1.6 \times 10^{-19}} = 1.04 \times 10^{-6} \text{ kg/C.}$$

$$m/e(101) = 101 \times 1.66 \times 10^{-27} \times \frac{1}{1.6 \times 10^{-19}} = 1.05 \times 10^{-6} \text{ kg/C.}$$

Now, in time of flight analyser

$m/e = 2Et^2/d^2$ if E = potential, d = distance of grid and detector, t = time

Since, $E = 2500$, t = unknown, d = 1000 mn or 1.0 meter. We have

$$m/e = 1.04 \times 10^{-6} = \frac{2(2500)t^2}{(1)^2} \text{ so } t^2 = 2.08 \times 10^{-10} \text{ kg m}^2/\text{C.V.}$$

Since tc = IAS, V = 1 kg m²/S³A. Hence, if S = 14.4 μS

$t^2 = 2.08 \times 10^{-10}$ so $t = 1.44 \times 10^{-5}$, S = 14.4 μS

For $$m/e = 1.05 \times 10^{-6} = \frac{2(2500)t^2}{(1)^2} = (14.5)^2. \text{ So, } t = 14.5.$$

42.6 Calculate the resolution for a mass spectral peak at *m/e* 447 amu with a peak width at 5% the peak height being 0.34?

Ans: Since $R = \frac{M}{W_{0.05}}$ so M = 447 amu, $W_{0.05}$ = 0.34

So, $$R = \frac{447}{0.34} = 1.3 \times 10^3.$$

42.7 What should be the resolution (R) of mass spectrometer if mass spectral peak centred at *m/e* = 352 with peak width at 5% of the peak height being 0.41?

Ans: Since $R = \frac{M}{W_{0.05}}$ we use this formula

$$R = \frac{352}{0.41} = 860.$$

42.8 Calculate the resolution of the mass spectrometer if the mass spectral peaks of m/e was 1257 and 1259 with the two peaks were having equal heights?

Ans: $$\frac{M}{\Delta M} = R \qquad \frac{H}{\Delta H} = 0.1$$

Now H_1 and H_2 is same

Since $(M_2 - M_1) = \Delta M = (1259 - 1257) = 2$

as $M_1 = 1257$ $M_2 = 1259$, $\Delta = 2$, if M = average mass = 1258

So $$R = \frac{M_{av}}{\Delta M} = \frac{1258}{2} = 629.$$

42.9 Calculate the resolution of the mass spectrometer if with mass spectral peaks at *m/e* of 748 and 749 exhibits 10% valley between the two peaks?

Ans: If $\Delta M = (M_2 - M_1) = (749 - 748) = 1$; M average = 748.5; ΔM = 1.0

$$R = \frac{M_{av}}{\Delta M} = \frac{748.5}{1} = 748.5.$$

42.10 Calculate the magnetic flux density essential to focus with *m/e* 245, on the exit slit with increasing potential of 7500 V and radius of curvature of 25 cm in typical mass spectrometer?

Ans: We can use any one of the following formulas as

$$m/e = B^2r^2/2E \qquad ...(1)$$

let us use the second formula amu/e = $4.824 \times 10^7 \times (B^2r^2/dE)$

Since E = 7500, v = 25 cm = 0.025 meter, we have *m/e* = 245.

$$245 = 4.824 \times 10^7 \times B^2 \times \frac{(0.025)^2}{7500} \text{ on solving, we have}$$

$B^2 = (0.78)^2$. So, magnetic flux = 0.78 tesla.

42.11 What should be the kinetic energy of singly charged ion ($Z = 1$) if it was accelerated through potential of 10^3 V in an electron impact source?

Ans: The kinetic energy added to the ion is due to accelerating potential (V) its kinetic energy = qv = Zev if $e = 1.6 \times 10^{-19}$ charge $Z = 1$

So kinetic energy = $1 \times 1.6 \times 10^{-19}$ $C \times 10^3$ V = 1.6×10^{-16} J.

42.12 What is the energy that electrons need as result of if it being accelerated by potential of 7V?

Ans: The kinetic energy = charge of electron × eV if V was potential.

We get energy per mole = $e \times V \times N$ if N = Avagardo's number = 6.02×10^{23}

Kinetic energy = eVN = (1.6×10^{-19}) (70) (6.02×10^{23}) = 6.7×10^6 Joule/mol.

42.13 Can you predict the resolution needed to separate the first two ions such as $C_2H_4^+$ and CH_2N^+?

Ans: We first calculate mass of each species or ion as

$$C_2H_4^+ (28.0313);\ CH_2N + (28{,}0187),\ N_2^+ (28.0061) \text{ and } CO(27.9949) \text{ Dalton}$$

$$\text{Diff. in mass } \Delta m = \left(C_2H_4^+ - CH_2N^+\right) = (28.03013 - 28.087) = 0.0126$$

$$C_6H_5\ CH_2CH_3 + e^- \rightarrow C_6H_5CH_2CH_3^+ + 2e^-, \text{ we have}$$

$R = M/\Delta M = 28.025/0.0126 = 2.22 \times 10^3$ as $M_{av} = 28.025$.

42.14 Calculate the accelerating potential (V) which was necessary to direct single charged water molecule through slit with magnetic field of 0.240 testa with radius of 12.7 cm?

Ans: Charge per ion = 1.60×10^{-19} $r = 0.127$ m

$$\text{mass } (m) = \frac{18.02\, H_2O/\text{mol}}{6.02 \times 10^{23}\, H_2O/\text{mol}} \times 10^{-3} \text{ kg/g} = 2.99 \times 10^{-26} \text{ kg } H_2^+O.$$

magnetic field B = 0.240 tesla = 0.240 W/m² in SI units.

$$V = \frac{B^2\gamma^2 ez}{2m} = \frac{(0.240)^2\,(0.127)^2\,(1.6 \times 10^{-19})}{2 \times 2.99 \times 10^{-26}} = 2.49 \times 10^3 \text{ V}.$$

42.15 What should be the ratios of the $(M + 1)^+$ to M^+ peak heights for the dinitrobenzene ($C_6H_4N_2O_4$) with mass of 168?

Ans: For every 100 C^{12} atoms there are 1.08^{13} C atoms so 6C atoms have 1.0813 × 6.0 = 6.48 mole of dinitrobenzene. Thus effect alone $(M + 1)^+$ peak will be 6.48% the M^+ peak.

42.16 Calculate the voltage range for scanning mass spectrometer ranging from 18 to 200 with field strength of 2400 gauss and radius of curvature of 12.7 cm? (Given $e = 1.6 \times 10^{-20}$).

Ans: $$m/e = \frac{H^2r^2}{2V} \text{ if } H = \text{field strength } r = \text{radius } V = \text{Volts.}$$

Since $H = 2400$, $r = 12.7$ M = 18 to 200, $e = 1.6 \times 10^{-20}$ we have

$$V = \frac{(2400)^2 (12.7)^2 (1.602\times10^{-20})}{2\times18\times6.023\times10^{23}} = 2478 \text{ to focus mass 18 on detector.}$$

$$\frac{(m/e)_1}{(m/e)_2} = \frac{V_1}{V_2} \text{ or } \frac{m_1}{m_2} = \frac{V_1}{V_2} \text{ i.e., } \frac{18}{200} = \frac{V_2}{2478} \text{ so } V_2 = 223\,\text{V}.$$

PROBLEMS FOR PRACTICE

42.17 What should be the accelerating voltage to focus ions with *m/e* = 350 when radius of curvature was 22.5 cms and field strength was 1.00 tesla? **[Ans:** 6860 V]

42.18 Calculate the time necessary for an ion with *m/e* = 425 if the ion with *m/e* = 298 demands time of 1.79 μs to colloide with the detector? **[Ans:** 2.09 μs]

42.19 Calculate the resolution (*R*) necessary to isolate the C_2H_8 – S doublet at mass of 200? **[Ans:** 1 part in 2210]

42.20 What is the range of mass if in time of flight analyser the beam as well as resolution were energised for 0.25 μs every 100 μs? **[Ans:** 5600 mass units]

42.21 Calculate the resolution of the mass spectrometer if a mixture of benzophenone and azobenzene with MW = 182.2214 and 182.2244 resp (ΔM = 0030) was analysed? **[Ans:** 182.2229/0.30 is resolution]

42.22 Calculate the probable composition of a molecule of mass 142 whose peak was 1.1% of the parent compound? **[Ans:** CH_3I one C-is indicated]

42.23 Can you calculate the empirical formula of unknown compound from given data:
(*i*) *m/e* (% base peak): 90P(100) 91(561) 92(4.69)
(*ii*) *m/e* (% Peak): 89P(17.12); 90(0.58): 91(5.36), 92(0.17)
(*iii*) *m/e* (% Peak) 230P(1.10); 232(2.12); 234(1.06)
[Ans: Mass 206 with two Sulphur atoms]

42.24 Calculate the time necessary for ions with *m/e* as 100, 150, 200 to collide with detector after emission in a time of flight mass spectrometer which had flight of 1,250 m and an accelerating potential of 2875 V? (A peak at *m/e* = 447 had 5% peak width)? **[Ans:** 16.8, 20.6, 23.8 and 26.6 μs]

42.25 Can you identify the ions corresponding to the lines appearing in spectrum of amyl acetate at mass number when *m/e* is 39, 50, 51, 77, 78, 103, 104? **[Ans:** 14 CH_2; 28CO; 42 – CH_3COO; 44 – CO_2; 57.5 $C_5H_{11}COO$, 59CH_3COO; 70C_5H_{10} and 71 C_5H_{11}]

42.26 Identify the particles and probable ions with respect to lines if the mass spectrum of (CO_2 + CH_4) with lines were for *m/e* 8, 13, 14, 15, 16, 28, 44? **[Ans:** 8CH_4/2; 13-CH; 14CH_2, 15CH_2; 16CH_4; 28CO; 44CO_2]

42.27 Can you calculate the atomic mass of inert gas if a mass spectrometric analysis data was *m/e* (% abundance) [mass]: 36(0.337) 35.698; 38(0.063) [37.963]; 40(99.60) [39.962]? **[Ans:** 39.947 Argon]

42.28 Find out mole percentage of each individual species in a unknown sample in a mixture if the unknown compound had following mass spectral data?

m/e	*x*	*y*	*z*	*w*	Unknown compound
67	100	16.0	17.0	12.3	100
84	0	25	0	0	14.9
97	0	7	0	44	80.8
105	5.4	0	42	8.5	26.1
109	4.8	4.2	100	1.5	26.6
sensitive	1.42	0.82	0.94	1.08	

[**Ans:** $x = 15.7$, $y = 24.2$, $z = 6.5$, unknown = 53.6 mole %]

42.29 What should be the accelerating voltage which would be needed to migrate singly charged ions of mass 10,000 through an instrument which was similar to the conventional model? [**Ans:** 4.48 V]

42.30 Calculate the time for singly charged C_6H_6 to migrate at a distance of 15 cm with accelerating voltage of 5V in quadrupole mass instrument? [**Ans:** 42.7 Ms]

42.31 Find out the resolution needed to resolve the peaks for $C_2H_4^+$ (MW = 28.0313) and CO^+ (M.W. = 27.9949)? [**Ans:** 2.22×10^3]

42.32 What resolution is necessary to resolve peaks for atomic mass number 115 (with = 115.90) and thorium mass number 232 with atomic weight as 232.03? [**Ans:** 992]

42.33 Determine the ratio of the $[(M + 2)]^T$ to M^+ peak height for a compound like $C_{10}H_6Br_2$? [**Ans:** 1.96]

42.34 Ascertain the ratio of the $[(M + 2)]^+$ to M^+ peak height for compound dichlorobenzene? [**Ans:** 71.50]

42.35 At constant field, calculate the range accelerating of voltage for scanning the mass between 16 and 250 for an charged ion when a magnetic sector model was operated with accelerating voltage of 3×10^3 V with field of 0.126 tesla in order to focus the CH_4^+ ion on the detector? [**Ans:** to 192 V to 3000 V]

REFERENCES

42.1 S.M. Khopkar, *Basic Concepts of Analytical Chemistry*, 3rd edn., New Age International Publishers (P) Ltd. (2007).

42.2 R.D. Braun, *Introduction to Instrumental Analysis*, McGraw-Hill and Co., pp. 664, 671, 693 (1996).

42.3 H.H. Willard, L.L. Merritt and J.A. Dean, *Instrumental Methods of Analysis,* 2nd edn., D. Van Nostrand and Co., pp. 436-37; 453-54 (1990).

42.4 G.W. Ewing, *Instrumental Methods of Chemical Analysis,* 3rd edn., McGraw-Hill Co., p. 427 (1985).

42.5 K. Biemann, *Mass Spectrometry Applications to Organic Chemistry*, McGraw-Hill and Co. (1962).

42.6 R.W. Kiser, *Introduction to Mass Spectrometry and Applications,* Prentice Hall and Co. (1965).

42.7 J. Rubo'z, *Introduction to Mass Spectrometry,* Wiley Interscience Publisher (1968).

42.8 S.R. Shrader, *Introductory Mass Spectrometry,* Allen and Bacon, Boston (1971).

42.9 F.W. McLafferly, *Interpretation of Mass Spectra,* 2nd edn., Benjamin (1973).

42.10 D.A. Skoog, F.J. Holler and T.A. Niemen, *Principles of Instrumental Analysis*, 5th edn., Harcourt Brace Publishers (1998).

Chapter 43

Thermal Methods of Analysis

INTRODUCTION

There are several methods which are related to change in the temperature or heat contents of sample with time mainly used to characterise a compound. Such methods include: TGA, DTA, DSC, DMA and thermometric tritrations. In TGA, *i.e.*, thermogravimetric analysis which is most widely used, we plot graph of change in weight with respect to the temperature. Such technique provides valuable information not only on the composition of substance but also on optimum temperature for drying or decomposition of sample in gravimetric analysis. However in DTA, *i.e.*, differential thermal analysis, we plot a graph of difference in temperature (ΔT) between sample and inert material (like alumina) against successive rise of temperature. The upward deflection shows exothermic reaction and downward deflection shows endothermic reaction. DTA is used for quantitative analysis. DSC, *i.e.*, in differential scanning calorimetry, we record a plot of amount of heat added as the function of time for isothermal operations. DSC is used to measure the enthalpimetric changes associated with particular process; or to measure the activation energy, specific heat, heat of fusion, etc. A glass transition temperature can also be ascertained by this method. In TMA, physical dimension of material is monitored as the function of the time. Linear variable differential transition are recorded. While in the DMA, *i.e.*, dynamic mechanical analysis, a mechanical property of test material is monitored as the function of temperature, when the material is subjected to periodic stress. DTG-derivative thermogravimetry is offshoot of TGA, wherein the ratio of the change in mass to change in heating time ($w/t \; \alpha T$) is plotted as a function of sample temperature. The nature of DTG and TGA curve is more or less identical in pattern.

However, the thermometric titrations have maximum application in the analytical chemistry. In such titration we plot a graph of change in temperature (t°C) against the volume of titrant (increament) added. The equivalence point is recorded. We get two straight lines intersecting at equivalence point of the titration on account of change in free energy $\Delta G = -nFE$ there is change in the heat (ΔH) contents of the sample. This change depends upon the enthalpy change in the chemical reaction and amount of reactants used. Such titrations can be used for quantitative analysis if ΔH is reasonably large causing measurable changes in the temperature. It is used in acid base and non-queous titrations in quantitative analysis.

Finally the enthalpimetry technique deals with the measurement of change in temperature with addition of excess of chemical reactant. If (H) for chemical reaction is known the change in temperature of the reaction is used to determine the amount of limiting chemical reactant in the sample. DTE, *i.e.*, direct injection enthalpimetry is technique in which a single part of the excess of one chemical reactant is added to a fixed volume of a sample that contains other reactant. We can calculate concentration of sample from the knowledge of heat evolved (or absorbed). DIE, *i.e.*, is also called as the 'Enthalpimetry' titration. In continuous flow enthalpimetry excess of one reactant is successively added to a flowing stream of sample and its temperature is monitored. It is related to the concentration of the limiting reactant in the flowing sample. We use DIE in acid base as well as in complexometric titration's for such purpose.

Thus the term 'Thermal analysis' is designated to group of techniques in which property of the sample is monitored against time or temperature. Other less commonly used technique involve change in sound (thermosonometry); magnetic properties (thermomagenetometry), light emitted (thermoluminscene), gas released (emanation thermal analysis). The simultaneous thermal analyser (STA) and controlled rate thermal analysis (CRTA) are rarely used in practice. The purging gases used in thermal analysis are He, N_2, CO_2 and air. They effect mainly heat transfer. In DTA we actually plot graph of $(T_s - T_R) = \Delta T$ if T_s = temperature of sample, T_R = reference material temperature. The simultaneous applications of two techniques like TGA-DTA or TGA-DSC are used for the study of hydrated sodium tungstate fire retarded wood, PVC, etc. Similarly decomposition of $Sr(NO_3)_2$; KNO_3, $BaClO_4$ with TGA-DTA have been studied. One must know history of the sample as well as running condition. Apart from two thermal methods many a times spectral as well as thermal methods of analysis are coupled together in simultaneous analysis, e.g., mass spectrometry—TGA, (i.e., MS-TGA); or (TGA-GC-MS); (IR-DTA); etc.

Thermal methods (like TGA) can be used for the quantitative analysis of mixtures of alkaline earths. They are used for study of polymers to know the decomposition process. In exothermic reactions ΔH is negative. The composition of naturally occurring material like silicate clays, ceramics, glass can be analysed. In the titration we have $\Delta H = \Delta G + T\Delta S$, if S = entropy change G = free energy and H = is heat contents. We have $\Delta T = \frac{n}{K}$, if K = heat capacity and n = number of moles of reactant used in titration.

In this chapter several solved and unsolved problems are provided. They pertain to calculations of composition of the mixture; study of % composition of a constituent in a mixture; kind of transition involved in thermal decomposition; and evaluation of the several other parameters.

SOLVED PROBLEMS

43.1 In thermogravimetric analysis of 0.250 gm of calcium hyroxide, the loss in weight at different stages of temperature was as follows:

(*a*) 0.018 gm at 100-150°C is loss of hygroscopic water

(*b*) 0.038 gm at 500-560°C is dehydration and

(*c*) 0.022 gm at 900-950°C is dissociation

what should be the composition of calcium hydroxide?

Ans: We determine the composition assuming reaction as

$$Ca(OH)_2\,H_2O \xrightarrow[800]{\Delta} Ca(OH)_2 \xrightarrow[\substack{\Delta\\800}]{} CaO \text{ if } \Delta = \text{represent process of heating}$$

$$\text{amt. of hydroscopic water} = \frac{0.018 \times 100}{0.250} = 7.2\%\ H_2O$$

Since mol. wt. $Ca(OH)_2 = 74$, $H_2O = 18$,

$$\text{amt. } Ca(OH)_2 \text{ dehydrated} = \frac{0.038 \times 74}{0.250} = 0.1562 \text{ gram of } Ca(OH)_2 \text{ in 250 g}$$

$$50\% \text{ weight is } \frac{0.1562 \times 100}{0.250} = 60.40\% \text{ of } Ca(OH)_2 \text{ technical grade.}$$

43.2 The thermal decomposition of calcium oxalate (CaC_2O_4) and magnesium oxalate (MgC_2O_4) occurs between 160-870°C in following manner:

(*a*) 0.125 gm at 160-250°C is loss of water molecule

(*b*) 0.100 gm at 450-600°C is decomposition of oxalate to carbonate

(*c*) 0.057 gm at 600-700°C is decomposition of carbonate to oxide

(*d*) 0.041 gm at 750-800°C an account of formation of gas.

What is the % composition of the given mixture?

Ans: The possible decomposition reaction are as follows:

$$CaC_2O_4 \xrightarrow{\Delta} CaCO_3 \xrightarrow{\Delta} CaO + CO_2\uparrow,\ CaO\ SiO_2 \rightarrow CaO + SiO_2$$

$$MgC_2O_4 \xrightarrow{\Delta} MgCO_3 \xrightarrow{\Delta} MgO + CO_2\uparrow,\ MgO\ SiO_2 \rightarrow MgO + SiO_2$$

$$CaC_2O_4 \cdot MgC_2O_4 \cdot H_2O \xrightarrow{\Delta} CaC_2O_4 + MgC_2O_4 + H_2O$$

Mol. wts. of $CaC_2O_4 = 112$, $CaCO_3 = 100$, $CaO = 56$, $H_2O = 18$,

So $$\frac{112 \times 0.125}{18} = x \text{ water molecule.}$$

So $$\frac{112 \times 0.100}{112} = y \text{ amt. of } CaC_2O_4$$

So $$\frac{100 \times 0.057}{56} = x \text{ amt. of } CaCO_3 \text{ also.}$$

43.3 Can you calculate the percentage composition of calcium oxalate if the thermogram of 0.1257 gm of the sample containing mixture of $CaC_2O_4\ H_2O$ (m.w. = 146.12) and stable salt had a mass loss due to water being 6.98 mg at 140°C?

Ans: The reaction proceeds as follows:

$CaC_2O_4\ H_2O \rightarrow CaCO_3(S) + H_2O$ from Δm, we can get original weight.

$$\frac{6.98 \times 1}{18.015} \times \frac{1 \text{ m mole } CaC_2O_4\ H_2O}{\text{m mole } H_2O} \times \frac{146.12 \text{ mg of } CaC_2O_4H_2O}{\text{m mole of } CaC_2O_4H_2O}$$

$$= 56.6 \text{ mg } CaC_2O_4$$

Now % $CaC_2O_4\ H_2O$ in the sample is

$$\frac{56.6 \text{ mg}}{125.70} \times 100 = 45.0\%$$

43.4 What should be the concentration of silver nitrate if 50 ml of it was titrated with 0.9758 M hydrochloric acid which needed 1.284 ml of acid for end point?

Ans: 50 ml of $AgNO_3$ X - moles of $AgNO_3$ = 1.284 × 0.9728 M of HCl

$$x = \frac{1.284 \times 0.9728}{50} = 0.0256 \text{ M of } AgNO_3$$

0.0256 M moles/Litre $x \times 20 = x = 20 \times 0.0256 = 0.05120$ M of $AgNO_3$.

43.5 What should be the concentration of ethylenediamine (en) if 25 ml of it was titrated with 1.046 M of hydrochloric acid with end point at 1.048 ml. The reaction is

$$NH_2,—CH_2,—CH_2,—NH_2 + HCl = (NH_2)—CH_2—CH_2(NH_3^+Cl)?$$

Ans: 25 ml *en* X ml. of HCl = 1.048 × 1.046 M.

So $$x = \frac{1.048 \times 1.046}{25} = 0.4385 \text{ M of ethylene diamine.}$$

43.6 Account for the following differences when the dissociation plot of dolomite substantially differs from that of magnesium carbonate?

Ans: Dolomite is supposed to be calcium salt of carbonate and magnesium carbonates, viz., $Ca[Mg(CO_3)_2]$ while $MgCO_3$ was independent which showed decomposition to oxide as $MgCO_3 \rightarrow MgO + CO_2$ but $Ca[Mg(CO_3)_2]$ dissociated differently hence there is variation of the thermogram of both salts.

43.7 Indicate the kind of transition involved in decomposition of hydromagnesites if thermal decomposition of $3MgCO_3$. $Mg(OH)3H_2O$ occurred when heated in atmosphere of CO_2?

Ans: Between 25-300°C, the 3 moles of water of hydration was liberated and 1 mole of water from $Mg(OH)_2$ leaving an intermediate product with 3:1 mole ratio of $MgCO_3$ to MgO. Then there was formation of 2:1 of carbon dioxide and ultimately the liberation of 2 moles of CO_2. The exotherm matches to crystal lattice transition of amorphous MgO to cubic form.

43.8 Indicate the type of transformation taking place for a microcosmic salt $[Na(NH_4)HPO_4]\ 4H_2O$ upon heating initially liberated $4H_2O$ then another H_2O and molecule of NH_3 at end to give $NaPO_3$ as the stable product?

Ans: The salt $Na(NH_4)\,HPO_4 \cdot 4H_2O \xrightarrow{\Delta} Na(NH_4)\,HPO_4 + 4H_2O$ first stage, then $Na(NH_4)\,HPO_4 \xrightarrow{\Delta} NH_3 + NaH_2PO_4 + H_2O$ second stage finally we have $NaH_2PO_4 \xrightarrow{\Delta} NaPO_3 + H_2O$ in final transition so we note end product $NaPO_3$ which was most stable in first transformation, $4H_2O$ molecule were lost, in second one molecule of NH_3 was liberated and finally we had one molecule of water along with thermally stable $NaPO_3$.

43.9 What was the kind of error present in following set of operations, e.g., Sample of complex compound was heated in platinum crucible till 500°C and thermal analysis curve was recorded?

Ans: For the sample formula or source was not specified. The kind of the apparatus used was also not stated whether it was TGA, DTA or DSC. The information on rate, kind of atmosphere and mass of sample was missing. These were glaring omissions in work.

43.10 Indicate the kind of following transformation which ordinarily cannot be detected by TGA such as (*a*) loss of moisture; (*b*) sublimation; (*c*) melting; (*d*) gas adsorption? Specify reactions which are detected by TGA.

Ans: Moisture loss sublimation and gas adsorption operations involve change of mass hence can be recorded but melting was not detected by TGA. The reactions are:

(*i*) $CaCO_3 + SO_2 = CaSO_4 + CO\uparrow$

(*ii*) $CaCO_3 + SiO_2 = CaSiO_3 + CO_2\uparrow$

(*iii*) $CaCO_3 + Na_2SO_4 \rightarrow CaSO_4 + Na_2CO_3$

(*iv*) $MgCO_3 + K_2SO_4 \rightarrow K_2CO_3 + MgSO_4$.

First reaction will be detected with increase in mass with $CaSO_4$ second reaction would be also detected as CO_2 would loose in reaction but third and fourth reactions would not be detected as they involve double decomposition with no change in mass.

43.11 Magnesium carbonate was dissolved in oxalic acid and crystalline product was obtained when it was heated in air, 9.20 mg of it lost mass in two stages as 2.23 mg were lost up to 220°C and 4.49 mg lost at 500°C. What kind of transformation is expected?

Ans:

$$MgCO_3 + H_2C_2O_4 \rightarrow MgC_2O_4 \cdot H_2O + CO_2\uparrow$$

$$MgC_2O_4 \cdot 2H_2O \rightarrow MgC_2O_4 + 2H_2O \text{ (Loss of 2.23 mg at 220°C)}$$

$$MgC_2O_4 \rightarrow MgO + 2CO_2 \text{ (loss of 4.49 Mg at 500°C)}$$

43.12 Can you identify change which was not detected by DTA as the loss of moisture or sublimation, or desorption of vapour or polymer surface or softening of polymer?

Ans: Loss of moisture, sublimation and desorption were endothermic reactions and hence were detected by DTA however polymer softening could not be detected in absences of dynamic mechanical analyser (DMA).

43.13 State whether it is possible to find out the difference between DTA or DSC peak without using any instrument for structural inversion or loss of mass due to the chemical reaction?

Ans: The peaks due to mass loss occur only on heating while inversion appears on inverted cooling curves. If the cooling was not possible then a second heating show inversion with no peak.

43.14 Name the specific thermal method you would prefer to use if you are asked to characterise a small sample of material?

Ans: The small sample could be easily analysed by modern methods like IR or electron microscopy could be tried for non-destructive analysis. DSC should give information on polymorphism, melting purity and thermal stability. DSC-FT-IR would furnish most of the useful information on characterisation.

43.15 A mineral feldspar contains impurities which on heating gave Al_2O_3, SiO_2. The DTA curve gave three peaks as endothermic peaks at 600°C and 750°C and exothermic peak at 99°C. If first and third peak was due to Al_2O_3, SiO_2 and $2H_2O$ indicate kind of impurities which were present?

Ans: A mineral of ideal composition was difficult to procure. Mixtures of kaolin (Al_2O_3 $2SiO_2 \cdot$ $2H_2O$) and feldspar (K_2O Al_2O_3 SiO_2 x H_2O) were made by mixing in various proportions of kaolin and feldspar and their DTA and TGA were recorded. For there mass loss at 600°C a graph of loss in weight against kaolin concentration was plotted. The % of the suspected material then was read from the graph. The height or area of the peak at 600°C may be also correlated to %K, *i.e.*, feldspar.

PROBLEMS FOR PRACTICE

43.16 What should be the composition of magnesium sulphate heptahydrate if thermogram was obtained after heating $MgSO_4 \cdot 7H_2O$. The sample weighted 0.0028 gm and rate of heating was 50°K/mt with air as purging gas. At temp 378°K single step in graph was noted with the formation of $MgSO_4 \cdot H_2O \cdot$ with mass of 0.59 mg?

[**Ans:** 47%]

43.17 Calculate the percentage composition of the sample if thermogram of $Al(OH)(HCOO)_2O$. $5H_2O$ showed stepwise loss of temperature 200°C, 260°C, 350°C in air corresponding to the decomposition to Al_2O_3, with 0.025 gm sample had total weight loss of 22.5%.
[Ans: 78.4%]

43.18 Draw important inferences from the following observation, viz., Differential thermal analysis contains an endothermic bond, which was not changed even by increasing partial pressure of oxygen in furnace, however it was shifted to higher temperature in a CO_2 from atmosphere?
[Ans: Sample is a carbonate]

43.19 Identify the original substance present in the sample weighing 0.067 gm if an unstable compound suffered a weight loss at 70°C and further loss of 0.014 gm of liquid at 100°C with no residue? **[Ans:** $(NH_4)_2$ CO_3 originally present]

43.20 Calculate the % composition of the mixture if gypsum dolomite and calcite together when heated showed change at 180-200°C due to the dehydration, at 720-740°C due to dissociation of the dolomite and finally at 880°C-920°C dissociation of calcite. The loss in weight was 5.2% (at 180-200°C), 3-5%, (at 720-740°C) and 16.3% (at 880-920°C)?
[Ans: $CaSO_4$ $2H_2O$ = 24.85%, $CaCO_3MgCO_3$ = 11.5%, $CaCO_3$ = 37.15%]

43.21 Find out the composition of sample containing SiO_2 if 0.1 gram of chloride of lime had weight loss of 0.012 gram at 100°C due to loss of water molecule, and weight loss of 0.0163 gm was due to decomposition of chloride of lime with liberation of oxygen at 300°C and loss of 0.035 mg for $CaCO_3$ formation of 500°C?
[Ans: $Ca(ClO_4)_2$ = 17.4%; $CaOCl_2$ = 12.9%; $Ca(OH)_2$ = 20%; $CaCO_3$ = 7.9%]

43.22 Calculate the composition if a $MgNH_4PO_4$ which was heated with activated carbon to eliminate ammonia and calcium hydride. The product was reacted with water to give hydrogen gas. The volume of hydrogen gas generated was 12.45 ml at 80°C due to loss of water, 102 ml at 105°C due to elimination of water of crystallisation and 8.5 ml at 220°C due to removal of chemically bound water?
[Ans: H_2O = 5% water of crystal = 40.95%, chemically bound 3.46%]

43.23 Calculate % potassium iodate in the sample if an oxide of potassium was formed by thermal decomposition of salt? If 0.254 gm of sample was heated at 70°C for one hour the weight loss was 0.42 gram? **[Ans:** 91.56%]

43.24 What is the amount of KO_3 in the sample if in TGA analysis of potassium compound showed following transformation:

$$KO_3\,(s) \rightarrow KO_2\,(s) + 1/2O_2(g)$$

If 0.2546 gm of sample when heated at 70°C for 60 mts weighted 0.213?
[Ans: 91.56%]

43.25 Can you calculate the amount of copper if copper solution was thermometrically titrated with 100 ml of solution containing excess of KI against 1.03 M sodium thiosulphate as the titrant. To 10 ml solution of potassium iodide was added and liberated iodine was titrated in 0.840 ml with hypo giving final reading of 0.280 ml/mt? **[Ans:** 2.42×10^{-2} Mcu]

43.26 Can you calculate the % weight of nitrogen in the zeolite, if DTA method was used to analyse the nitrogen in synthetic zeolites DTA was run under oxygen atmosphere and in exothermic region of 460°C, was noted from 0-4% by wt. nitrogen with the linear increase in height was from 0-88 mm. An unknown sample showed height of 27 mm? **[Ans:** 1.23%]

43.27 Will it be possible to estimate Na_2O in the sample of ore if in analysis of feldspar (0.4150 g) a mixture of (KCl + NaCl) was obtained with weight 0.0715 gm. From these chlorides 0.1548 gm of K_2PtCl_6 was obtained? **[Ans:** 3.07% Na_2O]

43.28 Evaluate the magnitude of '*n*' in the following process when the palladium complex was decomposed at temperature of 231°C to give polymer containing an average of $(n + 2)$ palladium atom. For each molecule of the polymer. TGA loss was 19% while nitrogen was purge gas? **[Ans:** 4.6]

43.29 What is the concentration of $AgNO_3$ if 50 ml solution of silver nitrate was thermometrically titrated with 0.9758 N hydrochloric acid with the end point of the titration was 1.284 ml? **[Ans:** 0.02506 M]

43.30 What should be amount of pyridine if one ml of 1M hydrochloric acid was added to 99 ml of pyridine? The temperature of the reaction vessel was raised by 4.55×10^{-3} K. Specific heat of the solution was 1.0 cals/g K and density of solution was 1.0089/ml? (absorbed heat = –4.97 kcals/mole). **[Ans:** 9.33×10^{-4} M]

REFERENCES

43.1 S.M. Khopkar, *Basic Concepts of Analytical Chemistry*, 3rd edn., New Age International Publishers (P) Ltd. (2007).

43.2 R.D. Braun, *Introduction to Instrumental Analysis,* McGraw-Hill and Co., pp. 37; 946-47 (1987).

43.3 H.H. Willard, L.L. Merritt and J.A. Dean, *Instrumental Methods of Analysis,* D. Van Nostrand and Co., pp. 473-76 (1978).

43.4 G.W. Ewing, *Instrumental Methods of Chemical Analysis,* McGraw-Hill and Co., p. 443 (1985).

43.5 L.F. Hamilton, S.G. Simpson and D.W. Ellis, *Calculation in Analytical Chemistry,* McGraw-Hill and Co., 7th edn., pp. 380-82; 129 (1978).

43.6 C. Duval, *Inorganic Thermiogravimetric Analysis,* 2nd edn., Elsevier Publishers (1963).

43.7 W.W. Wendlandt, *Thermal Methods of Analysis,* Interscience, New York (1964).

43.8 W.B. Guenther, *Quantitative Measurement and Equilibrium,* Addison Weslay (1990).

43.9 A. Musekin, *Problems in Quantitative Analysis,* Mir Publishing House (1984).

43.10 P.J. Haines, *Thermal Methods of Analysis: Principles, Applications and Problems,* Blackie Academic (1995).

Chapter 44

Radio Analytical Methods

INTRODUCTION

The radioactive disintegration consists of emission of α, β, γ or X-rays from the irradiated sample. When an element emitts α (*i.e.*, He_2^4) rays we have $M_Z^A \rightarrow He_2^4 + M_{Z-2}^{A-4}$ daughter with mass number less by 4-units and atomic number less by two units. However, when an element emitts β-radiation *i.e.*, $\left(e_{-1}^0\right)$ electrons, atomic number increases by one unit with no change in mass number while if there is emission of γ-rays, there is no change in either the mass number or the atomic number. Neutrons are used as projectiles to bombard and form unstable nucleus which in turn give out γ-radiation thereby attaining stability. e.g.,

$$M_Z^A + n_0^1 = \left[M_Z^{A+1}\right]; \; \left[M_Z^{A+1}\right] \rightarrow \gamma + M_Z^{A+1} \text{(stable form)}.$$

In radiochemical reaction activity is given by an expression $A = -dN/dt = \lambda N$ if N = number of parent nuclei at time 't' and λ is first order rate constant. It is called as nuclear decay constant. From this, we get $N = N_0 e^{-\lambda t}$, if N_o = is number of nuclei at time 't_o'.

So half life ($t^{1/2}$) of radiochemical reaction is given by an expression $t^{1/2} = \left(\dfrac{0.693}{\lambda}\right)$, it is constant for a particular radionuclei. In problems dealing with counting data we use standard deviation (σ*N*) it is given by equation $\sigma N = \sqrt{N}$ but relative standard deviation is given by an expression as,

$\left(\sigma N\right)\text{rel} = \dfrac{\sigma N}{N} = \dfrac{\sqrt{N}}{N} = \dfrac{1}{\sqrt{N}}$. If we use R = counts per minute then, we have $R = N/t$ and

$\sigma_{\text{rel}} = \dfrac{\sqrt{N}}{t} = \sqrt{\dfrac{R}{t}}$ is relative terms $\left(\sigma_R\right)\text{rel} = \sqrt{\dfrac{1}{R_t}}$ the r-ray spectrometer is used for measuring r-radiation.

Most popular techniques used in the quantitative analysis are neutron activation analysis and isotope dilution method. In the activation analysis an analytical as well as standard samples are subjected to irradiation by slow neutrons to give unstable nuclei (because slow neutrons are more effective than fast neutrons during transmutations. The newly formed nuclei has mass number increased by one unit with no change in its atomic number so rate of production is $R = N\,\phi\,\sigma$, if ϕ = slow neutron flux $cm^{-2}\,sec^{-1}$ and σ = neutron capture cross-section (cm^2).

Now an expression for newly formed nuclei with activity N^* is given by an expression as:

$$N^* = \left(\frac{Rt^{1/2}}{0.693}\right)\left(1 - \exp\frac{-0.693t}{t^{1/2}}\right)$$

If $t^{1/2}$ = half life period of nuclei we can also use an expression for activity A as $A = N\phi\sigma\, fs$ (s = saturation factor, f = fractional abundance of nuclide) so, we write as

$$A = N\sigma\phi - \exp\left(\frac{-0.693t}{t_{1/2}}\right) \text{ since } s = \left(1 - e^{-\lambda t}\right) = \left(1 - 10^{\lambda t/2.303}\right).$$

In terms of weight of element is given expression as

$$W_1 = \frac{A_1 M}{6.02\times 10^{23}\,\phi\sigma sf};\ W_2 = \frac{A_2 M}{6.02\times 10^{23}\,\phi\sigma sf}$$

W_2 = wt. of unknown, A_2 = activity of unknown
W_1 = wt. of compd, A_1 = its activity, If Avogadro's number (N) is 6.02×10^{23}
Their proportionality ratio is

$$\frac{W_1}{W_2} = \frac{A_1}{A_2}\left(\text{when all other factors, viz., } \phi, \sigma, s, f \text{ cancel}\right) \text{ so we get}$$

$$W_2 = \left(\frac{A_2}{A_1}\right)W_1,$$ if W_2 = is weight of unknown sample with its activity as A_2 while W_1 is weight of same element with the activity of A_1. This method is useful for the non-destructive analysis of the sample. The only requirement is need of stable isotope with reasonable half life ($t^{1/2}$) period.

In the isotopic dilution analysis; a sample is taken which is (W) not active and to it is added a known weight of (W_1) of the active sample with activity (A_1) of same element of compound. The mixture is chemically inseparable. Then small fraction of a compound is taken from the mixture and its specific activity (A) is then measured. The extent of dilution of the radiotracer shows the amount (W) of inactive element present. It is given by equation $W = W_1\left(\frac{A}{A_1} - 1\right)$ where

W = wt. of inactive material $\quad$ W_1 = wt. of active material in mixture.
A = activity of diluted sample, A_1 = original activity of known compound, we can evaluate (W) if we know W_1, A and A_1 in the analysis. We give $A_1 = \frac{A_0 W_1}{(W_0 + W_x)}$ if W_0 = wt. of radioisotope with activity as A_o is mixed with W_x = wt. of sample of inactive material if W_1 = wt. of species with activity as A_1 we have equation which is rewritten as $W_x = \frac{A_0}{A_1}(W_1 - W_0)$. This simplifies to $W_x = \left(\frac{A_0}{A_1}\right)W_1$. These are alternative equations.

In substochiometry isotope dilution analysis, we use an expression as $A_r^1 = \frac{A_0 W_r^1}{W_0}$, *i.e.*, which finally leads to $W_x = W_0\left(\frac{A^1 r}{A_r} - 1\right)$ is weight of the unknown sample.

In radiometric titrations, a radioactive substance is used in a chemical reaction with non-radioactive substance for analysis. In titration, titrant is radioactive against non-radioactive substance in flask. An end point is traced from titration curve of plot of radioactivity against volume of the titrant. The result are highly sensitivity. The method is most effective in precipitation titration.

As regards applications of neutran activation analysis, it is used for analysis of several elements like Re, Ir, Sm, As, Sb. However elements like Ca Pb Bi Na cannot be analysed with satisfaction in absence of the suitable isotopes. The isotope dilution method is used for the analysis of mercury or to find out chloride contents of biological sample with Cl^{36} with specific activity of 185 cpm.

The Instrumental neutron activation analysis (INAA) is simple useful at the level of 10^{-10} g/g for elements like Dy Eu In and Mn. INAA has better detection limit. This is direct method while in RNAA, *i.e.*, radiochemical neutron activation analysis, where the sample is chemically manipulated after bombardment but before counting technique. In INAA sample and standard are bombarded with same projectile for same period of interval. This is non-destructive analysis. Therefore in this chapter large number of problems both solved and unsolved are provided on analysis by neutron activation analysis (NAA) or isotopic dilution method (IDA). In addition some problems on calculations of actvity, standard deviation, counting error, etc., are also provided. For more details the reader should refer to classical references cited at the end of this chapter.

SOLVED PROBLEMS

44.1 A 5 mg containing (2 × 10%) chromium was irradiated for 360 mts with a neutron flux of 2×10^{19} n/cm/sec the half life of chromium was 667.2 hr and cross-section of thermal neutrons absorption was 17 barns for Cr^{50}. What activity was produced in iron alloy sample containing microgram concentrations of chromium?

Ans: If $A = N\phi\sigma f s, N^* = \left(\frac{Rt^{1/2}}{0.693}\right)\left(1 - \exp\frac{-0.693t}{t^{1/2}}\right)$

If N = nos. of nuclie, ϕ = slow neutron flux; N^* = nos. of radioactive atoms, R = rate of production, σ = slow neutron cross-section, λ = decay constant, A = Activity, $t^{1/2}$ = half life period.

Further $A = N\sigma\phi\left(1 - \exp\frac{-0693t}{t^{1/2}}\right)$, we substitute the values as $\phi = 2 \times 10^{19}$, $\sigma = 17 \times 10^{-24}$, $t^{1/2}$ = 27.8 *ds* = 667.2 hr. t = 60 hr, A = Avogadro's number 6.023×10^{23},

We have value on substitution is

$$\frac{(5\times10^{-3})(2\times10^{-4})(6.023\times10^{-23})x(17\times10^{-24})(1-e^{-0.693})^{60/66.7}}{52}$$

$A = 2.28 \times 10^3$ disintegrations per second.

44.2 A technique of isotope dilution was used for the analysis of mercury in the sample of catalyst. To a 1.6 gm sample of catalyst was added. One gm of the mixture containing 1% Hg^{203} with specific activity of 2400 cpm/g. Then 0.0100 gm of mercury was separated and indicated activity of 30 cpm. What is the % of mercury in the catalyst?

Ans: We use an expression

$$W_x = W^1\left(\frac{A^1}{A}\right).$$

If W_x = wt. of unknown compd., W^1 = wt. of compd. with radio active activity as 'A' A = specific activity, W^1 = 0.0100 gm, A^1 = 2400 cpm/g, A = 30 cpm/0.1 = 300 cpm/g, w = wt. of unknown mercury on substituting the values in above equation, we have W_x

$$W_x = 0.0100\left(\frac{2400}{300}-1\right) = 0.01(8-1) = 0.01\times 7 = 0.0700$$

so, 100 gram of sample contained 0.7% of mercury.

44.3 Calculate the energy of γ-rays which were formed during position annihilation? [as $e^- + e^+ \rightarrow 2r$ (0.5 mev)]

Ans: Mass of electron = 5.486×10^{-4} amu in terms of kg the mass is

$$(2e^- \times 5.486 \times 10^{-4}) \times (1.666 \times 10^{-27}) = 1.822 \times 10^{-30} \text{ kg}$$

$$E = mc^2 \text{ so } E = (1.822 \times 10^{-30})\,(2.9979 \times 10^8)^2 = 1.638 \times 10^{-3}\text{ J}$$

Velocity of light (c) = 2.9979×10^8 ms^{-1}.

Mass of e$^-$ = 1.822×10^{-30} kg.

$$\frac{1.638\times 10^{-3}\times 1}{1.60219\times 10^{-19}} = 1.022\times 10^6 \text{ mev.}$$

This is for two photons

energy for single photon $\dfrac{1.022\times 10^6}{2} = 0.511$ mev.

44.4 Find out counting rate as well as standard deviation of a number of counts if in a particular analysis, 225 counts were recorded during 75 seconds?

Ans: Since $\sigma = n^{1/2} = \sqrt{n}$ we have value of standard deviation

$\alpha = \sqrt{225} = 15$ counts was standard deviation.

Now, $R = n/t$ = 225/75 = 3 counts per second of counting rate was

$\alpha_R = (R/t)^{1/2} = (3/75)^{1/2}$ = 0.20 counts per second.

44.5 Find out the activity of the sample having $(1 \times 10^{-3}$g) of Na^{23} when irradiated for one hour with neutron flux from Apasra reactor was 1×10^{11} n/cm^2/sec. the same on measurement showed 3250 count/second as $\gamma Na_{11}^{23} + n_0^1 \rightarrow Na_{11}^{24} + \gamma Na_{11}^{23} \rightarrow Mg_{12}^{24} + \rightarrow \beta(1.4 \text{ Mev})\,(t^{1/2} = 15\text{hr for } Na_{11}^{24}$, with 100% abundance and cross-section 0.54)?

Ans: The activity was calculated as

$A = \frac{W}{M}\phi\sigma$ of $(1-e^{-1})$ in this problem, we have value for $W = 1 \times 10^{-3}$g; $M = 22.99$ for sodium,

ϕ = fractional abundance with cross-section, $\sigma = 1 \times 0.59$; $t = 15$ hr, $f = 1 \times 10^{11}$,

$$A = \left(\frac{1\times10^{-3}}{22.99}\right)(1\times0.54)(1\times10^{11})(1-e^{-0.693/15}),$$

$$= \frac{3250}{11,700} = 0.278, \text{ because}$$

A = activity, W = wt. of sample, M = mol. wt. of sample, t = time of irradiation, f = neutron flux, σ = cross-section for activation and ϕ = fractional abundance of isotopes so $A = 0.278$.

44.6 Find out the concentration of the sodium if 1.6502 gm sample of sodium was to be analysed by activation analysis, with 5200 counts/sec. was activity after irradiation. Given (cross-section σ) = 0.54 fractional abundance ϕ = 100%, $t^{1/2}$ = 15 hr for sodium)?

Ans: 52,000 counts/second = 52.00 × 0.278 dps = 187 dps wt. of the sample was therefore,

$$W = \frac{AM}{\phi\sigma f(1-e^{-xt})} = \frac{(187)(22.99)}{(1.0\times0.54)(1\times10^{11})(1-e^{-0.003(1)(15)})}$$

$$= 1.59\times10^{-5}\text{ g}$$

amount of sodium 1.650 gm sample was

$$\frac{(1.59\times10^{-5})\times100}{1.650} = 9.6\times10^{-4}\text{ \% of sodium.}$$

44.7 Estimate the likely error (%) if 8100 counts were timed in terms of standard deviation and relative standard deviation (%)?

Ans: The standard deviation $(\sigma) = \sqrt{8100} = 90$ counts

Then the relative standard deviation (%) was

$$100\sigma = \frac{100}{\sqrt{N}},\ i.e.,\ 100\sigma = \frac{100}{8100}\times90 = \frac{100}{\sqrt{8100}} = 1.11\%\text{ error.}$$

44.8 Find out the number of counts which must be measured, so that deviation is 95% of determination (2σ) will not exceed 2%?

Ans: $\frac{2\sqrt{N}}{N} = 0.02,\ \sqrt{100} = \sqrt{N}$

so, $N = 10,000$ counts.

44.9 Find the activity of sample of 1×10^{-3}g of an alloy containing Mn^{56} being 4.1×10^{-2} % was after 1800 seconds irradiated with flux from CIRUS reactor of 5×10^{11} n/cm²/sec.? [if ϕ = 13.4 = 100%, $f = 5 \times 10^{11}$, $t = 2.58$ hr].

Ans: We use fundamental equation of neutron activation analysis.

$A = N\phi\,\sigma f s$ as $s = (1 - e^{-\lambda t}) = (1 - 10^{-\lambda t/2.3})$.

Therefore, $W = (A \times M)/6.02 \times 10^{23} \times \phi \times f \times s \times \sigma$.

$A = N\phi\sigma fs = N\phi\sigma f\,(1 - e^{-\lambda t}) = N\phi\sigma f(1 - 10^{-\lambda t/2.3})$ or

$$A = \frac{M \times W}{(6.02 \times 10^{23})\phi\sigma fs}$$ we substitute these values in equation as

$$A = \frac{(0.0041)(0.00100)(1.0)(6.02 \times 10^{23})(5 \times 10^{11})(13.4 \times 10^{-24})s}{54.94}$$

Where $s = (1 - e^{-\lambda t}) = (1 - 10^{-\lambda t/2.3}) = (1 - e^{0.693 \times 30/258 \times 66}) = (1 - 0.878) = 0.122.$

So $A = 3680$ dps.

Now after 2 hr discharge at A (2.3 mt) and most of other short-lived isotopes will be decayed to insignificant activity.

So $A = 3680^{-\lambda t} = 2320$ dps.

44.10 Find out at what activity gamma radiation (from Na_{11}^{24}) will not interfere if measured in analysis of Fe in alloy A decay period of one week prior to chemical processing permitted the decay of 15hr Na^{24} by following the reaction:

$$Al_{13}^{27} + n_0^1 \rightarrow Na_{11}^{24} + He_2^4$$

Ans: $\frac{A}{A_0} = e^{-\lambda t}$. Now $\lambda = 0.693$, $t = 15$ hr.

$$A = A_0^{e-\frac{(0.693 \times \phi 68x)}{15.0}} = 0.00235 A_0$$

Thus after an interval of 7 *ds* the Na_{11}^{24} activity will be reduced to 0.00235 of its original activity (A_o). Thus γ-radiation from Na_{11}^{24} at 1.368 and 2.754 Mev would no longer show an interference in the quantitative analysis.

PROBLEMS FOR PRACTICE

44.11 Calculate the counting rate and standard deviation of counting of 15 minutes, 1148 count were measured? **[Ans:** 76.5, 2.26 mintues]

44.12 Find the activity of Y^{90} formed from 0.05 mg of Y^{89} during Y^{89} (n, γ) Y^{90} reaction after an irradiation for 72 hr the cross-section for the reaction was 1.31 barns, the half life of Y^{90} was 64.3 hr with 2×10^7 n/cm/s as neutron flux.? **[Ans:** 4.8 sec^{-1}]

44.13 Can you find out the concentration of chloride in the original sample weighing 3.2051 gm was irradiated for a period of 60 minutes with neutron flux of 1×10^{12} n/cm/sa activity. Other portion of sample (1.5029 gm) containing chloride was added 1 μg of chloride was treated similarly and showed activity of 3569 count? **[Ans:** 0.248 ppm]

44.14 Calculate the half life period of copper (Cu^{64}) if the activity of Cu^{64} sample was 3250 cps if after 10 hr of the count rate for same sample was 2230 dps.? **[Ans:** 18.4 hr]

44.15 Estimate time of cooling a sample prior to chemical reaction or counting if 0.010 gm sample of aluminum sheet was irradiated for 1800 seconds in neutron flux of 5×10^{11} n/cm/s. (It is presumed that the strontium aluminium activity had subsided to less than 1 count/mt.)? **[Ans:** 69 minute]

44.16 Find the chloride concentration in the sample of the plant if an isotope dilution method was used to determine in plants with Cl^{36}. If to 0.890 gm sample of plant, 1 ml of 0.001 M of $NaCl^{36}$ with the activity of 185 cpm was added and if 0.1 gm of hydrochloride was separated and counted to 400 cpm? **[Ans:** 0.561 %]

44.17 Ascertain the specific activity of the phosphorus in the sample if 0.55 gm of tissue contains P^{31} 6.2% irradiated with neutrons with flux of 6×10^{10} n/c/s for 1200 minute and the cross-section was 0.23 barns for P^{32} and half life of 340.8 hr?

[Ans: 6.65×10^6 C/hr/g, 3.04×10^6 C/h/gm.]

44.18 Ascertain the concentration of selenium from unknown sample if a neutron activation analysis was carried out with standard sample with 0.75 mg/g of selenium and unknown sample were bombarded with slow neutrons for 600 hrs with γ-radiation 0.05–0.2 mev. the data obtained was

E mev:	0.04	0.06	0.08	0.10	0.12	0.16	0.20
dpm standard:	400	400	500	500	1600	800	400
dpm unknown:	600	600	900	1700	2800	1200	600

The background intensity was 400 and for an unknown sample intensity was 600 dpm.

[Ans: 1.57 μg/g of se]

44.19 Find the concentration of silver in the sample if in radiometric titration of 25 ml solution with 1.49×10^{-3} M of Cl^{36} was used and then counted after silver chloride was precipitated. The data was:

Vol. of Cl^{36}:	0	1	2	4	6	8	10	11
Counts/mt:	105	115	127	148	180	340	620	780

[Ans: $4.26 + 10^{-4}$ M]

44.20 Ascertain the concentration of cobalt in sample of an alloy weighing one gram which was dissolved in acid. Then 5 ml of an aliquat counting Co^{60} was added to solution (250 ml); after elimination of iron (III) by solvent extraction, cobalt was anodically deposited at Pt electrode as Co_2O_3 which on measuring its activity showed to be 847 c/mt. With one gram of alloy process was repeated 0.0105% Co^{60} in second sample, the activity of deposited Co from it was 758 c/mt? **[Ans:** 0.0117%]

44.21 Find optimum weight of the sample to be taken if activity was 1500 cpm. When an alloy containing 5% copper was irradiated in a neutron flux of 10^6 n cm^2s^{-1} for 10 hr and cooled for 5 hr? ($t^{1/2}$ of Cu^{64} = 12.7 hr, σ = 4.5 b, *r* = 69.1% and counting efficiency = 10%)

[Ans: 0.0269, Cu = 0.525 g sample]

44.22 Find requisite weight of the sample after cooling for five hours if 0.1 gm manganese powder was irradiated in a neutron flux of 10^6 n cm^{-1} s^{-1} for 90 mt? (Assume *v* = 100%, σ = 13.3 b, $t^{1.2}$ of Mn^{50} = 2.58 hr) **[Ans:** 75654 dpm]

44.23 Ascertain the amount of germanium in the unknow sample, if it was irradiated in a neutron flux of 10^{12} ncm^{-2} s^{-1} for one hour when ^{76}Ge was formed, ^{77}Ge of a half life of 1 minute. Supporting the activity measured 1 mt after 1 hr irradiation was 2500 dps? (Assume cross-section of reaction was 3.28 mb and isotopic abundance of ^{76}Ge was 7.8%) **[Ans:** 2.47 mg of Ge]

44.24 What should be the amount of chromium in a ruby which on weighting 0.59 was irradiated in a neutron flux of 10^{12} n cm^{-2} s^{-1} for exactly 24 hr and Cr^{51} activity (*J* = 277 d) and was counted immediately thereafter. It was found to give 35,000 c/s. (Given for Cr^{50} = 15.9 b the counting efficiency = 10% and count of natural chromium Cr^{50} to be 4.35%) **[Ans:** 0.34% Cr]

44.25 Find the amount of the antibiotic in mixture if a 2 liter mixture was assesed for its contents 5 mg of the labelled antibiotic with an activity of 20,000 counts per five mt was added to this mixture. After thorough mixing 0.6 mg of pure crystalline antibiotic was isolated which gave an activity of 3000 count per mt? **[Ans:** 0.15% penicillin in mixture]

44.26 Ascertain the concentration of an iodide in the mixture if one liter of iodide content 2 ml labelled iodine having an activity of 5000 counts per 2 mt were added to the mixture. After through mixing 3 ml or per iodide were separated which was found to give an activity of 6000 counts for 10mt? **[Ans:** 1.066% Iodide]

44.27 Ascertain the quantity of aluminium in the alloy if a 1.40 gm sample of Devadras alloy containing 1.24% Al was irradiated with neutrons with flux (ϕ) of 1.0×10^9 n cm^{2-} sec^{-1}, if active Al^{28} has $t^{1/2}$ 2.27 and after half an hour irradiation? (Given r = 0.21 Barns) **[Ans:** 0.85 gm]

44.28 Find the amount of aluminium if it was present after 30 mt. irradiation when 1.4 gm sample of rock containing 1.24% Al was placed in container and irradiated with neutrons with flux of $1 \times 10^{8-}$ n/cm^2/sec. The Al^{27} was converted to Al^{28} with $t^{1/2}$ = 2.27 mt (assume cross-section of neutrons by Al^{27} was 0.21 barns)? **[Ans:** 6.88×10^{-15}%]

44.29 Ascertain the amount of phosphorus taken by the soil if about 5 gm of soil was crushed and mixed with 500 ml of 0.02 M Na_3PO_4 to this mixture 10 ml Na_3PO_4 soln (2×10^{-2} M) containing P^{32} had an activity of 7450 cpm/ml. after equilibration sample was filtered and one ml of solution was counted yielding a rate of 35 cpm? **[Ans:** 0.80 3 gm]

44.30 Find out the amount of zirconium in the steel if after 30 days activity was 19 cpm in nickel base high alloy steel was done by the activation analysis Zr^{95} ($t^{1/2}$ = 65 ds) was obtained by NAA with a flux of 8×10^{12} ncm^2/sec. The cross-section for Zr^{94} was 0.08 barns with isotopic abuandance of 17.4%. If a one gram sample was activated for 25 hr for steel? (If counter was 100% efficient for x-rays produced by Zr^{95})? **[Ans:** 0.70%]

REFERENCES

44.1 S.M. Khopkar, *Basic Concepts of Analytical Chemistry,* 3rd edn., New Age International Publishers (P) Ltd. (2007).

44.2 R.D. Braun, *Introduction to Instrumental Analysis,* McGraw-Hill Co., pp. 617, 637, 645-46 (1987).

44.3 D.J. Pietrzyke and C.W. Frank, *Quantitative Analytical Chemistry,* Alloy Bacon, pp. 638-40 (1979).

44.4 H.H. Willard, L.L. Merritt and J.A. Dean, *Instrumental Methods of Analysis,* D Van Nostrand and Co., pp. 257, 267-68, 278 (1980).

44.5 G. Friedlander, J.W. Kennedy and J.M. Miller, *Nuclear and Radio Chemistry*, 2nd edn., Wiley (1964).

44.6 W.S. Lyon, *Guide to Activation Analysis*, Van Nostrand and Co. (1964).

44.7 G.R. Choppin, *Experimental Nuclear Chemistry,* Prentice Hall, NY (1961).

44.8 H.J. Arnikar, *Essentials of Nuclear Chemistry*, 4th edn., New Age International Publishers (P) Ltd. (1995).

44.9 H.J. Arnikar and N. Rajurkar, *Nuclear Chemistry Through Problems and Solutions,* New Age International Publishers (P) Ltd., pp 135-137 (1997).

44.10 L.F. Hamilton, S.G. Simpson and D.W. Ellis, *Calculations in the Analytical Chemistry*, 7th edn., McGraw-Hill and Co. (1970).

44.11 S. Brewster, *Solving Problems in Analytical Chemistry,* John Wiley Co. (1980).

Chapter 45

Electroanalytical Methods —Potentiometry

INTRODUCTION

Electroanalytical methods are as effective as the spectroscopic methods for the quantitative analysis. In such studies efforts are made to measure either potential, current or resistance (reciprocal of resistance, *i.e.*, conductance) to correlate them with the concentration of the constituents. The amount of material deposited at electrode is largely governed by what is known as Faraday law of electrolysis'. The subject dealing with correlationship of current against amount of product generated at electrodes, is called coulometry or electrodeposition. In this chapter we shall deal with the measurement of potential and its correlation with concentration. This technique has numerous applications as in ion selective electrodes, in potentiometric titrations and in pH metric studies.

The electrochemical redox reaction is represented as $A_{red} + B_{ox} = A_{ox} + B_{red}$. The equilibrium constants is given by an expression

$$K = \frac{\left[A_{ox}\right]_{eq}\left[B_{red}\right]_{eq}}{\left[A_{red}\right]_{eq}\left[B_{ox}\right]_{eq}}$$ after considering their activity coefficient we have equation as

$$Q = \frac{\left[A_{ox}\right]_{act}\left[B_{red}\right]_{act}}{\left[A_{red}\right]_{act}\left[B_{ox}\right]_{act}}$$ if 'act' represent actual values, eq = equivalent concentration so thermodynamically.

The expression for the free energy ΔG is given as:

$$\Delta G = \left[\frac{RT}{nQ} - \frac{RT}{nK}\right]$$ if R = universal gas constant (8.136 J/mole/deg)

T = absolute temp, ΔG = free energy which is related to electrical parameter as

$\Delta G = -nFE_{cell}$ if F = Faraday constant (96,500 columbs); K = equilibrium constant.

E = cell potential, n = number of electrons transferred

If G_0 is free energy in standard state we also have an equation as

$$\Delta G_0 = -nFE_0 \text{ if } E_0 = \text{standard potential for a typical reaction,}$$

So $\Delta G = \Delta G_0 + \frac{RT}{n}\frac{[B]_{red}}{[B]_{ox}}$. On substituting the value in this equation for

$$\Delta G_0 = -nFE \text{ we have } \Delta G = -nFE_0 + \frac{RT}{n}\left[[B]_{red} \div [B]_{ox}\right]$$

$E = E^0 - \frac{RT}{nF}\ln\frac{[B]_{red}}{[B]_{ox}}$. This is called 'Nernst equation', with natural logrithm it is

An $E = E^0 - \frac{2.303RT}{nF}\log\frac{[B]_{red}}{[B]_{ox}}$ (If $R = 8.13$, $T = (273 + 25) = 298°K$, $F = 96{,}500$, n = electrons transfered).

$E = E_0 = \frac{0.05917}{n}\log\frac{[B]_{red}}{[B]_{ox}}$ (Simplified Nernst equation) (At room temp. if $T = 300°K$)

The Nernst equation is most important equation as it facilitates the calculation of potential of the cell the E_0 represents standard electrode potential for hydrogen.

$$Zn^{2+} + 2e^- = Zn; \quad E^0_{Zn} = -0.763\,V$$

$$2H^+ + 2e^- = H_2(g); \quad E^0_H = 0\,V$$

$$Cu^{2+} + 2e^- = Cu; \quad E^0_{Cu} = +0.337$$

The signs are as per IUPAC (1953) convention; with reducing reagent have negative sign like zinc. The details of values for various metals are listed in Appendix 3 for standard electrode potential.

Now, the activities used are related to concentration (C) by equation $a = fC$, if f = dimensionless activity coefficient, C = concentration for solution with ionic strength (I) (less than 0.001 M) the activity coefficient is $-\log f = Az^2 I^{1/2} = \frac{Az^2 I^{1/2}}{J + 1.5I^{1/2}}$. If A = constant, Z = charge with t_c^0 in dilute solutions the Nernst equation is valid as following equation:

$$E = E^0 - \frac{RT}{nF}\ln\frac{[\text{red}]b}{[\text{ox}]a} \quad \text{(In transformation } a_{Ox} + e^- = b_{red} \text{ species in action)}$$

The ionic strength of a solution is dependent upon all ions in the solution so

$$I = \frac{1}{2}\Sigma C_1, Z^2, \text{ if } C_1 = \text{con. of each ion in solution and } Z \text{ is charge of ions.}$$

In calculations involving ion selective electrodes with potential E_m external reference electrode (E_r) net potential is $E = (E_m - E_r)$, *i.e.*,

Since $E_m = \frac{2.303RT}{nF}\log\frac{\alpha i}{\alpha j}$ if αi = activity of ion in sol., αj = activity internal ref. solution.

We have an expression for $E = (E_m - E_r)$

$$E = \left[\frac{2.303RT}{nF}\log a_i - \frac{2.303RT}{nF}\log a_i\right] - E_{\text{ref}}$$

$$= \text{constant} + \frac{2.303RT}{nF}\log a_i = \text{constant} + \frac{0.05917}{n}\log a_i$$

A change in activity of a univalent ion by a factor often should reflect change of 0.05917 V at 25°C for divalent ion it will be $\frac{0.05917}{n} = 0.02959\,\text{V}, (n = 2)$.

If voltage is expressed in terms of millivolts we have an equation

$$E = \text{Constant} + \frac{0.1984T}{n}\log a_i \quad (\text{If} = a_i \text{ is activity of ion in solution})$$

An error caused by electrode in anion analysis with known activity is obtained by following expression as:

$E = \text{Constant} + \frac{0.0591}{Z_A}\log\left[\alpha_A + K_B^A(\alpha_B)^{Z_A/Z_B}\right]$. In this equation K_B^A is selectivity coefficient A = ion and B is interfering ion n = charge, α = activity of ion.

In ion selective electrodes the developed potential is given by an expression

$E_{\text{cell}} = \text{constant} + \frac{0.0591}{n}\log M^{n+}$ if M^{n+} is ion being measured with 'n' valancy

Now in problems dealing with calculation of concentration with ISE we use following equation as

$$C_x = \frac{C_s V_s}{(V_x + V_s)}\left[10^{\Delta E/S} - \frac{V_x}{V_x + V_s}\right]^{-1}$$

Where C_x = concentration of unknown solution, V_x = volume of unknown solution, Vs = volume of standard solution added, C_s = concentration of standard solution

If ΔE change of potential $\left(S = \frac{RT}{nF} = \frac{-0.0591}{n}\right)$ one can also use simplified form of equation as

$$C_x = C_s\left(\frac{V_s}{V_x}\right)\left(10^{\Delta E/S} - 1\right)^{-1} = C_x\left[\frac{V_s}{V_x}\right]\left(10^{\Delta E/\frac{0.0591}{n}}\right).$$

Alternatively one can write equation for ISE for its potential as follows:

$E = \text{Constant} + 2.303\frac{RT}{nF}\log a$ if a = activity, n = change since activity is (a)

$a = F \int C$ if F = uncomplexed ion, a = activity coefficient, C = Concentration we have, then

$E = \text{Constant} + 2.303\,\frac{RT}{nF}\log(EfC)$. Now if C is standard solution of the analytical ion with concentrations C_s and volume V_s is added to have total volume of V_t then the total activity (αt) of ion is given as follows provided Ft = Fraction of uncomplexed ion

$$a_T = F_t f_r \left(\frac{V_t C_t + V_s C_s}{V_t + V_s} \right)$$, if f_s = activity coefficient so

$$E_T = \text{constant} + 2.303 \frac{RT}{aF} \log F_\tau F_R \left(\frac{V_t C_t + V_s C_s}{V_t + V_s} \right)$$ in solution, F_τ = activity coefficient so net ISE potential is

$$\Delta E_\tau = \text{constant} + 2.303 \frac{RT}{ZF} \log \left(F_\tau F_\tau \frac{V_t C_t + V_s C_s}{FF(V_t + V_s) C_E} \right)$$. If s = slope of graph of E_{vs} log activity ΔE is change of potential say 'ΔE is then and further (assume $E = E_\tau$ and of f_τ) we have an expression

$$\Delta E = 2.303 \frac{RT}{nF} \log \frac{(V_t C_t + V_s C_s)}{(V_t + V_s) C_t} = \frac{2.303 RT}{nf} \log \frac{V_t C_t - V_s C_s}{(V_t + V_s) C_t}$$

$$\Delta E = 2.303 \frac{RT}{nF} \log \frac{(V_t C_t + V_s C_s)}{(V_t + V_s) C_t}$$ this is formal equation used, for calculation for change.

If s = slope of graph of E against log (analytical activity).

$$\Delta E = 2.303 \frac{RT}{nF} \log \frac{(V_t C_t + V_s C_s)}{(V_t + V_s) C_t}$$ this is formal equation used, for calculation for change in potential.

In such problems the values of $\frac{C_t V_t}{C_s V_s}$ is given for corresponding values of R in Tables in Appendix 12 (e.g., 1–4, $C_t/C_s V_s$ = 0.325 If R = 1.5 = 0.618, = $C_t\ V_t/C_s\ V_s$ R 1.6 = 1.086 × 1.7 = 1.894 and 1.8 = 3.536) these values of R can be used in normal calculations with ($C_t\ V_t/C_s V_s$).

So far we have considered the Nernst equation the activity and the potential of ion selective electrode, evaluation of error in measurement in the calculation of concentration and unknown potential with ISE.

We would conclude this discussion with a brief account of various kinds of electrodes used for pH measurement. The most versatile is glass electrode also called as indicator electrode. It is filled with hydrochloric acid with mixture of (Ag + AgCl) with thin glass membrane. The expression with glass with electrode against SCE reference electrode for $\text{pH} = \frac{(E_{\text{cell}} - E_g)}{0.0591}$ if E_{ref} = potential of reference electrode. If glass electrode is filled with buffer of pH = 7.0 or pH of buffer of pH = 4.0 then at that pH we get an expression as $\text{pH} = \frac{E_{\text{cell}}}{0.0591} + 7$ at pH = 7.0 and $\text{pH} = \frac{E_{\text{cell}}}{0.0591} + 4$ at pH of 4.0. Thus for glass electrode the potential is given by an expression as $E = \text{constant} + \frac{2.303\,RT}{n,\ F} \log a_i$. If a_1 = activity of ion in solution provided the concentration of solution is fixed E = constant + 0.0591 log a_1, [for univalent ions as Na, K, H]. Now for the hydrogen ion sensitive glass electrode, E = constant –0.05 pH. As per modern theory, glass electrode works like ion exchange membrane. From this fundamental equation for ISE is

$$E = \text{constant} + \frac{2.303\,RT}{nAF}\log\left[aA + K_B^A(aB)Z_A/Z_B\right]$$

is universal equation. The liquid membranes electrodes are usually filled with internal solution with Ag/AgCl internal reference electrode with organic liquid ionexchanger acting as organophilic membrane instead of glass surface. In such liquid membrane electrode the potential is expressed as

$$E = \pm\frac{2.303RT}{nAF}\log\left[-a_A + \left(\frac{U_A U_B}{U_A U_B}\right)aB^-\right],$$

when $U_A U_B$ are mobiltes of ions and D_A, D_B are distribution rates of ions A and B. We also have solid state electrodes or precipitation electrode in which, an epoxybody solid is in contact with sensing element (e.g., for Fluoride LaF_3 is sensing element). The electrode potential is

$$E = \text{con} + \frac{2.3\,RT}{nF}\log a_M^{n+},\ \left(\text{if } a_M^{n+}\right) \text{ is activity of ion } M \text{ with valancy } 'n'.$$

Finally in enzyme electrode we have enzyme which shows catalytic reaction (e.g., urease electrode for urea). It has internal reference electrode, with ion selective membrane with enzyme matrix layer as enzyme substrate.

Finally let us consider the reference electrode. It is called as saturated calomel electrode (SCE). It consists of insulated wire with platinum and mercury with glass seal filled up with solid KCl + Hg_2Cl_2 as with porous plug and fiber tip. The potential of SCE against NHE is + 0.241 V with saturated KCl. It is + 0.286 V with 1M KCl and 0.336 V with 0.1 M KCl solution. The half cell is Hg (1) | Hg_2Cl_2 (Solid), KCl (aq) || with any other suitable electrode. The reaction proceeds $Hg_2Cl_2(s) + 2e^- \rightarrow 2Hg(e) + 2Cl^-$.

The SCE is very useful in potentiometric titration technique. In such methods we plot a graph of E_{cell} (volts) versus titrant volume. The inflection shows end point. The potentometric end point is more accurate however reaction is time consuming to attain an equilibrium. For precipitation titration indicator electrode is metal for which reacting cation is derived or platinum electrode. For $AgNO_3$ as titrant silver electrode is best. With SCE it is better to use agar bridge. It is useful for analysis of halides. In complex formation titration, both metal and membrane electrodes are useful, most of the EDTA titrations one which uses mercury as second order electrode. In acid base titrations we use glass and SCE electrode. Finally in redox titration platinum electrode is best. It is responsive to both oxidation and reduction reactions. The liquid membrane electrode differs from glass electrode where the solution of known and fixed activity is separated from analytical solution by thin layer of immovable organic liquid instead of thin glass membrane. In gas sensing electrode, thin replaceable gas permeable membrane is attached to one end of tube thus separating internal solution from analytical solution. With glass electrode if alkaline error ($pH > 9.0$) acid error ($pH < 0.5$), dehydration error is encountered) then one uses unbuffered neutral solution. Finally the variation in junction potential is commonly encountered as error.

In this chapter large number of solved and unsolved problems are given for calculation of EMF of half cell, concentration of cation, activity of ion, concentrations, calculations, with ISE measurement, error in analysis, measurement of pH, etc.

SOLVED PROBLEMS

45.1 What is the e.m.f. of the following cells $\left(\text{Given } E^0_{Zn} = -0.763\,\text{V}, E^0_{Cu} = +0.337\text{V}\right)$?

(*a*) Zn | Zn^{2+} (1×10^{-6} || Cu^{2+} (0.01 M) | Cu.

(*b*) Cu | Cu^{2+} (2M) || Cu^{2+} (0.01 M) | Cu.

Ans: (*a*) We use Nernst equation which states that

$E_1 = E^0_{Zn} + RT/nF \log\left(Zn^{2+}\right)$. If $RT/F = 0.0591$ when $n = 2$.

So, similarly for copper potential.

$$E_2 = E^0_{Cu} + \frac{RT}{nF}\log\left(Cu^{2+}\right),\ i.e.,\ E_2 = E^0_{Cu} + \frac{0.0591}{2}\log(0.010) = 0.277\,\text{V},\ n = 2$$

Hence, $E = (E_1 \sim E_2) = 0.277 + (-0.940) = 1.217$ Volts.

(*b*) For second part of the problem the cell of copper Since $E^0_{Cu} = +0.337$.

$$E_1 = E^0_{Cu} + \frac{RT}{nF}\log\left(Cu^{2+}\right) \text{ so } E_1 = 0.337 + \frac{0.0591}{2}\log(2) = 0.3458 \text{ V as } n = 2$$

$$E_2 = 0.337 + \frac{0.0591}{2}\log(0.01\text{M}) = 0.2779 \text{ V.}$$

$$E_1 = \left(E_1 \sim E_2\right) = (0.3458) - (0.2779) = 0.9679 \text{ volts.}$$

45.2 What was the half cell potential at end point in the titration of Fe (II) with $KMnO_4$ in H_2SO_4 at pH = 2.3? $\left(E^0Fe^{3+} = 0.77, E^0_{MnO_4^-} = -1.51\text{V}\right)$?

Ans: The possible redox reaction proceeds as follows:

$$5FeSO_4 + KMnO_4 + 4H_2SO_4 = 5Fe\left(SO_4\right)_3 + MnSO_4 + 4H_2O.$$

$$\text{i.e., } 5Fe^{2+} + MnO_4^- + 8H^+ \rightarrow 5Fe^{3+} + Mn^{2+} + 4H_2O$$

$$E_1 = E^0_{Fe} + \frac{RT}{nF}\log\left[\frac{Fe^{3+}}{Fe^{2+}}\right] \text{ or } 0.771 + \frac{0.0591}{1}\log\left[\frac{Fe^{3+}}{Fe^{2+}}\right], \text{ as } n = 1.$$

$$E_2 = E^0_{Fe} + \frac{RT}{nF}\log\left[\frac{MnO_4^-\left[H^+\right]^8}{Mn^{2+}}\right] = +1.51 + \frac{0.0591}{5}\log\left[\frac{MnO_4^-\left[H^+\right]^8}{Mn^{2+}}\right] \text{ as } n = 5.$$

Adding two equations

$$6E = \left(E_1 + 5E_2\right) = 0.771 + 0.0591 + \log\left[\frac{Fe^{3+}}{Fe^{2+}}\right]$$

$$+5(1.51) + 0.0591\log\left[\frac{Fe^{3+}}{Fe^{2+}}\right]\left[\frac{\left[MnO_4^-\right]\left[H^+\right]^8}{\left[Fe^{2+}\right]\left[Mn^{2+}\right]}\right]$$

Since $[Fe^{3+}] = 5\ Mn^{+2}$ and $[Fe^{2+}] = 5[MnO_4]$

$$E = \frac{0.771 + 5(1.51)}{6} + \frac{0.0591}{6}\log\left[H^+\right]$$ other terms cancel out as pH = 2.3 –log H^+.

E = 2.031 + 0.07883 (2.3)

E = 2.031 – (0.1812) = + 1.8498 Volts.

45.3 If 30 ml of 0.1 N $FeSO_4$ was diluted to 100 ml and titrated with 0.1 N ceric sulphate with calomel as reference electrode, what was the emf of the cell when 20 ml of $Ce(SO_4)_2$ solution was added?

(Given $E^0_{Fe} = +0.771$, $E^0_{SCE} = +0.286$).

Ans: The redox reaction proceeds follows

$Fe^{2+} + Ce^{4+} = Fe^{3+} + Ce^{3+}$

When 30 ml of solution was neutralised with 20 ml of solution, ratio of concentration was 2:1,

$$E_1 = E^0_{Fe} + \frac{RT}{nF}\log\frac{\left[Fe^{3+}\right]}{\left[Fe^{2+}\right]}, \ i.e., \ E_1 = 0.771 + \frac{0.0591}{1}\log\left[\frac{2}{1}\right] = 0.7887 V_n$$

E_2 = was SCE. So $E^0_{SCE} = +0.286$V

$E = (E_1 \sim E_2)$ = (+0.7887 – 0.286) = 0.5V

45.4 Find out the concentration of silver if electrolysed with combination of silver electrode and SCE? (Given $E^0_{SCE} = +0.246$ v; $E^0_{Ag} = 0.799$ if $E_{Cell} = 0.450$).

Ans: By Nernst equation

$$E_1 = E^0_{Ag} + \frac{RT}{nF}\log\left[Ag^+\right]; \ E_{SCE} = +\ 0.246 \text{ v}$$

$$E = \left(E_1 - E_2\right) = E^0_{Ag} + \frac{0.0591}{1}\log\left(Ag^+\right) - E_{SCE} \text{ as } n = 1$$

or $$\log\left[Ag^+\right] = \frac{E_{cell} - E^0_{Ag} + E^0_{SCE}}{0.0591} = \frac{0.450 - 0.799 + 0.246}{0.0591} = -1.74$$

$\log\left[Ag^+\right] = -1.74$, antilog of $\left[Ag^+\right] = 1.8 \times 10^{-2} M^-$.

45.5 Ascertain the activity of calcium ion in solution of strength 1.50 × 10^{-4} M calcium chloride and 5 × 10^{-4} M concentration of sodium chloride solution?

Ans: Ionic strength was calculated as $I = \frac{1}{2}\left|\Sigma C_1 Z_1^2\right|$ if I = ionic strength

$$I = \frac{1}{2}\left(1.50 \times 10^{-4}\right)(z)^2 + \left(3 \times 10^{-4}\right)(-1)^2 + \left(5 \times 10^{-4}\right)\left(+1^2\right) + \left(5 \times 10^{-4}\right)(-1)^2$$

= 9.50 × 10^{-4} M.

$-\log f = Az^2 I^{-1/2}$ = 0.51 $(2)^2$ $(9.5\ 10^{-4})^{1/2}$ = 6.31 × 10^{-2}

if f = 0.85, then afc = 0.865 (1.5 × 10^{-4}) = 1.3 × 10^{-4} M

45.6 Find out the redox potential of ferrisoferrocyanide in given cell.

$$\text{Pt}\left|\begin{matrix}0.01\\ Fe^{3+}\end{matrix}\right|\left|\begin{matrix}0.02\\ Fe^{2+}\end{matrix}\right|\text{SCE with } E_{\text{cell}} = 0.30\text{V}, E_{\text{SCE}} = 0.243?$$

Ans: Redox reaction proceeds as follows:

$$[Fe(CN)_6]^{3-} + e^- \rightarrow [Fe(CN)_6]^{4-}$$

$$E_{cell} = (E_1 - 0.240)\ (E_1 - E_{SEC})\ E_1 \text{ be cell potential to be determined}$$

$$E_1 = E^0 + \frac{RT}{nF}\log\left[\frac{Fe^{2+}}{Fe^{3+}}\right] \quad i.e., E_1 = \frac{0.0591}{1}\log\left(\frac{0.020}{0.010}\right) \text{ if } n = 1.$$

$$E_{\text{cell}} = 0.30 = (E^1) - \left(\frac{0.0591}{1}\log\frac{0.020}{0.010}\right) - 0.240$$

So, $E^1 = 0.30 + -240 + 0.059\ (0.30) = 0.56$ V.

45.7 Find out the potential of the instrument if the real potential between an indicator and reference electrode was 1.0V with internal resistance of cell as 100 MΩ = 108 with resistance of 10^9 ohms?

Ans: Since Ohm's law $C = \frac{E}{R} = \frac{10}{(10^8 + 10^9)} = 9 \times 10$ amps current.

The voltage difference was $eR_{cell} = (9 \times 10^{-10}) - (1 \times 10^8) = 0.09$ V.
Since voltage was opposite in polarity to the cell emf. Therefore net voltage of voltmeter was 0.91 V.

45.8 Find out the concentration of analytical solution when ion selective electrode was immersed into 50 ml of analyt solution to show electrode potential 113.5 mv. After 5 ml addition of 5×10^{-3} M solution of analytical ion the potential had changed to 132 mv. A second addition of 5 ml of the solution further changes it to 144.4 mv?

Ans: $\Delta E_1 = S\log\frac{C_v + C_sV_s}{C(V+V_s)}$ and $\Delta E_2 = S\log\frac{C_v + 2(C_sV_s)}{C(V+2V_s)}$

$$R = \frac{\Delta E_2}{\Delta E_1} = \frac{\log[CV + 2C_sV_s]/C[V+2V_s]}{\log[CV + C_sV_s]/C[V+V_s]}$$

Now in a given problem $\Delta E_1 = (132 - 113.5) = 18.5$ mV; $\Delta E_2 = (144.4 - 113.5) = 0.30.9$ mV.

$R = \frac{\Delta E_2}{\Delta E_1} = \frac{30.9}{18.5} = 1.67$. The value for $\frac{CV}{C_sV_s} = 1.598$ see from Tables in Appendix 17.

So $$R = \frac{CV}{C_sV_s} = 1.598 = \frac{C(50)}{(5\times10^{-3}\,\text{M})(5)} = 7.99\times10^4\,\text{M}.$$

45.9 What is the potential of cell after addition of 5 and 50 ml of 0.01 M ceric ion solution with SCE and Pt electrodes to 25 ml of 0.0100 M Fe (II) solution?

Given: $\left(E^0_{\text{Fe}} = 0.77\,\text{V}, E^0_{\text{Ce}} = 1.61\,\text{V}, Fe^{2+} + Ce^{4+} \rightarrow Fe^{3+} + Ce^{3+}, E_{\text{SCE}} = 0.243\right)$.

Ans:

$$E_{cell} = (E_1 - E_{SCE})$$

$$E_1 = E^0 + \frac{RT}{nF}\log\left[\frac{Fe^{2+}}{Fe^{3+}}\right];$$

$$E_1 = 0.771 + \frac{0.0591}{1}\log\left[\frac{Fe^{2+}}{Fe^{3+}}\right] \text{ (if } n = 1)$$

$$E_{cell} = (E_1 - E_{SCE}) = 0.771 + 0.0591 \log\left[\frac{Fe^{2+}}{Fe^{3+}}\right] - 0.243 = 0.527 - 0.0591 \log\left[\frac{Fe^{2+}}{Fe^{3+}}\right]$$

After addition of 5 ml of cerric ion the concentration is

$$[Fe^{3+}] = \frac{5 \times 0.01}{(25+5)} = 1.67 \times 10^{-3} \text{ M strength of solution}$$

$$[Fe^{2+}] = \frac{(25 \times 0.1) - (5 \times 0.01)}{30} Fe^{2+} = 6.67 \times 10^{-3} \text{ M}$$

$$E_{cell} = 0.527 - 0.591 \log\left[\frac{6.67 \times 10^{-3}}{1.67 \times 10^{3}}\right] = 0.491 \text{ V}$$

After addition of 50 ml of solution we have

$$Ce(IV) = \frac{(50-25) \times 0.01}{(25+50)} = 3.33 \times 10^{-3} \text{ M; } Ce(III) = \frac{25 \times 0.01}{75} = 3.33 \times 10^{-3} \text{ M.}$$

$$E_{cell} = 1.61 - 0.0591 \log\left(\frac{3.33 \times 10^{-3}}{3.33 \times 10^{-3}}\right) - 0.244, \textit{ i.e.}, \left(E^0_{Ce} - E_1 - E_{SCE}\right) = E_{cell}$$

$$E_{cell} = 1.37 - 0.0591 \log\left(\frac{3.33 \times 10^{-3}}{3.33 \times 10^{-3}}\right) = 1.37 \text{ Volts.}$$

45.10 What is the potential with SCE at end point at pH 3.5 for the titration of 25 ml of 0.01 M ferrous sulphate with 0.01 M potassium dichromatic solution?

Ans: The redox reaction was as follows:

$$6Fe^{2+} + Cr_2O_7^{-2} + (14H^+) = 6Fe^{3+} + 2Cr^{3+} + 7H_2O.$$

The potential E_1 of indicator electrode w.r.t. SEC was

$$E = E_{Fe} + \frac{RT}{nF}\log\left[\frac{Fe^{2+}}{Fe^{3+}}\right] = 0.771 - \frac{0.0591}{n}\log\left[\frac{Fe^{2+}}{Fe^{3+}}\right] \text{ as } (n = 1)$$

$$E = E^0_{Cr} + \frac{RT}{nF}\log\frac{[Cr^{3+}]^2}{[Cr_2O_7^{2-}][H^-]^{14}} \text{ as } n = 6$$

Since $E_1 = 0.771 - 0.05917\log\left[\frac{Fe^{2+}}{Fe^{3+}}\right]$

$$6E_2 = 7.98 - 0.05917\log\frac{[Cr^{3+}]^2}{[Cr_2O_7^-][H^+]^{14}}$$ adding together two equation.

$$(E_1 + 6E_2) = 7E = 8.75 - 0.0591\log\frac{[Fe^{2+}][Cr^{3+}]^2}{[Fe^{3+}][Cr_2O_7^{2-}][H^+]^{14}}$$ but $[Fe^{2+}] = 6[Cr_2O_7^{2-}]$.

$$7E = 8.75 - 0.0591\ \log\frac{6(Cr_2O_7^{2-})(Cr^{3+})^2}{(Fe^{2+})(H^+)^{14}}$$

Volume of added $K_2Cr_2O_7$ at end point was

$$\frac{0.01MFe^{2+} \times 250 \times 1m\ mol\ K_2Cr_2O_7}{6m\ molFe^{2+}} \times \frac{1\ ml}{0.010\ ml\ Cr_2O_7^{2-}} = 4.2\ ml.$$

$$[Fe^{3+}] = \frac{0.01 \times 25}{(25 + 4.2)} = 8.6 \times 10^{-3}\ M.$$ Now at end point.

$$[Cr^{3+}] = [Fe^{3+}] \times \frac{2}{6} = 8.6 \times 10^{-3} \times \frac{2}{6} = 2.9 \times 10^{-3}\ M.$$

$pH = -\log[H^+] = 3.50, [H^+] = 3.2 \times 10^{-4}\ M$ as $Cr^{3+} = 2.9 \times 10^{-3}$, $H^+ = 3.2 \times 10^{-4}$,

$Fe^{2+} = 8.6 \times 10^{-3}$.

$$7E = 8.75 - 0.05917\log\frac{6[2.9 \times 10^{-3}]^2}{[8.6 \times 10^{-3}][3.2 \times 10^{-4}]} = 0.855\ V = E.$$

The net potential was

$$E = (E - E_{SCE}) = (0.855 - 0.244) = 0.611\ V.$$

45.11 Calculate the potential of the electrode if sodium selective electrode and SCE were immersed in solution of sodium ion with activity of 1.25×10^{-3} M?

($E_{Na} = -0.2034$ v, assume [Na] = 1.50×10^{-3} M and [K⁺] = 1.20×10^{-3} M if $K_k^{Na} = 0.24$.)

Ans: $$E = \text{constant} + \frac{0.05917}{n}\log a_1 \text{ if } a_1 = \text{activity coeff.}$$

$$E = \text{constant} + \frac{0.05917}{n}\log a_i \text{ if } E = -0.2034,\ a_i = 1.25 \times 10^{-3}$$

The $E = \text{constant} + \frac{0.05917}{n}\log(1.25\times10^{-3}) - 0.2034$ so constant $= -0.0316$ but solution concentration was calculated from the equation.

$$E = \text{constant} + \frac{2.303RT}{nF}\log\left(a_1 + k_p a_j^{n/a}\right)$$

$$E = \text{constant} + \frac{0.05917}{n}\log\left(a_1 + K_j \cdot a\right)a^{n/a}$$

$$= -0.0316 + \frac{0.05917}{n}\log\left(1.5\times10^{-3} + 0.24\left(1.20\times10^{-3}\right)\right)$$

$$= -0.1942\,\text{V}.$$

45.12 Evaluate the probable error likely to occur for nitrite electrode if nitrite with activity of 0.010 M was measured?

Ans: $0.010 = a_{NO_3}$

$$E = \text{constant} + \frac{0.0591}{nA}\log\left[a_A + K_{AB}\left(a_B\right)^{nA/nB}\right] \text{ so, we have}$$

$$E = \text{constant} + \frac{0.0591}{1}\log\left[0.0001 + (0.06)(0.010)\right]$$

$$E = \text{constant} + (0.0591) - (-2.80) = \text{constant} + 0.165\,\text{V} \text{ so in above equation.}$$

$$\text{Constant} + 0.165 = \text{const.} - 0.059\ \log a_{NO_3}$$

$$\log^a_{NO_3} = -2.80 a_{NO_3} = 0.0016.$$

$$\text{Error} = \left(\frac{0.0016 - 0.0010}{0.00}\right)\times100 = +60\%.$$

45.13 Predict the possible interference by 0.2 M of sodium in measuring calcium (II) $a^+_{Ca} = 0.0010$ other ionic strength effect? ($K_{ca/Na} = 0.016$)?

Ans: $a^+_{Ca} = 0.0010$, $K_{Ca/Na}$ = selectivity coeff. = 0.016 given

So, $$E = \text{constant} + \frac{0.0591}{2}\log\left[a^+_{Ca} + K_{Ca/Na}\left(a_{Na}\right)^{a/b}\right] \text{ if } n = 2.$$

$$E = \text{constant} + \frac{0.0591}{2}\log\left[0.001 + 0.016(0.20)^2\right] \text{ as } a/b = 2 = Ca^{2+}/Na^+.$$

Since $a^{2+}\ Ca = 1.06\times10^{-3}$ instead of 1×10^{-3} true value.

The error is $\frac{0.06\times10^{-3}}{1.0\times10^{-3}}\times100 = 6\%.$

45.14 What is the concentration if cadmium selective electrode when it was used with method of standard addition. The volume of unknown sample was 100 ml with potential of 379 mv. After addition of one ml of one molar cadmium potential rose to 414 mV?

Ans: With usual equation was $C_x = \frac{C_s + V_s}{(V_x + V_s)}\left(10 - \frac{V_x}{(V_x + V_s)}\right)$ whereas

C_x = Con. of unknown soln., V_x = Vol. of unknown soln. Before addition.
V_s = Vol. of standard soln. Added, C_s = concentrate of the standard solution.

ΔE = change in the potential, $S = \frac{RT}{nF} = \frac{0.0591}{n}$ (since $n = 2$).

Now in given problem, C_s = 1M, C_x = unknown, V_s = 1 ml, V_x = 100 ml.

$\Delta E = (414 - 379)/1000 \text{ volts} = 0.035$; if $S = \frac{0.0591}{2}$

On substituting these value in above equation we have

$$C_x = \frac{(1)(1)}{(100+1)}\left\{10^{0.035/0.0519/2} - \frac{100}{100+1}\right\}^{-1}$$

$$= 6.9 \times 10^{-4} \text{ M.}$$

45.15 Find out the amount of fluoride if fluoride ion selective electrod was used with standard addition method when the volume of unknown solution was 100 ml when 10 ml of 0.1 sodium fluoride was added, then it changed its potential to 70 mv assuming the original potential was 30 mv?

Ans: Here C_s = 0.1, V_s = 10 ml, C_x = unknown, V_x = 100 ml.

$$\Delta E = \frac{(70-30)}{1000\,j} = 0.040 \text{ volts}, \; S = \frac{0.0591}{1} \text{ as } n = 1 \text{ for fluoride ion.}$$

On substituting the values in first expression given above, we have

$$C_x = \frac{(0.1)(10)}{(100+10)}\left\{10^{0.035/0.0591/2} - \frac{100}{100+1}\right\}^{-1}$$

$$= (0.00909)\,(4.76 - 0.009)^{-1} = 2.4 \times 10^{-3}\text{M}$$

We can also solve using simplified equation as

$$C_x = (0.1)\left(\frac{10}{100}\right)\left(10^{0.40/0.059} - 1\right)^{-1}$$

$$= 2.6 \times 10^{-3} \text{ M.}$$

45.16 What is the pH of a solution if Pt electrode immersed into the solution to which quinhydrone was added and it showed 0.189 V against SCE = 0.247 V; $\left(E^0_{\text{quinhydron}} = 0.703\text{ V}\right)$?

Ans: Now pH was calculated from equation as

$$\text{pH} = \frac{0.703 - (0.247 + 0.189)}{58} = 4.6.$$

45.17 What was the pH of a solution if the potential of the glass electrode against Ag/AgCl electrode was –0.133 V? (E^0_{glass} = –0.033 V at pH 3.28)?

Ans: We calculate pH as given above.

$$pH = \frac{0.133 - 0.033}{58.1} + 3.28 = 5.0.$$

45.18 Ascertain the content of an impurity of ferric ion in solution of $FeSO_4$, if the potential of the platinum electrode against SCE was 0.440 V? $\left(E^0_{Fe^{3+}} = 0.77 \text{ V and } E^0_{SCE} = 0.247 \text{ V}\right)$

Ans: We see ratios of ferric to ferrous ion as

$$\frac{Fe(III)}{Fe(II)} = \frac{0.440 + 0.248 - 0.770}{58} = -1.41 = 2.59$$

$$\frac{Fe(III)}{Fe(II)} = 0.039 = \frac{x}{(100 - x)}. \text{ So, } x = 3.7\%.$$

45.19 What was the potential of silver electrode against SCE electrode in 0.1 N solution of sodium chloride when titrated with 0.1 N of silver nitrate by 99%?

Ans: We know that

p Ag + pCl = 10.0

Now E = (800 – 58)

p Ag = 222 + 58 pCl.

Before equivalence point was reached.

$$pCl = -\log \frac{0.1(100 - p)}{(100 + p)}$$

After equivalance point

$$p\,Ag = \frac{0.1(100 - p)}{(100 + p)}$$

To potential against SCE is was

$$E^1 = (E - 250) \text{ mV} = 164 \text{ mV}$$

45.20 Evaluate the resistance of a solution of copper sulphate solution if in the electrolysis by a current of 0.1 A the voltage at the terminal was 1.9 volts with polarisation voltage of 1.5 V?

Ans: We have general equation

$$E_{appt.} = (P_a - P_c) + IR = P + IR$$

If P_a and P_c were anodic and cathodic polarisation and P = difference in polarisation e.m.f. I = current and r = resistance and E_{app} was applied voltage. We have an expression as

If P = 1.9, I = 0.1 R, we have

$$E_{appl} = (P + IR);\ 1.9 = 1.5 + 0.1\ R$$

as $\quad E_{apl} = 1.9$ and $P = 1.5$

So $\quad R = 4.0\ \Omega.$

PROBLEMS FOR PRACTICE

45.21 Calculate the solubility product of cadmium chloride if in the typical cell was having potential of 0.893 V in cell Cd|CdX_2 (Std) (0.02 M) |SCE|. **[Ans:** 1.6×10^{-2}]

45.22 Calculate the potential of mercury cathode with SCE after the addition of 5 and 50 ml of 0.05 M mercuric nitrate to 50 ml of 0.08 M sodium chloride?
(K_{sp} = 1.3 × 10^{-2}) **[Ans:** 0.89 V and 0.479 V]

45.23 Calculate the pH of the solution with glass and SCE electrode which gave a potential of 0.0620 V at pH = 7.0 assuming the potential for the unknown solution was 0.2794 V? **[Ans:** pH = 10.67]

45.24 Ascertain the probable error if it was caused by $Mg^{2+}\left(a^{2+}_{Mg} = 0.01\right)$ at time when calcium $\left(a^{2+}_{Ca} = 0.01\right)$ was measured with this electrode $\left(K^{Mg}_{Ca} = 0.01\right)$? **[Ans:** 8 × 10^3]

45.25 What was the e.m.f. of the cell of silver and mercury having concentration of cell in two compartments as 0.40 M and 0.001 M if $E^0_{Ag} = 0.799$ V and $E^0_{Hg} = 40.920$ V and concentration of Hg was 0.10 M and 0.0071 M? **[Ans:** 0.15 V for Hg 0.740 V for Hg]

45.26 Calculate the e.m.f. of the hydrogen gas and SCE electrode [H^+] = 3.6 × 10^{-3} M? and ($E^0_{SCE} = 0.285$, $E^0_{H} = 0$ V). **[Ans:** 11.56 and 0.429 V]

45.27 Calculate the total strength of a solution containing 0.015 M of calcium chloride and 0.01 M of copper nitrate? **[Ans:** 0.075 M]

45.28 Calculate the cell potential at pH 4.0 of a solution containing 1 × 10^{-3} M $K_2Cr_2O_7$ and 2 × 10^{-3} M of chromium (III) with platinum and SCE electrode? **[Ans:** 0.56 V]

45.29 Calculate the potential of cell at pH 4.0 containing SCE and Pt electrode in a solution of 2 mM manganese (II) and 5 mM $KMnO_4$? **[Ans:** 0.89 V]

45.30 Calculate the potential of cell containing SCE and Ag/AgCl electrode in solution of 50 ml of 0.015 M sodium chloride after the addition of 5 ml of 0.1 M silver nitrate in the mixture? **[Ans:** 0.117 V]

45.31 What should be the concentration of the fluoride ion if 50 ml of fluoride solution was analysed by electrode. The potential of electrode was –102.5 mV after addition of 5 ml of 1 × 10^{-2} M fluoride the potential changed to –112.7? **[Ans:** 3.24 × 10^{-3} M]

45.32 With copper selective electrode find out concentration of copper if the potential of 25 ml was + 247.2 mV and after addition of 5 ml of (2 × 10^{-3} M) copper the potential changed to 261.8 mv, with second addition of 5 ml of copper (2 × 10^{-3} M) solution it changed to 271.9 mv? **[Ans:** 7.15 × 10^{-4} M]

45.33 Calculate the voltage of the solution, if SCE and hydrogen electrode were used to measure the pH of the unknown solution as 4.6? **[Ans:** 0.512 V]

45.34 If the voltage of cell was 0.518V, what was the value of K_a for weak acid for a cell Pt/H_2 (atm) HA (1 × 10^{-2}) (A^-) (1 × 10^{-2}) | Hg_2Cl_2 | Hg, A = anion. **[Ans:** 7.55 × 10^{-7} is K_a]

45.35 Calculate the e.m.f. of the following platinum tin cell with SCE as reference electrode

$\left|Pt/Sn^{4+}(0.001M)\right|\left|Sn^{2+}(0.06M)\right|$ with SCE in half cell?

If $E_{SCE} = 0.285$, $E^0_{Sn} = -0.136$. **[Ans:** 0.188 V]

45.36 Calculate the voltage reading at the end point in potentiometric titration of stannous chloride with $KMnO_4$ with SCE as reference electrode?

(pH at end pt was 2.0; $E^0_{SCE} = 0.285$ V, $E^0_{Sn} = -0.136$, $E^0_{Mn} = +1.51$ V). **[Ans:** –0.98 V]

45.37 Calculate solubility product (K_{sp}) for AgCl at 18°C, if the potential was 0.518 V with Hydrogen electrode (NHE)? Given: $E^0_{Ag} = +0.800$ V). **[Ans:** $K_{sp} = 1.79 \times 10^{-10}$**]**

45.38 What was the potential of following half cells
(*a*) || $Pb(NO_3)_2$ (0.001 M) Pb (*b*) | $CdCl_2$ (0.01) 1 Cd
(*c*) || $FeCl_3$ (0.001 M) $FeCl_2$ (0.010 M)/Pt. (*d*) || $HClO_4$ (1 M) Cc^{3+} (0.10 M) (e^{4+}(0.05)/Pt.
[Ans: (*a*) −0.214 V, (*b*) −0.462 V, (*c*) +0.712 V, (*d*) +1.68 V**]**

45.39 What was the voltage of the following cell if there was no complexation or hydrolysis of Ag with oxalic acid to form soluble species.

$$\text{SCE}\left|\text{Na}_2\text{C}_2\text{O}_4\left(1.0\times10^{-3}\text{M}\right)\text{Ag}_2\text{C}_2\text{O}_4\left(\text{satd}\right)\right|\text{Ag}.$$

$\left(E^0_{\text{Ag}} - 0.799,\ K_{sp} = 1.1\times10^{-11}\right),\ \left(E^0_{\text{SCE}} = 0.246\,\text{V}\right).$ **[Ans: 0.318 V]**

45.40 Calculate the magnitude of K_1 of orthophosphoric acid in cell as the potential of the cell was 0.323 V.
SCE | Q (Satd.) H_2Q(Satd.), H_2PO_4 (0.1 M) NaH_2PO_4 (0.1 M) | Au. Assuming no reaction of phosphate quinhydrone or with gold in the cell occurred? **[Ans:** 7.1×10^{-3}**]**

REFERENCES

45.1 S.M. Khopkar, *Basic Concepts of Analytical Chemistry,* 3rd edn., New Age International Publishers (P) Ltd. (2007).

45.2 G.W. Ewing, *Instrument Methods of Chemical Analysis,* McGraw-Hill and Co., p. 278 (1985).

45.3 R.D. Braun, *Introduction of Instrumental Analysis,* McGraw-Hill and Co., pp, 703, 707, 710, 720, 738, 742, 754 (1987).

45.4 H.H. Willard, L.L. Merrite and J.A. Dean, *Instrumental Methods of Analysis,* D. Van Nostrand and Co., p. 548 (1980).

45.5 J.S. Fritz and G.H. Schenk, *Quantitative Analytical Chemistry,* Allen and Beaccon, pp. 336, 338, 346 (1979).

45.6 S. Brewster, *Solving Problems in Analytical Chemistry,* John Wiley and Co., pp. 336-8, 346 (1980).

45.7 D.A. Skoog and D.M. West, *Analytical Chemistry,* 3rd edn., Holt Reinhard and Winston, p. 387 (1979).

45.8 H. Rosseti, *Chemical Applications of Potentiometry*, Van Nostrand and Co. (1969).

45.9 P. Delalay, *New Instrumental Methods in Electro Chemistry,* Interscience (1954).

45.10 I.M. Kolthoff and H.A. Laitinen, *pH and Electro Titration,* 2nd edn., Wiley (1941).

45.11 G. Sevhla, *Automatic Potentiometric Titrations,* Pergamon Press (1978).

45.12 A. Musekin, A. Kharopkorsky, S. Shaikini and S. Efors, *Problems in Quantitative Analysis,* Mir Publishers, p. 148 (1984).

Chapter 46

Voltammetry and Polarography

INTRODUCTION

The change of diffusion current against time or variation of diffusion current against potential is plotted on the graph. For such purpose, we need two electrodes out of which one electrode should be polarisable while another electrode should be inert. In a technique in which dropping mercury electrode (DME) is polarisable and acts as cathode with SCE as anode. It is used to study current time or current voltage graphs. It is called polarography. In case the polarisable electrode is not DME but platinum or other suitable electrode such technique is termed as 'voltammetry'. The simplest example of the latter is cyclic voltammetry or triangular wave voltammetry which is generally used in study of co-ordination complex in inorganic chemistry. The rate of movement of species on account of diffusion is directly proportional to the concentration gradient, the difference between the concentrations at two points divided by distance between electrodes. Hence we have an expression as follows:

$\frac{dc}{dt} = D\frac{dc}{dx}$ if c = concentration, t = time in seconds, x = distance between two electrodes.

By application of Fick's law, the diffusion current (id) is given as $id = nFA$ if i_d = diffusion current, F = Faraday's constant, A = area of electrode, c = concentration, n = number of electrons transferred, t = time in seconds, D = Ficks diffusion coefficient. The study of current against time at fixed potential is called chronoramperiometry. The study of potential against time at fixed current is called chronopotennometry. By using DME and SCE in analysis we have $id = 708.2\ nD^{1/2}\ m^{2/3}\ t^{1/6}\ c$ (since Fick's law states

$$id = nFA\left(\frac{D}{\pi_t}\right)\frac{1}{2}c \text{ on substituting } id = \int_t^d id\,dt = \frac{6}{7} = id.$$

Now i_d is average diffusion current, we have $i_d = 607\ nd^{1/2}\ m^{2/3}\ t_d^{1/6}\ c$. This is called as an "Ilkovic equation where in m = rate flow of mercury, t_d = time between two successive drops, d = diffusion current. Now in quantitative chemical analysis if we can keep $n, D, m\ t_d$ all constant then id would vary directly with "c" that is on plotting a graph of id against c we would get straight line graph. When we use unknown solution (c_x) with idx as diffusion current and if current for standard solution with concentration c_s is known we can easily evaluate the magnitude of c_x easily. Apart from (I) Ilkovic equation in polarographic analysis we use another term ($E^{1/2}$) called as half wave potential which is typical of particular element in the presence of specific supporting electrolyte. For this we have an expression as:

$E^{1/2} = E_{DME} + \frac{RT}{nf}\ln\left(\frac{c}{id-i}\right)$. From which we can evaluate 'n'. Now if we plot $\log\left(\frac{i}{id-i}\right)$ against E_{DME}, we get slope $\left(\frac{0.0591}{n}\right) = \left(\frac{2.303RT}{nF}\right)$ where Kc = instability constant, X = complexing ligand, p = moles of X used for metal complexation (provided $D = D^1$). In above expression (i) stands for limited diffusion current while (id) is normal diffusion current with same supporting, electrolyte like say NaOH (1 M) $E^{1/2}$ for different metals will be different as Cd^{2+}(−0.78 v), Cs^{+}(−1.46 v), Cu^{2+}(− 0.41 v), Fe^{3+}(1.12 v), Pb^{2+}(− 0.76 v), Zn^{2+}(−1.53 v) while same metal with different supporting electrolyte $E^{1/2}$ varies with nature of electrolye, e.g., *Cd*, $E^{1/2}$ is (− 0.6 v in KCl), (− 0.81 v in NH_4Cl), (− 0.78 v in NaOH), (− 1.88 v in KCN). In the presence of oxygenation cell the reaction is $O_2 + 2H^+ + 2e^- \rightarrow H_2O_2$ zf; $H_2O_2 + 2H^+ + 2e \rightarrow 2H_2O$.

In order to prevent this we must flush inert gas like nitrogen through DME further the hump in polarographic maxima is eliminated by use of surfactants like Triton X −100, gelatin, methyl red. This provides smooth curve in polarogram.

In amperiometric titrations, we plot graph of diffusion current (id) against volume of titrant added (ml). In amperiometry, both electrodes are polarisable. We need to give correction to volume (v) as $\left(\frac{\nu+\gamma}{\gamma}\right)$ if γ is increment added each time to (v) volume of sample solution. In chronopotentiometry, a study of potential with fixed current, we plot graph of current against time with three electrode assembly. We use an expression for finding out concentration as $t^{1/2} = \frac{\pi^{1/2} nFAD^{1/2} c}{2I}$ if n = nos. of electrons transferred, A = area, c = concentration, F = Faraday constant, d = Fick's diffusion coefficient, I = diffusion current passed, T = transition time.

Apart from DC polarography which is most commonly used we have other techniques like AC polarography, pulse polarography, differential pulse polarography and square wave polarography. In case of AC polarography the peak height (ip) is expressed in terms of concentration and frequency as

$$ip = \frac{2.5n^2F^2AD^{1/2}\Delta EN^{1/2}c}{4RT} = \text{peak height.}$$

Where ΔE = amplitude of superimposed alternating potential. A = area of electrode N = frequency of AC potential D = diffusion current c = Concentration i_p = peak current. In case of pulse polarography the diffusion current is measured as i_p. Various equations are shown herewith

$$i_p = \frac{nFAD^{1/2}}{\pi^{1/2}S^{1/2}}\left[\frac{\exp(nF\Delta E/2RT-1)}{\exp(nF\Delta C/2RT-1)}\right] \text{ as } ip = \frac{nFAD^{1/2}C\Delta\psi(Es_1E_{sum})}{(\pi tp)^{1/2}}$$

Cyclic voltammetry or triangular wave voltammetry uses triangular voltage ramp with DME as indicator electrode, entire scan is taken in fraction of the drop time the peak current (ip) is given by an expression as; $ip = 0.27n^{3/2}\ AD^{1/2}\ C^{1/2}$ where A = area of electrode, D = diffusion coefficient, c = concentration and v = the scan rate. This technique though useful for quantitative analysis is mainly used to understand reaction mechanism and kinetics. A repetative forward reverse scans when performed sequentially without a pause between the scan technique is designated as 'cyclic voltametry'.

The amperiometry technique envisages use of both polarisable electrodes. In this method diffusion current generated during titration is plotted against the volume of titrant when point of intersection indicates equivalance end point in titration, e.g., Karl Fischer titration when for moisture determination can be amperiometrically analysed. Finally in anodic stripping voltametry it is used for the determination of small concentration of metal in the range of 10^{-5} to $10g^{-10}$. Stripping voltammetry is carried in two steps. In first stage metal (M) is deposited on electrode in 10-30 mt. Then potential of said electrode is varied. The metal is oxidised, we have peak height which is proportional to the concentration. The height which is proportional to the concentration in a plot of anodic current against potential of mercury thin film is used. In this electrode is anode during plating step and at cathode during stripping stage it is called cathodic stripping voltammetry.

Finally in organic polarography reactions at microelectrode are slow, but it is useful for elucidation of structure. The electrode reaction are pH sensitive, so good buffering is necessary Glycol, dioxane, alcohol, celloslove, acetic acid are the solvents used. Lithium is invariably used as the supporting electrolyte. Carbonyl, carboxylic and peroxides, nitro, nitrasamine, and azo compounds, haloorganic entities containing double bond or hydroquinone mercaptans leads to lower value of half wave potential.

This chapter provides large number of numerical problems involving calculation of unknown concentration using Ilkovic or chronopotentiometric equation. The calculations involving half wave potentials or polarogram in the presence of complexing ligands are also provided. The problems on evaluation of concentration with peak height of polarogram are quite interesting. Some other miscellaneous problems are also included for exercises for practice. The reader should also refer to books described in the references at end of this chapter.

SOLVED PROBLEMS

46.1 Calculate the diffusion coefficient of cadmium if the transition time for 5×10^{-2} M solution was 22.1 seconds. The area of the working electrode was 2.4 cm^2, when 144 μa current was flowing with total number of electrons transfered were two?

Ans: $t_{1/2} = \dfrac{\pi^{1/2} nFAD^{1/2C}}{2l}$ as $c = 5 \times 10^{-2}$ M, $A = 2.4$, $I = 144$, $n = 2$, $F = 96500$

as Transition time $(t_{1/2}) = \dfrac{\pi^{1/2} nFAD_C^{1/2}}{2I}$

$$22.1 = \frac{\left((3.14^{1/2} \times 2)(96500)(2.4)D^{1/2}\right)5 \times 10^{-2}}{2 \times 144}$$ D is 6.0×10^{-5} is the value.

46.2 What was the value of diffusion current if $c = 3 \times 10^{-3}$ moles/liter, $D = 7.2 \times 10^{-5}$ cm^2/sec, $m = 3$ mg/se, $t = 4$ seconds and $n = 2$?

Ans: We use Ilkovic equation to know value of (id) diffusion current, $id = 607\ nD^{1/2}\ m^{2/3}\ t^{1/6}\ c$

we have $n = 2$, $c = 3 \times 10^{-3}$, $t = 4$, $m = 3$

$id = 607 \times 2 \times (3 \times 10^{-3})^{1/2} \times (7.2\ 10^{-5}) \times (3)^{2/3} \times (4^{1/6})$ on substitution

or $id = 3.642 \times (3)^{2/3} \times (4)^{1/6} \times (0.000072)^{1/2}$

$id = 3.642 \times 2.08 \times 8.4 \times 10^{-3} \times 2.5 \times 10^{-7} = 25.4$ A.

46.3 If the diffusion current constant (I) for zinc was 8.25, when m = 32.5 mg/sec, t = 4.3 second, with diffusion current of unknown sample polarographed under similar condition was 4.3 μa? What is concentration of unknown sample?

Ans: $I = 607\, D^{1/2}\, id = c \times m^{2/3} \times t^{1/6}$

Now I = 8.25, n = 2, m = 32.5, t = 4.3, id = 4.3 on substituting these values

$$8.25 = \frac{4.3}{0 \times (32.5)^{2/3} \times (4.3)^{1/6}}$$

$$c = \frac{4.3}{8.25 \times (32.5)^{2/3} \times (4.3)^{1/6}} = 4.02 \times 10^{-2}\,\text{M}.$$

46.4 An organic compound (4.20 mg) was reduced in solvent. The drop time was 34.9 seconds for 10 drops and 20 drops had mass of 0.1296 gm with diffusion current of 14.25 μg from following data calculate half wave potential ($E^{1/2}$) n (nos of electrons) and diffusion coefficient?

E versus Ag/Ag NO_3 (0.01) V:	−2.400	−2.440	−2.460	−2.480
I (diffusion current), μ a	1.00	3.5	6.0	9.0

Ans: The equation used is

$$E = E^{1/2} - 0.0591 \log\left(\frac{i}{id - i}\right).$$ We calculate value $\left(\frac{i}{id-i}\right)$ from data as

$$-2.400 = \left(\frac{1}{14.5 - 1.00}\right) = 0.0755,\ i.e.,\ \log(-1.222)\ \text{with}\ -2.4\,\text{V}$$

$$-2.400 = \left(\frac{1}{14.5 - 3.5}\right) = 0.325,\ i.e.,\ \log(-0.488)\ \text{with}\ -2.4\ \text{V}$$

$$-2.460 = \left(\frac{1}{14.5 - 6.0}\right) = 0.727,\ i.e.,\ \log(-0.138)\ \text{with}\ -2.460\ \text{V}$$

$$-2.480 = \left(\frac{1}{14.5 - 9.0}\right) = 1.71 \log 0.233\ \text{with}\ -2.480\ \text{V}$$

We plot graph of E versus $\log\left(\frac{i}{id - i}\right)$, we get $E^{1/2}$ = −2.466 v slope was − 0.059 a^4 1/n v

since n = 1. So, slope was −0.0594 v. Now $m = \frac{129.6}{20} \times \frac{1}{3.49} = 1.86$ mg/s, $t^{1/6} = 3.49^{1/6}$ = 1.23$s^{1/6}$ as $m^{2/3}$ = 1.51

so on substituting these values in equation

$id = 708 n D^{1/2}\, m^{2/3}\, t^{1/6} c$

14.5 = (708) (1) ($D^{1/2}$) $(1.51)^{2/3}$ $(1.23)^{1/6}$ (4.20)

as i_a = 14.5 n = 1, $m^{2/3}$ = 1.51, $t^{1/6}$ = 1.23 and c = 4.20

$D^{1/2} = 2.58 \times 10^{-3}$ or $D = 6.66 \times 10^{-6}$ cm²/sec.

46.5 Find out the value of the half wave potential ($E^{1/2}$) for zinc versus background of 0.1 M ammonium chloride with NHE if the half wave potential measured with reference to the SCE was 60 volts?

Ans: $\left(E^{1/2}\right)c = E^{1/2} + \frac{0.059}{n}\log K_{in} - m\frac{0.059}{n}\log Ca$

We use an expression for $E^{1/2}$ as shown above.

Where K_{in} = instability constant, m = co-ordination number, Ca = concentration of anion and $\left(E^{1/2}\right)$ half wave potential in the presence of complexing ligand.

The $E^{0}_{\text{SCE}} = 0.246\,\text{V}$; if $E^{1/2}$ = 60 v. The actual potential will be – 0.600 + 0.246 = – 0.345 v, *i.e.*, for zinc $E^{1/2}_{\text{Zn}} = -0.40\,\text{V}$.

46.6 What should be height of the polarographic wave if the following data was obtained in the polarographic analysis:

$-E(v)$:	0.40	0.50	0.60	0.65	0.70	0.75	0.80	1.0
G Reading:	0	0	0.5	10	27	35	36	38

Ans: We plot the graph of potential against galvenometer reading (G).

The wave height was assessed from the point of intersection ($y = ax + b$)

(where $a = \tan\alpha = \frac{\Delta i}{\Delta E}$, x = potential b = current at E = 0)

For first line (E = 0.4 – 0.5 b) (a = 0) pressuming y = 0

For second line $(a_2) = \left(\frac{27-10}{0.75-0.70}\right) = 0.340$ if $b_2 = -211$; $(27 = 340 \times 0.70 + b_2)$

For third line $(a_3) = \left(\frac{38-35}{1-0.75}\right) = 12$, $(35 = 12 \times 0.75 + b_3$ if $b_3 = 26)$

We have two simultaneous equations as

$y_2 = 340x - 211$, $ys = 12x + 26$ if $x = 0.724$, $y = 35$

As (y_1 = 0) we have h = 35 mm is height of polarographic wave.

46.7 Calculate the concentration in polarographic analysis of unknown solution for thallium when following data was obtained:

C(mg/ml):	0.4	0.8	1.0	Unknown
H(mm):	30	61	74	50

Ans: We know following equation:

$\frac{c_x}{c_0} = \frac{hx}{ho}$ if $(x = \text{unknown solution})$ $\qquad$ c_o = con. of known solution

hx = ht of unknown $\qquad$ h_o = ht of known solution

Therefore, we have

$$\frac{c_x}{c_0} = \frac{hx}{ho} \quad c_x = c_0\left(\frac{hx}{ho}\right) \text{ value of } c/n \text{ for known solution was}$$

$$\frac{0.4}{30} = 0.01333, \ \frac{0.8}{61} = 0.01313 \text{ and } \frac{1.0}{74} = 0.01352. \text{ The mean} = 0.01313 \text{ mgl}$$

Now if $hx = 50$ and $c_o/h_o = 0.1313$, then

Now $c_x = \left[\frac{c_a \times hx}{ho}\right]$, *i.e.*, $0.01333 \times 50 = 0.67$ mg/ml.

46.8 Find out the concentration of the unknown solution if a standard solution of zinc contained 10 mg/ml. The various solution were made with total volume of 50 ml with varying volume of standard solution when following heights were noted:

Soln (ml)	1	2	4	8	unknown zinc solution
H-(height) (mm)	17	35	67	136	71 mm (2.5 ml)

Ans: We first calculate ratio of $\frac{(v+10)}{h}$ for all solutions, *i.e.*, $\frac{10+4}{67}, \frac{10+8}{136}$ average of 0.568 m.

Now since $C_o/C_x = h_o/hx$ so $\left[\left(\frac{v+10}{h}\right)xh\right] \div 2.5 = \frac{0.568 \times 71}{2.5} = 16.7$ mg/ml.

46.9 Find out the concentration of the unknown solution if in polarographic analysis of thallium aliquat of 2 ml of solution were taken, reagent added and made up to 50 ml. The corresponding wave height was 30 mm to other solution with same volume one ml of standard solution was added (strength 10 mg/ml) with height 55 mm.

Ans. We use relation $c_o/c_x = ho/hx$, $hx = 30$ mm, $ho = 55$ mm, $c_x = x$, while $c_o = 10$ mg/ml on substituting these values
If x = original solution

$$\frac{2x}{50} = \frac{(2x+1)10}{(50+1)} = 30{:}\ 505 = h_1/h_2. \text{ If } x = \text{original solution}$$

On solving this equation we have $x = 5.75$ mg/ml.

46.10 Evaluate the magnitude of the diffusion current constant of lead with following data $id = 4.2$, $m = 1.41$ mg/sec, $t = 1.85$ sc $C = 1$ mole/1.

Ans: The diffusion current constant (I) was equal to

$$1 = 607nD^{1/2} = \frac{id}{Cm^{2/3} \times t^{1/6}}, \text{ we substitute the value}$$

$$I = \frac{4.2}{1 \times (1.41)^{2/3} \times (1.85)^{1/6}} = 3.2\ \mu\text{A was diffusion current constant}$$

Since, $m^{2/3}$, $t^{1/6} = 1.40$ was capillary constant for lead.

46.11 Ascertain the concentration of cadmium in solution from following data for polarographic analysis. A 2 ml soln. to be analysed was made up to 50 ml., when polarographed the diffusion current was 5.1, if ($m^{2/3}$ $t^{1/6}$) = 1.40, i = 3.2 μa. What is value of I?

Ans: $I = \dfrac{id}{Ca}$. This was reduced equation for (I) if

$$C = \frac{5.1 \times 112.4 \times 50}{1.40 \times 3.2 \times 1000 \times 2}$$

at wt of Cd = 112.4, volume = 50 ml

If i = 3.2, a = 1.40; id = 5.1, I = 3.20 mgs/ml.

46.12 Evaluate the magnitude of the half wave potential for a solution of zinc in 0.005 M solution of ammonium oxalate (as complexing ligand) if in the absence of it the $E^{1/2}$ = –0.60 V, If K_{ins} = 4.2 × 10^6 was instability constant for zinc oxalate complex?

Ans: $\left(E^{1/2}\right)c = E^{1/2} + \dfrac{0.059}{n}\log K_{ins} - m\dfrac{0.059}{n}\log c[a]$

In above equation the numerical values were

$E^{1/2}$ = –0.60 V, n = 2, K_{ins} = 4.2 × 10^{-6}, m = 2 co-ordinate number Ca = 0.005 we substitute the

$$\left(E^{1/2}\right) = -60\,\text{V} + \frac{0.059}{2}\log\left(4.2\times10^{-6}\right) - (2)\times\frac{0.059}{2}\log(0.005)$$

$$= 0.6\,\text{V} + \frac{0.059}{2}\times 5.37 - 2\times\frac{0.059}{2}\times\log 0.05$$

$$= -0.90\,\text{V}.$$

46.13 Find out the magnitude of instability constant and the co-ordination number in a complex compound of bivalent metal with EDTA, on the basis of following information, the concentration of EDTA = 0.001 N, $E^{1/2}$ = –1.40, $E^{1/2}$ of complex = 1.43 V?

Ans: We use as expression as

$$\left(E^{1/2}\right)c = E^{1/2} + \frac{0.059}{n}\log K_{ins} - m\frac{0.059}{n}\log ca$$

Where in above expression

$E^{1/2}$ = –1.40, ($E^{1/2}$) comp. = 1.43, K_{ins} = x, n = 2, Con. of EDTA = 0.001N.

Conversely the values of (n) and K_{in} from the values of

Δ($E^{1/2}$)/log Ca can be also used for such calculations

$$p = \frac{\Delta E}{\Delta \log C}\times\frac{2}{0.059} = \frac{0.03\times 2}{1\times 0.06} = 1$$

$$-\log K = \left[\left(E^{1/2}\right)c - \left(E^{1/2}\right)\right]\times\frac{2}{0.06} + p\log Ca = (-1.43 + 1.0)\times\frac{2}{0.06} - 1\times 2 = -163$$

$$(-1.40 - 1.0)\times\frac{2}{0.06} - 1\times 3 = -16.3.$$

$K = 10^{-16.3}$ or pK = 16.3.

46.14 Find out the concentration of lead in a solution if the titration of 50 ml of this solution with 0.03 M solution of potassium dichromate gave following readings:

S.N.	0	2	4	6	7	8
ml	80	45	12	25	60	93

Ans: We calculate corrected micro current as shown in Table

V	0	2	4	6	7	8
T $\Delta i/\Delta r$	80	45	12	25	60	93

→ 16.5 ←→ 17.0 →← 20 →← 20 →

So *ie*, $= -16.7$ V + 80, $ia = (20 + b)$

The value of (b) was calculated at $V = 7$ so $68 = (20 \times 7) + b$ which had $b = -72$

$ia = (20\text{ V} - 72)$ At this point of intersection $e_1' = e_2'$. Which gave $v_{eq} = 4.16$ as lead was

$$x = \frac{416 \times 0.05 \times 2}{50} = 0.0083 \text{ M}.$$

46.15 Depict the curve for the titration of 50 ml of 0.01 M solution of lead with 0.05 M solution of sulphate at 0.8 V_s against V (SCE).

At this potential electrode reaction was $Pb^{2+} + 2e^- \rightarrow Pb^0$, $id = 10^4$ μa. Calculate corrected current (t) with data given as:

V	0	2	4	6	8	10	12
t-corrected	100	80	60	40	20	0	0

Ans: Taking into account the correction for the increase in the volume on titration

$$t = \frac{(0.01 \times 50 - V \times 0.05) \times 10^4}{(50 + V)}.$$

From this we set-up two equation one can easily get the magnitude of concentration.

46.16 Find out the concentration of the unknown solution if the diffusion current for 0.5 M solution was 45 μa and for unknown 90 μa?

Ans: $\frac{id_1}{id_2} = \frac{c_1}{c_2}$. If $id_1 = 45$, $id_2 = 90$, $c_1 = 0.5$, c_2 = unknown

$$\frac{45}{90} = \frac{0.5}{x} = \frac{0.5 \times 90}{45} = 1.0 \text{ M}.$$

PROBLEMS FOR PRACTICE

46.17 Estimate the flow rate of the mercury when the mass of its ten drops was found to be 0.0660 gm and drop time was 2.47 second? **[Ans:** 1.34 mg/s]

46.18 Calculate diffusion coefficient of divalent metal if polarogram of 1.25 mM of solution had diffusion current of 7.12 μa if the capillary properties were as t = 3.47 seconds, m = 1.42 mg/sec? **[Ans:** 6.70 × 10^{-6} cm/sec]

46.19 Ascertain level of dissolved oxygen in polluted sample of water when the effluent had the diffusion coefficient of oxygen at room temperature was 2.65×10^{-5} em/s and droping mercury electrode had ($m^{2/3}$ $t^{1/6}$) of 1.86 mg/sec with, the diffusion current as 2.3 µa?

[**Ans:** 0.17 mm]

46.20 Find out the concentration of the unknown solution of an organic compound which gave diffusion current of 40.3 µA. If 10 ml of solution was mixed with 5 ml of 2.5 mg/ml of the organic compound in standard addition method the diffusion current was 65.2 µa?

[**Ans:** 0.876 mg/ml]

46.21 Find out the concentration of a drug when a solution had a diffusion current of 24.3 µa. A standard solution containing 600 µg/ml gave corresponding diffusion current of 40.3 µa?

[**Ans:** 362 µg/ml]

46.22 What should be the concentration of antimony if analysed with ten ml of 0.2 M perchloric acid in 5 ml of sample with 5 ml of 7.82×10^{-3} M of its oxide with 5 ml of water gave a current of 66.2 µa when voltage was –1.0 V?

Ans: 7.77×10^{-3} M]

46.23 What was the concentration of germanium in solution which when analysed by standard addition method gave height of 20.5 mm. Now after adding 2 ml of 0.053 germanium solution gave height of 24.3 mm with volume as 15 ml?

[**Ans:** 0.020 N]

46.24 Calculate strength of the cobalt solution in unknown sample with following data?

Co content %	0.1	0.32	0.54	Unknown
h-mm	6	20	35	15 mm

[**Ans:** 0.23% cu]

46.25 Find out concentration of unknown solution of metal halide which was analysed by polarography. The reading for 0.045 N metal halide were as follows:

Metal halide ml	2	4	8	16	Unknown
h-mm	7	13	25	42	37

[**Ans:** 3.3×10^{-4} g]

46.26 In polarographic analysis if $m = 9$ second for 2 drops (*i.e.* $m = 4.5$ sec per drop) and for each drop weight was 6.1×10^{-3} g, what was the product of last two terms (viz. $m^{2/3}$ $t^{1/6}$) in Ilkovic equation?

[**Ans:** 1.53×10^{-3} M]

46.27 In a polarographic determination of copper with concentration of 1×10^{-4} M the diffusion current was 17.5 µa while unknown solution had a diffusion current of 27.9 µa. Calculate the concentration of unknown solution?

[**Ans:** 1.59×10^{-3} M]

46.28 In a polarographic estimation of mercury with equal volume of 5.50×10^{-4} M solution and unknown solution gave diffusion current of 58.5 µa and 47.4 µa resp. Calculate the concentration of unknown solution?

[**Ans:** 3.75×10^{-4} M]

46.29 In chronoamperiometric anlaysis the various parameters were as follow with usual notations: $n = 6$, $D = 1.4 \times 10^{-5}$ cm²/sec $C = 41$ sec, $id = 1.50$ µA, $A = 1.60$ cm². Calculate the concentration of this solution?

[**Ans:** 0.810 M]

46.30 What was the diffusion current which flowed through a cell at –1.0 V if capillary was 1.92 with $c = 2 \times 10^{-3}$ and $D = 0.72 \times 10^{-5}$?

[**Ans:** 12.5 µA]

46.31 What was the concentration of cadmium solution if in polarographic analysis the diffusion coefficient was 0.72×10^{-5} cm^2/sec, with rate of flow = 2 mg/sec, with time between successive drop was 4.4 and magnitude of average diffusion current was 10 μA?

[Ans: 1.5×10^{-3} M]

46.32 Find out the concentration of lead and barium in solution if the titration of 100 ml of this solution with 0.05 M $K_2Cr_2O_7$ solution (E = 1 V) gave following readings.

V(ml)	0	2	4	6	8	10	11	12	13
i(μA)	100	60	20	10	10	10	35	30	114

[Ans: 0.0045 M Pb^{2+}, 0.0057 M Ba^{2+}]

46.33 In micro determination of a radioactive element by amperiometry with 0.05 N solution of ferrous sulphate following readings were obtained:

$V \times 10^{-6}$ (ml)	0	50	100	125	150	170
i (μa)	0	0.05	0.04	0.06	0.13	0.20

What was the concentration of radioactive element in the given solution?

[Ans: 0.73 μg]

46.34 What was the concentration of arsenic in a solution if an amperometric titration of 100 ml of this solution (with KBr + HCl) with 0.01 M potassium bromate at 0.3 V (vs SCE) gave following readings:

V(ml)	5	9	9.3	9.6	9.9
i(μA)	0	0	1.4	–4.1	7.0

[Ans: 4.07×10^{-4} M]

46.35 In analysis of zinc by polarography by standard addition method the unknown sample showed diffusion current of 47.4 μA. When equal volumes of unknown and standard solution (5.50×10^{-4} M) were mixed the diffusion current was 58.5 μA. What was the concentration of zinc in unknown sample? **[Ans:** 3.75×10^{-4} M]

46.36 Calculate the change of current in the electrolytic oxidation of 0.005 g of Uranium of 830 mv if t_0 = 200 mA? **[Ans:** 2.301 – 0.0018077]

46.37 What should be the concentration of the Vitamin B (cynocobalamine) in unknown solution if at pH 11.5 it gave following data?

Vitamin B con (mg/ml)	0.40	0.90	1.20	1.60	Unknown
Diffusion current (i_d)	32.0	73.0	95.8	1.60	63.2

[Ans: 0.790 mg]

46.38 Find out the strength of the unknown solution of zinc with data given as:

Zn^{2+} (1×10^3 M)	0	0.40	0.60	1.0	1.2	Unknown
Diffusion current (i_d)	2.6	12.9	19.6	20.5	24.6	10.2 μA

[Ans: 4.8×10^{-4} M of zinc solution]

46.38 Find out the concentration of unknown solution of divalent metal if samples data was as follows:

Divalent metal (con.)	1.16	3.11	4.19	5.32	Unknown
Diff. current (i_d)	5.5	12.8	17.7	22	14.7

[Ans: 3.5 mM]

46.39 In a typical polarographic set-up, the value of the diffusion current was 25.4 microamperes of solution of the concentration of 0.003 moles/liter for particular metal, when polarographed at rate of flow of mercury was 3 mg/second when the time between two successive drops was 4 seconds. Assuming the unknown metal was divalent (*i.e.*, $n = 2$), what was the magnitude of the diffusion coefficient? **[Ans:** 7.2×10^{-6} cm^2/sec]

46.40 In a typical chronopotentiometric analysis of mercury the following data was obtained: $n = 2$, $m = 22.1$, $A = 2.4$, $I = 144$ μa and $D = 6.5 \times 10^{-5}$.
What should be concentration of mercury? **[Ans:** 0.05 M]

REFERENCES

46.1 S.M. Khopkar, *Basic Concepts of Analytical Chemistry,* 3rd edn., New Age International Publishers (P) Ltd. (2007).

46.2 A. Musekin, *Problems in Quantitative Analysis,* Mir Publishers, p. 106, 165-69 (1984).

46.3 R.D. Braun, *Introduction to Instrumental Analysis*, McGraw-Hill and Co., p. 815-16 (1987).

46.4 D.J. Pietrzyke and C.W. Frank, *Analytical Chemistry An Introduction,* Academic Press, p. 627-28 (1979).

46.5 D.A. Skoog and D.M. West, *Analytical Chemistry*, 3rd edn., Holt Rienhart and Winston, p. 431 (1979).

46.6 Y. Lyalikov, *Physico-chemical Analysis,* Mir Publishing House, Moscow p. 429, (1964).

46.7 R.C. Kapoor and U.C. Agarwal, *Polarography,* Wiley Eastern Ltd. (1980).

46.8 I.M. Kolthoff and J.J. Lingane, *Polarography,* 2nd edn., Interscience, NY (1952).

46.9 L. Meites, *Polarographic Technique,* Interscience, NY (1955).

46.10 J. Herrovsky and J. Kuty, *Principles of Polarography*, Academic Press (1966).

46.11 H. Schmidt and M. Stackberg, *Modern Polarographic Methods*, Academic Press (1963).

46.12 P. Zeman, *Organic Polarography Analysis*, MacMillion, NY (1964).

Chapter 47

Conductometric Methods

INTRODUCTION

The Ohm's law states that current is directly proportional to the potential and it is inversely proportional to the resistance *(i.e.,* $C = E/R$). We have seen in electroanalytical methods involving measurement of the current in voltammetry, or potential in potentiometric methods. The methods involving actual measurement of resistance are few however those involving the measurement of reciprocal of resistance, viz., conductance are many. The units for measurement of resistance are Ohms, while reciprocal of this, viz., mhos or Sieman is the unit for conductance. There is distinct difference between terms conductometry and conductivimetry. The former pertains to the titration's, while latter pertain to absolute measurement of conductivity of the solution.

The conductance is expressed in terms of area and length of the conductor as

$$L = \frac{a}{d}\Sigma c_1\lambda_1 = \frac{a}{d}\left[c_1\lambda_1 + c_2\lambda_2 + c_3\lambda_3 \ldots\ldots c_n\lambda_n\right]$$

Where d = distance between two electrodes, a = area of cross-section of electrode, c = concentration of ion per unit volume of solution and λ_1 = equivalent of conductance of the particular ion. Σ represent summation of several such ions in solution. Now the ratio of $d/a = \theta$ is called 'Cell Constant' so we have conductance as $L = \dfrac{\Sigma c_1\lambda_1}{1000\,\theta}$, if L represent conductance per liter of solution.

We can replace c_1 and λ_1 by terms as $\Sigma\lambda t$. Another term viz. specific conductance is most frequently used. It is $K = L\left(\dfrac{d}{a}\right) = L\theta$. This provides an experimental measure for evaluation of Θ if we can evaluate (K) specific conductance, if L = equivalence conductance. The specific conductance at fixed temp. varies with concentration (e.g., 71.13 g/kg. KCl = 0.11134, 7.419 g/kg = 0.01265 and 0.749/kg = 0.0014). It varies to the extent of approximately 2% with rise of 1°C of temperature. The equivalent ionic conductance gives quantitative information of their relative contribution to conductance mesurements.

The conductance is measured on platinum plate electrodes with large area with AC as power supply, as DC is of no use for such measurements. The conductance is also refered to 'G'. It is given in terms of ions as $G = \frac{\Sigma \lambda m_1 c_1}{1000\,\theta}$ if θ = cell constant, c_1 = concn. of ion and λm_1 is the equivalent conductance of ions. (For instance (λ°+) the equivalent conductance of ([H^+] = 3498, [K^+] = 73.5, [Na^+] = 50.1, [Li^+] = 38.7, [OH^-] = 198.6 [NO_3] = 71.4). The methods used for conductometric titration of sodium hydrodoxide against hydrochloric acid gives two st. lines intersecting at point of equivalence. Such methods are best for acid base titration though they were also used for the precipitation titration of silver sulphate with barium chloride. This technique is rarely used for redox titration of ferrous ion with permanganate as there is not much change in magnitude of conductance. It is an excellent method for the titration of weak acid and weak base, e.g., the solution of boric acid is best titrated by conductometric technique in the presence of the mannitol with weak base. The other nomenclature normally encountered in calculations in titration are as follows:

$$G = \frac{1}{R} \text{ if } G = \text{conductance, } R = \text{resistance, } R = \frac{L}{A} \text{ (if } L = \text{dist. between electrodes; } A = \text{area)}$$

$$G = \frac{A}{L} = \frac{KA}{L} \text{ if (} K = \text{specific conductance) (Here } L/A = \Theta \text{ is cell constant) } = \frac{K}{\theta}$$

$$G = KL = \frac{\lambda mc}{1000\,\Theta}, \text{ Since } G = \frac{\lambda mc(A)}{1000(L)} \text{ as } \lambda_m = \frac{1000\,K}{c} \text{ for several ions}$$

$$G = \frac{\Sigma \lambda m_1 c_1}{1000\,\theta} \text{ sometimes term 'transference number' is used (} N \text{), viz., } N = \frac{\lambda m_j c_j}{\Sigma \lambda m_1 c_i}.$$

It is the ratio of conductance due to particular ion in solution to the total conductance of the solution.

The electrophoretic and relaxation effects influences conductance with change in the concentration e.g., 0.1 M NaCl = 106.7 (λ), 0.01 M = 118.5 (λ), 0.001 M = 123.7 (λ) for infinite dilution 126.4 (λ_0) usually $\lambda_0 = \left(\lambda_c^0 + \lambda_a^0\right)$ is sum of ionic conductance of both cation and anions is equivalent conductance at infinite dilution represented by λ_0.

Finally oscillometry or high frequency titration is actually an example of electrode less titration as the electrodes do not come in contact with solution under the influence of potential or current. The dipolar molecules orient to + or –ve form to exhibit oriental polarisation. This is temporary distortion or polarisation which develops a small current for short interval. This is caused by radiofrequency (10^5-10^7 Hz.) waves.

The impedance of a solution to flow a current varies with the dielectric constant, electrolyte concentration, and frequency employed. The glass cylinder is juxtaposed between two electrodes to be part of sinewave oscillator at high frequency resistive when impedance is low.

These high frequency measurement are used for analysis of the binary mixtures of non-ionic species. The success of such work depends upon the actual difference in dielectric constant for two species, e.g., in analysis of benzene and chlorobenzene or alcohol, water.

The efficiency of high frequency titration depends upon the change in frequency with, volume of the titrant. Such method is mostly used where one does not intend to use an electrode in contact with the solution.

This chapter covers large number of solved as well as unsolved problem pertaining to calculation of equivalent conductance (λ); dissociation constant (*K*); conductance of solution; evaluation of the transference number (N); solubility product; conductance at equivalent point in the titration; titration curve, determination of cell constant (θ) or determination of the specific conductance of the given solution (*K*).

In order to get clear picture of theory the reader is advised to refer to books listed at end of this chapter.

SOLVED PROBLEMS

47.1 A 5% solution of potassium sulphate with electrode area of 2.54 cm² with distance between two electrodes 0.65 cm had a specific conductance of 0.178, what was the equivalent conductance of solution if the density of electrolyte was unity?

Ans: The specific conductance was calculated from expression as:

$$\omega = x\left(\frac{A}{L}\right) \text{ if } x = \frac{WL}{A} \text{ so } \omega = \left(\frac{WL}{A}\right)\left(\frac{A}{L}\right) \text{ also } K = \left(\frac{d}{A}\right) L\Theta = \text{cell constant.}$$

A = area of electrode surface.

So, x = eq. and W = specific conducts, L = dist. between two electrodes, we have

$$x = \frac{WL}{A} = \frac{0.178 \times 0.65}{2.54} = 0.0458 \text{ ohm (equivalent conduct).}$$

The number of gm equivalent of K_2SO_4 in 1 cm² of solution (*i.e.*, 5%)

$$n = \frac{5}{95 \times 87.13} = 6.039 \times 10^{-4} \text{ gm/eq (equi. weight of } K_2SO_4 = 87.13)$$

The equivalent conductance shall be $\dfrac{4.58 \times 10^{-6}}{6.039 \times 10^{-4}} = 75.8 \text{ ohm/gm/cm}^2$ equi conduct.

47.2 The electrical conductivity of 0.0109 N ammonium hydroxide was 1.02×10^{-4}/ohm. What was the dissociation constant of the solution?

Ans: $\lambda = \dfrac{K \times 1000}{c}$; K = sp. conductance, c = concentration, λ = eq. conductance.

Since $K = 1.02 \times 10^{-4}$; $c = 0.0109$ g, we have equivalent conductance (λ) as

$$\lambda = \frac{1.02 \times 10^{-4} \times 1000}{0.01090} = 9.38 \text{ ohm/gm eq/cm}^2$$

The equivalent conductance at infinite (∞) dilution was evaluated as follows:

$$\lambda_\infty = (76 + 205) = 281 \text{ ohm/gm/cm}$$

as $$L = \frac{\Sigma c_1 \lambda_1}{1000} = \frac{\Sigma c_1 \lambda_1}{1000} \times \left(\frac{A}{d}\right)$$

so $$\alpha = \frac{\lambda}{\lambda_\infty} = \frac{9.38}{281} = 0.0334 \text{ was degree of dissociation}$$

$$K = \frac{\alpha^2 c}{(1-\alpha)} = \frac{(0.0334)^2 \times (0.0109)}{(1 - 0.0384)} = 1.26 \times 10^{-5} \text{ was the dissociation constant.}$$

47.3 In conductometric titration of 25 ml of sodium hydroxide against 1 M hydrochloric acid the following readings were obtained:

Conductance (y)	3.15	2.6	2.04	1.4	1.97	2.86	3.66
Vol. of titrant (x)	0	1	2	3	4	5	6

What is the number of gram equivalent of sodium hydroxide?

Ans: We plot graph of vol. of titrant (x) axis against conductance on (y) axis. The intersection is at point of 2.5 ml for acid therefore the calculation leads to.

$$2.5 \times 1 = 25 \times x \text{ so } x = \frac{2.5 \times 1}{25} = \frac{1}{10} = 0.1 \text{ N}(\text{NaOH} = 40 \text{ Mol. wt.})$$

Gm eq of NaOH = 0.1 × 40 = 4 gm eq. for liter.

47.4 If a 0.0001 M potassium chloride had conductance value of 2.83×10^{-6} S. What is the conductance of a solution containing 1×10^{-4} M hydrochloric acid and 2×10^{-4} M ferric oxalate in the cell?

Ans: Cell constant $\theta = l/A$ was fixed.

The specific conductance, κ = conductance of KCl (0.01 M) = 1.437×10^{-3}

$$\text{So } 283 \times 10^{-6} = \frac{1.413 \times 10^{-3}}{\Theta} \text{ so } \Theta = 4.99 \text{ cm}^{-1}.$$

The molar conductance value from standard tables was (H^+) = 0.0001 was 349.8, (Cl^-) = 0.0001 is 76.4, (Fe^{3+}) = 0.0004 was 2.4 and $\left(C_2O_4^{2-}\right)$ of 0.0006 M was 148 so we put in equation.

$$L = \frac{\lambda mc}{1000}\left(\frac{A}{l}\right) = \frac{349.8(0.0001) + 76.4(0.0001) + 204(0.0004) + 148(0.0006)}{1000 \times 4.9}$$

as $\quad \Theta = 4.99 = \frac{1}{A}$

$L = 4.27 \times 10^{-5}$ specific conductance.

47.5 What was the 'Transference number' (N) of chloride ion in above problem?

Ans: $N = \frac{\lambda m_j c_j}{\Sigma \lambda m_1 c_1}$ is equivalent conductance of chloride ion was 76.4.

$$N = \frac{76.4(0.0001)}{349.8(0.0001) + 76.4(0.0001) + 204(0.0004) + 148(0.006)}$$

$= 3.59 \times 10^{-2}$

47.6 Calculate the solubility product of silver chloride at 25°C if the cell constant $\theta = 1.20 \text{ cm}^{-1}$ and if the cell contained saturated solution of silver chloride at room temp. with conductance of 1.44×10^{-6}?

Ans: $L = \frac{\Sigma \lambda m_1 c_1}{1000\,\Theta}$ also molar conductance for each ion was known the ionic conductance of

Ag = 61.9, Cl = 76.4, θ = 1.20

So, $$L = \left[\frac{\lambda m^6 C_{Ag} + \Delta MCl\,C_{Cl}}{1000\theta}\right], \text{ we put } L = 1.44 \times 10^{-6}.$$

So, $$1.44 \times 10^{-6} = \frac{61.9 \times C_{Ag} + 76.4\ C_{Cl^-}}{1000(1.20)} = 1.25 \times 10^{-5}\ M$$

At equivalence $C^+_{Ag} + C^-_{Cl} = C$ so $AgCl = Ag^+ + Cl^- = (Cl^- + Cl^-) - (Ag^+ + Ag^+)$.

$$1.44 \times 10^{-6} = \frac{61.9C + 76.4C}{1.2 \times 10^3} \text{ as } 1.25 \times 10^{-5}\ M = C_{Cl} = C_{Ag} = C$$

The solubility product was calculated as

$$K_{sp} = [Ag^+][Cl^-] = (1.25 \times 10^{-5})(1.25 \times 10^{-5}) = 1.56 \times 10^{-10}.$$

47.7 What was the equivalence conductance of sulphuric acid and ionic conductance of ions brackets as |H(350)|; SO_4^{2-} (80) are given?

Ans: The molar conductance of sulphuric acid was

$$2[H^+] + [SO_4^2] = 2[350] + [80]2 = 860\ \text{ohm}^{-1}\text{cm}^2.$$

Now equivalent conductance (λ_ω) = (350 + 80) = 430 ohm^{-1} (Eq. Conductance (λ_w) = $\Sigma(\lambda^+)$ + (λ^-) of cation and anion).

47.8 Find equivalent conductance of the electrolytic cell 0.01 M solution of potassium chloride it had resistance of 161.8 and for 0.005 M sodium hydroxide the resistance was 190 ohm?

Ans: The cell constant (θ) was calculated from table the specific conductance of 0.010 M KCl = 0.01286 and 0.001 M KCl = 0.001409.

$$K = \frac{\Theta}{R}\ A/R \text{ if } K = \text{sp. conductance so } (K \times R) = \theta.$$

$$(0.001409) \times (161.8) = 0.2281\ \text{cm}^{-1}$$

$$\text{Sp. cond. of alkali} = \frac{\theta}{R} = \frac{0.2281}{190} = 0.00120\ \text{cm}^{-1}.$$

Hence, equivalent conductance of alkali hydroxide was

$$\lambda = 1000 \times \frac{K}{C} \text{ if } K = \text{specific conductance, } C = \text{concentration.}$$

$$\lambda = \frac{(1000) \times 0.00120}{0.005} = 240\ \text{cm}^{2-}\text{eq}^{-1}\text{ohm}^{-1}.$$

47.9 The conductivity bridge reads values from 0.005 – 2% sulphuric acid in approximately logarithm scale. Predict range of resistances involved and cell constant, if the specific conductance ranges from 0.00044 to 0.176 ohm cm^{-1}?

Ans: The resistance value ranges as:

$$R = \frac{\theta}{0.00044} = 22{,}800 \text{ as } R = \frac{\theta}{K};\ K = 0.0004$$

$$R = \frac{\theta}{0.176} = 56.9 \text{ as } R = \frac{\theta}{K} \quad K = 0.176$$

An appropriate cell constant was 10 cm^{-1} for resistance (57 – 22,800) ohms.

47.10 What was the conductance at equivalent point if 100 ml of 0.01 N hydrochloric acid was titrated against 0.1 N sodium hydroxide? (cell constant (θ) = 1 cm)?

Ans: $$R = \frac{\theta}{K} \text{ so } \frac{1}{R} = \frac{K}{\theta} = \frac{0.01(350+76)}{(1000\times 1)} \text{ as } H^+ = 350, Cl^- = 76.0$$

$1/R = 0.00426$ ohm^{-1} = R^{-1}.

At infinite dilution when titration was complete at 0.9.

$$\frac{1}{R} = \frac{0.01}{(1000)(1)}\left[350 + 76 + 0.9(50 - 350)\right] = 0.00156 \text{ ohm}$$

If we correct for the dilution caused during titration, then we have

$$\frac{1}{R} = \frac{(100+9)}{(1000)(1)}(0.00156) = 0.00170 \text{ ohm/cm}$$

At equivalence point $\frac{1}{R} = \frac{0.01}{(1000)(1)}(76+50)\left(\frac{110}{100}\right) = 0.00139$ ohm.

When end point exceeded by 10%.

$$\frac{1}{R} = \frac{0.01}{(1000)(1)}(126+0.1)(248)\left[\frac{110}{100}\right] = 0.00167 \text{ ohm.}$$

47.11 What was the conductivity of potassium chloride with concentration of (1 × 10^{-2} M)?

Ans: We use formula as

$$x = \frac{1}{R}\times\frac{1}{S} = \frac{0.01\times 0.94}{1000}(73.5+76.4) = 0.00410 \text{ ohms.}$$

If 0.94 was mean activity coeff. of KCl, (73.5 + 76.4) is mol wt. of KCl. 0.01 N is con. of KCl, S = electrode surface area $1/R$ = conductance value.

47.12 Predict the resistance of 0.001 N potassium chloride solution between electrodes of area 0.4 cm^2 with separating distance was one cm?

Ans: $$\frac{1}{R} = \frac{s}{1}\Sigma\frac{C_1}{1000}\lambda_{\pm}$$

Where $\frac{1}{R}$ = conductance, S = electrode surface area, d = distance between electrodes

$\frac{1}{S}$ = cell constant, $\left(\frac{1}{R}\times\frac{1}{S}\right)$ = electrical conductivity, C_i = ionic concent, λ = eq. conductance.

So $$\frac{1}{R} = 0.001\times 0.98(73.5+76.4)\times 0.4 = 0.0588 \text{ m ohms.}$$

So R = 17 ohms if 0.98 degree of dissociation, *i.e.*, activity coefficient of KCl (0.001 N) was con. of KCl; (73.3 + 76.4) mol. wt. of KCl and 0.4 is electrode area (s), d = 1 distance between two electrodes $\frac{1}{R} = K \times (\text{con. KCl}) \times (\text{Mol. wt. KCl}) \times \frac{\text{electrode area}}{\text{dist.}} = 170$ ohms.

47.13 What was the concentration of a solution of hydrochloric acid if 50 ml of this solution was titrated with 0.001 N alkali hydroxide furnished following results:

V(ml)	0	2	4	6	8	10
R(ohms)	664	915	1490	1580	1010	740

Ans: The equivalance point lies at the intersection of two lines on the graph.

$$r = a_r + r_o \text{ if } r = \frac{1}{R} \times 10^{14} \quad a = \text{slops of line } (i.e., \tan\theta\ v = \text{vol of NaOH}, r_o = r \text{ as}$$

We substitute these values and find out conductance as follows

V(ml)	Q	2	4	6	8	10
$1/R \times 10^4$	15.1	10.9	6.5	6.35	9.9	13.5
$\tan\theta = \Delta r/\Delta r$	–0	–2	–2.2	–0.08	+1.78	+1.8

We get intersection $a = 2.15$, $r_o = 15.1$ for second line $a = 1.79$ if $r_o = 0$

$$V_{eq} - 2.15\,V_{eq} + 15.1 = 1.79(V_{eq} - 6) + 6.35$$

$$V_{eq} = 4.95 \text{ so } \frac{4.95 \times 0.1}{50} = 0.00099.$$

47.14 Portray the titration curve for 50 ml of 0.01 N hydrochloric acid against 0.1 N sodium hydroxide if $\alpha = 0.90$? (cell constant = 0.4 cm) with following readings:

Vml of NaOH	0	1	2	4	6	8	10
$1/R \times 10^4$ (ohm)	15.3	13.5	11.23	6.7	6.36	9.9	13.5
$\Delta a/\Delta v = \tan\theta$	–1.8	–1.8	–2.25	2.25	–	+1.78	+180

Ans: We solve the problem as

$$\frac{1}{R} = \frac{0.9 \times 0.4}{1000}\left[(H^+) \times 350(Cl)76\right] + [Na^+]500 + [OH][198]$$

Thus for 2 ml sodium hydroxide,

$$[H^+] = 0.06\ [Cl^-] = 0.01\ [Na^+] = 0.004 \text{ where } \left(\frac{1}{R} \times 10^4 = 0.001125\right)$$

The slope of lines near end point was at 2.25 and after end point 1.8.

47.15 Depict the curve for the titration of 50 ml of 0.01 N solution of acetic acid with 0.10 N alkali hydroxide? (cell constant 0.4 cm^{-1}).

V of NaOH ml	0	2	4	6	8	10
$1/R \times 10^4$ (ohm)	0.66	1.48	2.88	5.64	9.60	13.6

Ans: Before eq. pt $[H^+]$ was calculated from diss. constant was 1.8×10^{-5}.

$$[H^+] = 1.8\times10^{-5}; \frac{[CH_3COOH]}{[CH_3COO^-]} = 1.8\times10^{-5} = \frac{[acid]}{[salt]}$$

e.g., when the amount of NaOH was 2 ml.

$$[H^+] = \frac{50\times0.01-2\times0.1}{2\times0.1}\times1.8\times10^{-5} = 2.7\times10^{-5}$$

$[CH_3COO^-] = 0.004$ $[Na^+] = 0.004$

$$\frac{1}{R} = 0.000148.$$

At eq. pt. $[OH^-]$ was calculated from volume of excess of alkali

The slope of st. line to the *x*-axis before the eq. pt. was 0.5 after eq. pt. it was $2.0\ ohm^{-1} \times 10^4\ ml^1$.

47.16 What was the resistance of 0.001 N chloride of potassium solution between electrodes of area of 0.4 cm^2 when placed at as distance of 1 cm? (Given 0.98 is degree of diss.)

Ans: Since we have $\frac{1}{R}$ = conductance

So $$\frac{1}{R} = 0.001\times0.98(76.4+73.5)\times0.4 = 0.0588\ ohm^{-1}.$$

So R = 17 ohms as

Here 0.98 was degree of dissociation of KCl.

47.17 What was the concentration of sodium hydroxide and ammonium hydroxide from following result if 0.01 N acid was titrated with 50 ml of the solution?

V(ml)	0	1	2	4	6	8	10
$\gamma = 10.1/R$	6.3	5.41	4.52	3.71	5.85	9.0	15.13
$\Delta a/\Delta_v = \tan\theta$	–0.91	–0.89	–0.89	1.08	1.08	3.08	3.05

Ans: The equivalent pt. (first) was 0.9 V eq. + 6.3 = 1.08 (V_{eq} – 4) + 3.71.

So V_{eq} = 3.49 ml. Now second equivalence point was

1.08 (V_{eq} – 4) + 3.71 = 3.08 (V_{eq} – 8) + 9.0 so V_{eq} = 7.2.

$$\text{NaOH concn. } \frac{3.49\times0.01}{50} = 0.000698,\ i.e.,\ NaOH = 6.9\times10^{-5}$$

$$NH_4OH\text{ concn.} = \frac{(7.52-3.49)\times0.01}{50} = 0.000805\ i.e.,\ NH_4OH = 8.5\times10^{-5}.$$

47.18 Evaluate the magnitude of cell constant (θ) for 0.1 M potassium chloride with conductance of 1.58?

Ans: $$G = \frac{K}{\theta}\left(\text{if } G = 1.58,\ c = 0.1\text{M},\ K = 1.288\times10^{-2}\text{ sp. cond.}\right)$$

Then $$\theta = \frac{K}{G} = \frac{1.288\times10^{-2}}{1.58} = 8.15\ cm^{-1}.$$

47.19 What was the conductance of solution containing 1.15×10^{-4} M common salt and 5.65×10^{-5} M of potassium chromate with cell constant $\theta = 8.15$? ($K = 1.46 \times 10^{-4}$ Given).

Ans: Given: $\theta = 8.15$, $K = 1.46 \times 10^{-4}$

$$L = \frac{K}{\theta} = \frac{1.46 \times 10^{-4}}{8.15} = 3.98 \text{ µs.}$$

4.20 What was the specific conductance (K) of 5% potassium sulphate with electrode area of 2.54 cm (A) and dist. between electrodes 0.65 cm with equivalence conductance of 75.8?

Ans: $w = x\left(\frac{A}{L}\right)$ or $\left(\frac{WL}{A}\right) = x;\ \frac{0.178 \times 0.65}{2.54} = 0.045$ ohms.

PROBLEMS FOR PRACTICE

47.21 A conductance at room temp. for 0.1 N potassium chloride was of 0.01315 and one with 0.105 M acetic acid had conductance of 5.58×10^{-3}. What was the dissociation constant of acetic acid? **[Ans:** 1.88×10^{-5}**]**

47.22 A 0.02 M sulphuric acid containing sodium iodide was titrated with 1N of potassium permanganate. The readings for volume versus conductance were as follows:

Volume (ml)	0.2	0.4	0.6	0.8	1.0	1.2	1.4	1.6	2.4
Conductance	943	872	803	737	670	622	608	609	614

What was the amount of sodium iodide in the original solution? **[Ans:** 2×10^{-45} M**]**

47.23 The specific conductance of 0.02 M hydrochloric acid was 7.92×10^{-3} ohm. What was the equivalent conductance of 0.02 M solution of same acid? **[Ans:** 396 ohm^{-1}**]**

47.24 What was the equivalent conductance at infinite dilution of a solution of silver sulphate? **[Ans:** 141.7 ohm^{-1}**]**

47.25 When a solution of 0.01N sodium hydroxide was titrated against hydrochloric acid conductometrically following titre values were obtained:

Volume (ml)	0	4	8	9	10	11	12	14	17
Conductance	218	230	243	247	256	269	285	323	380

Calculate the amount of common salt present in a solution? **[Ans:** 0.820 g**]**

47.26 Evaluate the magnitude of cell constant if conductance of 0.1 M potassium chloride was 1.58 at room temperature? **[Ans:** 8.15 cm^{-1}**]**

47.27 What was the conductance of solution with cell constant (θ) 8.15 containing 1.15×10^{-4} M sodium chloride and 5.65×10^{-5} M potassium dichromate in a cell? **[Ans:** 3.98 µs)

47.28 Calculate the transference number of iron in a solution having 1×10^{-3} M ferric chloride in decimolar hydrochloric acid? **[Ans:** 2.52×10^{-3}**]**

47.29 What was the solubility product of cell with cell constant (θ) as 2.38 cm^{-1} with conductance of saturated solution of baryata was 1.27 µs? **[Ans:** 1.10×10^{-10}**]**

47.30 What was the conductance of 1N potassium chloride solution in the titration cell, if the electrode area was 1.25 cm^2 and spacing between two electrodes was 0.84 cm?

[Ans: 0.143 cm^{-1}**]**

47.31 If 0.324 ohm resistance was showed by 10% sulphuric acid in cell, find out the specific molar and equivalent conductance if the area of the electrode was 5.25 cm^2 and space between two electrodes was 0.65 cm? (Given density of solution 1.079/cm). [**Ans:** 0.388 ohm^{-1} cm^{-1}]

47.32 In cell with cell constant of 20 measuring conductivity in concentration range of 1-18% hydrochloric acid with corresponding conductance varying from 0.0630 – 0.750 ohm. What would be span of resistance of the said cell? [**Ans:** 318 – 26.7 ohms]

47.33 If a solution of 20% hydrochloric acid had specific conductance of 0.85 mhos, calculate the resistance? (cell constant 1.0). [**Ans:** 1.17 Ω^{-1}]

47.34 If 100 ml of dilute hydrochloric acid was titrated with 0.02 N sodium hydroxide the titre value with conductance in brackets were 0(26552), 1(26036), 2(25328), 3(24506), 4(23536), 5(22585), 6(22940), 7(23631), 9(24930), 10(925445). Calculate strength of hydrochloric acid? [**Ans:** 0.00105 M]

47.35 In typical titration if 50 ml of acetic acid was titrated against 0.1 M sodium hydroxide the titer values were 0(55), 1(35), 2(62), 3(91), 4(170), 5(252), 6(337) as volume (conductance). Calculate the concentration of acetic acid? [**Ans:** 6×10^{-3} M]

47.36 If the specific conductance of N/50 hydrochloric acid was 0.00792 $ohms^{-1}$ what was the equivalence conductance of the acid at 25°C? What was the equivalence conductance as infinite dilution of a solution of silver sulphate? [**Ans:** 396 $ohms^{-1}$ 141.7 $ohms^{-1}$]

47.37 When a very dilute solution of sodium hydroxide was titrated conductometrically with hydrochloric acid following readings were obtained as:
0(3.15), 1(2.60), 2(2.04), 3(1.40), 4(1.97), 5(2.86), 6(3.66). Draw the titration curve and determine strength of sodium hydroxide. [**Ans:** 0.127 gm]

47.38 A solution of 0.001 N sodium acetate was titrated conductometrically with 1 N hydrochloric acid. From following reading, calculate no. of g of sodium acetate present in solution. The readings were as: 0(218), 4(230), 8(243), 9(247), 10(256), 11(269), 12(285), 14(323), 17(380). [**Ans:** 0.820 gm]

47.39 A 65 mg sample of organic compound containing iodide was dissolved in 50 ml of methanol and it was titrated with methanol solution of silver nitrate which was 2×10^{-1} M. The high frequency conductance was measured after each 10 ml of titrant added. The end point was 0.95 ml. Calculate the molecular weight of the organic compound? [**Ans:** 342]

47.40 A solution of 0.04 N in sulphuric acid containing potassium iodide was titrated with 1.1 N potassium permanganate when following results were obtained. Find out the nos. of gram of potassium iodide present in the original solution? The readings were 0.2(943), 0.4(872), 0.6(803), 0.8(737), 1(670), 1.2(622), 1.4(608), 1.6(609), 1.8(610), 2.4(614). [**Ans:** 0.214 gm]

REFERENCES

47.1 S.M. Khopkar, *Basic Concepts of Analytical Chemistry*, 3rd edn., New Age International Publishers (P) Ltd. (2007).

47.2 R.D. Braun, *Introduction to Instrumental Analysis*, McGraw-Hill and Co., pp. 810-12; 818-20 (1987).

47.3 H.H. Willard, L.L. Merrit and J.A. Dean, *Instrumental Methods of Analysis,* D. Van Nostrad and Co., pp. 718, 722-25, 742-43 (1980).

47.4 A. Musekin, *Problems in Quantitative Analysis,* Mir. Publishing House, pp. 176-180 (1984).

47.5 Y. Lyalikov, *Physico Chemical Analysis,* Mir. Publishers, Moscow, pp. 368 (1964).
47.6 H.T.S. Britton, *Conductometric Analysis,* Van Nostrand and Co. (1934).
47.7 E. Püngor, *Oscillometry and Conductometry,* Pergamon Press, N.Y. (1965).
47.8 J.J. Lingane, *Electroanalytical Chemistry,* 2nd edn., Interscience (1958).
47.9 P. Delahay, *New Instrumental Methods in Electrochemistry,* Interscience (1954).
47.10 G.H. Ayeres, *Quantitative Chemical Analysis*, Harper and Row Brothers (1980).
47.11 S. Brewster, *Solving Problems in Analytical Chemistry,* John Willey and Co. (1980).
47.12 M.A. Yoroslavesei, *Problem and Exercises in Analytical Chemistry,* Mir Publishing House, Moscow (1984).

Chapter 48

Coulometry

INTRODUCTION

Electrogravimetry, coulometry and coulometric titrations are most important analytical methods. In the first technique namely in the electrogravimetry a cathode is initially weighted, electric current is then passed through electrolyte to deposit metal in form of the film at cathode, then again it is weighted. The difference in weight indicates the amount of metal deposited by known quantity of electricity in fixed amount of time. The amount of electricity needed to generate one gram equivalent of the metal is called 'Faraday'. Thus 96,500 coulombs is equivalent to the one Faraday of electricity. The coulomb is defined as that amount of the electricity which flows during the passage with constant current of one ampere for one second. The quantity (Q) of electricity is given in terms of I-curent and t-time as $Q = It$ for variable current coulombs are given by an expression as $Q = \int_0^t dt$. One Faraday is equal to 6.02×10^{23} electrons Platinum in the form of cylinderical sponge is made as the cathode and pure platinum wire is made as the anode. The deposition of zinc or copper creates problem unless platinum gauge is coated with the thin layer of the copper. The metals like cadmium, cobalt, copper, iron, lead, nickel, arsenic tin and zinc are determined quantitatively by electrogravimetry techniques. During electrolysis the potential applied is measured in terms of potential at cathode and anode and the over potential $E = (E_c - E_a) + IR$, if IR represent over potential or ohms drop. This expression should account for W_a and W_e in voltage as $V = (E_a + W_a) - (E_c + W_c) + IR$. We impose the decomposition potential to start plating of metal. The ohmic drop (IR) is due to the three effects, viz., polarisation effect or the concentration polarisation or kinetic polarisation with increase in current density over voltage increases. However with rise of temp. potential decreases, it is more with softer metals, e.g., lead, zinc, mercury. It is also more if gases are evolved but its magnitude cannot be exactly predicted at any time.

The controlled potential electrolysis gives best result. In controlled potential cathode electrolysis we have as expression relating current with concentration as $\frac{c_t}{c_o} = \frac{I_t}{I_o} = 10^{-KC}$, if c_t and c_o are concentration at time t and initial concn. While I_t and I_o are current time 't' time KC is constant which can be defined as $K = \frac{DAS}{V}$, if D = diffusion coefficient, A = area of cathode, S = rate of stirring and V = volume of solution in ml.

In coulometry technique, it encompasses a method involving measurement of the quantity of electricity (in coulombs) required to convert analyt into the different chemical form. We express it mathematically as $W = \frac{QM}{nF}$, if W = wt. of substance with formula weight as 'M'.

Q = coulombs of electricity passed, n = number of equivalent per formula weight, F = Faradays constant. It permits quantitative determination, if we have 100% current efficiency. Such methods are accurate like conventional gravimetric and volumetric methods. These methods are rapid and adaptable for on line analysis. The method works both at constant current and constant potential but latter is most common in use. The technique at constant potential proceeds without involvement of less reactive species. In beginning current is high but it decreases to zero. While in second at constant current it is continued till reaction reaches to completion. The amount of electricity used is related to amount of material deposited. The latter method is also called as 'Coulometric titration'. It involves electrolytic generation of a reagent which in turn reacts with analyt sample. The electrode may just prepare reagent or analyt may be directly involved at generation of electrode. The net process must have 100% current efficiency. The current is kept constant, a quantitative oxidation or reduction of unknown species proceeds at electrodes with transfer of electrons with net rise in electrode potential. The reaction occurs away from working electrode. Such techniques have been developed for all kinds of titrations like acid base, redox, precipitation and complexometric titrations. Pt is used as anode. The possible source of error in coulometric tritration is due to variation in current imposed, absence of 100% current efficiency; or wrong measurement of current or time and error arising due to the location of the end at equivalent point. Coulometric titration at constant potential are known, e.g., reduction of picric acid or trichloroacetic acid at mercury cathode. It is used to monitor gas or liquids. In the atmosphere with this method the sulphur dioxide in the polluted atmosphere can be monitored.

We use Nernst equation to calculate concentration of species generated. At a times we use following relationship as if $W = \frac{i \times t \times eqwt}{F}$, if W = wt. of sub., I = current in amps., t = time in seconds and F = Faraday = 96,500 coulombs. During electrolysis at fixed potential current decreases with concentration of species as $-Kt = \log \frac{I_t}{I_t - 0}$, if I_t = current at any time t, I_o = current at time t_o and K = constant. A plot of log $\left(\frac{I_t}{I_0}\right)$ against c will be st. line graph with slope of K.

This chapter therefore provides solved as well as unsolved problems dealing with calculation of electrolysis normality of titrant, concentration of metal in a cell (*i.e.*, cathodic compartment). Calculation of potential of cathode or calculation of over potential; current efficiency, time needed for decomposition, etc. For detailed theoretical discussion the reader is advised to consult the various books cited at the end of this chapter.

SOLVED PROBLEMS

48.1 If a solution of 4.8×10^{-10} M silver and copper ions was electrolysed; what will be the concentration of copper before it would be deposited with silver in solution?

Ans: By Nernst eq. (Given $E^0_{\text{Cu}} = +0.337,\ E^0_{\text{Ag}} = +0.799$ v, over potential = O_v)

$$E = E^0_{\text{Cu}} + \frac{RT}{nF} = \log\left(\text{Cu}^{2+}\right) = E^0_{\text{Ag}} + \frac{RT}{nF}\log\left[\text{Ag}^+\right].$$ We substitute the values

$$0.337 + \frac{0.0591}{2}\log\left[\text{Cu}^{2+}\right] = 0.799 + \frac{0.0591}{1}\log\left(4.8\times10^{-10}\right)$$ as $n = 2$ (for Cu), $n = 1$ (for Ag)

on solving both the sides of the equations, we have

$$\log\left[\text{Cu}^{+2}\right] = \frac{0.799 - 0.551 - 0.337}{0.0295} - \left(\frac{0.089}{0.0295}\right)$$

Hence, Cu = 1.0×10^{-3} M.

48.2 How many grams of cadmium will be deposited in 240 minutes by a current of 2 amperes?

Ans: 240 minutes × 60 = 14400 seconds = *t*; *I* = 2 amps

Nos. of coulombs used = *I* × *t* = 2 × 14400 = 28,800 coulombs.

Since one Faraday, *i.e.*, 96500 coulombs give 56 of cadmium, 28,800 coulombs would give

$$\frac{28{,}800\times56}{96{,}500} = 16{,}712 \text{ gram of cadmium.}$$

48.3 In coulometric titration of 20 ml of potassium dichromate with ferric ion (III) which was self-generated in solution, took 25 mts to reduce when 200 μA of current was employed. Calculate the normality of potassium dichromate solution.

Ans: Since *Q* = (*I* × *t*) = (25 × 60) × 200 = 300 coulombs.

(if *i* = 0.200 and *t* = 25 × 60 = 1500 seconds)

1 Faraday = 96,500 coulomb $\frac{1\times300}{96500} = 3.01\times10^{-3}$ gm

equivalent for 20 ml of solution,

Therefore $\frac{3\times10^{-3}\times1000}{20} = 0.1554 \text{ N } K_2Cr_2O_7.$

48.4 In coulometric determination of zinc 35.4 ml of hydrogen and oxygen was generated after reduction of metal. Calculate the concentration of zinc in the cell. (Eq. wt. of zinc = 32.7).

Ans: In electrolysis of water 22.4 liter of hydrogen and 11.2 liter of oxygen was liberated. Hence, total volume generated was (22.4 + 11.21 = 33.62). Now one gm eqivalent of zinc (viz. 32.7) give 33.62 liter so 35.4 ml corresponded to $\frac{32.7\times3.54}{33.62}$ wt. of zinc in cell was 34.2 mg.

48.5 Find out the weight of deposited silver and copper when 96,500 coulombs of electricity was passed through solution assuming 100% current efficiency.

Ans: The reaction in cell is as follows:

$Ag^+ + e^- \rightarrow Ag^0$, $Cu^{2+} + 2e^- = Cu^0$, eq. wt. of Ag = 107.9/1 = 107.9 as n = 1.

eq. wt. of copper = 63.54/2 = 31.77 as *n* = 2.

Now $Q = (i\times t)w = \left[\frac{i\times t\times eqwt.}{F}\right]$. We substitute the values,

$$w = \frac{96,500 \times 107.1}{96,500} = 107.9 \text{ gm equivalent of silver.}$$

$$w = \frac{96,500 \times 31.77}{96,500} = 31.77 \text{ gm equivalent of copper.}$$

48.6 Zinc was electrolysed for half a minute with one microampere current. What was the weight of zinc deposited assuming 100% current efficiency?

Ans: The reaction was $Zn^{2+} + 2e^- = Zn$.

The eq. wt. of zinc = Mol wt./n = 65.38/2 = 37.69 as n = 2.

Current passed = 0.001 × 30 seconds = (i × t)

So $w = \left[\frac{i \times t \times eqwt.}{F}\right]$. If i = 0.001 amps, t = 30, eq. wt. = 37.69, F = 96,500.

$$w = \frac{0.001 \times 30 \times 37.69}{96500} = 1.17 \times 10^{-5} \text{ gm.}$$

48.7 When a 100 ml of chloride solution was titrated with silver nitrate with one microampere current, find out the concentration of the chloride when the end point was noted after 102 seconds when Ag and SCE electrodes were used.

Ans: $\text{wt. of Ag} = \dfrac{i \times t \times \left(\dfrac{at \cdot wt \cdot Ag}{n}\right)}{F}$ with usual notations.

$$\text{Moles of } Ag^{2+} = \text{moles of } Cl^- = \frac{\text{wt. of Ag}}{\text{at. wt. of Ag}} = \left(\frac{i \times t}{nF}\right) \text{ if } n = 1.$$

$$\text{Moles of } Cl^- = \frac{(0.001)(102)}{(1 \times 96,500)} = 1.1 \times 10^{-6} \text{ moles/100 ml of chloride.}$$

48.8 What should be the concentration of the iron in the sample if at constant coulometry if used to find out iron concentration, with 100% current efficiency, 0.1 M ceric sulphate solution was consumed. The electrode reaction was:

$Fe^{2+} \rightarrow Fe^{3+} + e^-$ and $Ce^{3+} \rightarrow Ce^{4+} + e^-$ and $Fe^{2+} + Ce^{4+} \rightarrow Ce^{3+} + Fe^{3+}$ and

at equivalence point of titration of 25 ml, 6.43 μA current was used for period of 223 seconds?

Ans: $Q = i \times t$ = (6.43 × 10^{-3}) (223) = 1.43 coulombs.

Since 96,5487 coulomb gave 1 equivalent 1.43 coulombs would give 1.48 × 10^{-5} Faraday.

$$\frac{1.43 \times 1}{96,487} = 1.48 \times 10^{-5} \text{ Faradays.}$$

Nos. of moles of Iron (II) in the solution was

$$1.48 \times 10^{-5} \times \frac{1\text{mole Fe}}{\text{Fe}} = 1.48 \times 10^{-5} \text{ moles of Fe.}$$

Now nos. of moles of iron is, since volume of titrant = 25 ml.

$$[Fe^{2+}] = \frac{1.48 \times 10^{-5}}{0.0250} = 5.92 \times 10^{-4}\,M.$$

48.9 If a constant current of 10 μa was passed through solution for 200 seconds. Calculate the amount chloride which had reacted with silver anode?

Ans: Since $Q = i \times t$; here $i = 10 \times 10^{-3}$ ampere, $t = 200$ second.

So $Q = 10 \times 10^{-3} \times 200 = 2.0$ coulombs.

As 96,487 coulomb was 1 gm eq. 2.0 coulombs gives 2.075×10^{-5} Faraday.

$$\frac{1 \times 2.00}{96{,}487} = 2.075 \times 10^{-5} \text{ Faraday or moles of chloride.}$$

Now as $Ag^0 + Cl^- = AgCl + e^-$.

2.075×10^{-5} was equivalent to chloride obtained from AgCl.

So amount was (2.075×10^{-5}) (35.45) $= 0.735 \times 10^{-3}$ as 35.45 is equivalent weight = 0.735 mg of silver deposited.

48.10 Find out the cathode potential when 0.01 M solution of copper was slowly electrolysed? ($E^0_{Cu} = 0.337$)

Ans: As per Nernst equation.

$$E_{Cu} = E^0_{Cu} + \frac{RT}{nF}\log\frac{1}{[Cu^{+2}]}. \text{ As } E^0_{Cu} = 0.337;\ \frac{RT}{F} = 0.0591,\ n = 2.$$

We have following expression.

$$E_{Cu} = 0.337 - \frac{0.0591}{2}\log\left[\frac{1}{0.01}\right] \text{ as } [Cu] = 0.01M.$$

$E_{Cu} = 0.278$ V versus normal hydrogen electrode.

48.11 In electrolysis of 0.01 M copper solution if concentration of copper was changed to (1×10^{-5}) M indicate exact potential of current needed?

(Given E^0_{Cu}=0.337, RT/F = 0.0591, n = 2).

Ans: Since $$E_{Cu} = E^0_{Cu} - \frac{RT}{nF}\log\frac{1}{[Cu^{+2}]} = 0.337 - \frac{0.0591}{2}\log\frac{1}{[Cu^{2+}]}$$

$$= 0.337 - \frac{0.0591}{2}\log\frac{1}{[0.000001]}$$

$$= +0.160 \text{ V against normal hydrogen electrode.}$$

48.12 A complex of $[Cd(NH_3)_4]^{2+}$ of (0.01 M) in 1 M ammonia at platinum cathode and 0.02 M of complex of was $[Zn(NH_3)_4]^{2+}$ was electrolysed. Calculate the potential at which cadmium as well as zinc will be plated with normal hydrogen electrode?

Ans: The various reactions in the cell are as follows:

$[Cd(NH_3)_4]^{2+} + 2e^- = Cd + 4NH_3$. $\quad E^0_{Cd} = -0.61\,V$

$[Zn(NH_3)_4]^{2+} + 2e^- = Zn + 4NH_3$.

$$E_{Cd} = E^0_{Cd} + \frac{RT}{nF}\log\frac{[NH_3]^4}{[Cd(NH_3)_4]^{2+}} \qquad [\text{Since } E^0_{Cd} = -0.6\,v\ [NH_3]_4 = 1]$$

and $[Cd(NH_3)_4]^{2+} = 0.01$ M

$$E_{Cd} = -0.61 - \frac{0.0591}{2}\log\frac{(1)}{(0.01)} \text{ against NHE; } -0.67\text{ V against NHE.}$$

similarly, $E_{Zn} = E^0_{Zn} + \frac{RT}{nF}\log\frac{[NH_3]^4}{[Zn(NH_3)_4]^{2+}}$ as $NH_3 = 1[Zn(NH_3)_4]^{2+} = 0.02$.

We have $E_{Zn} = -1.04 - \frac{0.0591}{2}\log\frac{(1)^4}{(0.02)} = -1.09$ V against NHE.

48.13 Zinc and cadmium were to be separated as amine complexes by controlled potential electrolysis with Pt cathode in 1 M ammonia solution containing 0.01 M cadmium and 0.02 M of zinc solutions. If $E^0_{Cd} = -0.67$ V and $E^0_{Zn} = 1.09$ V applied potential is –0.88 V. Calculate the potential for deposition of zinc which is required to keep complete plating in which solution?

Ans: By Nernst Equation, we have

$$E_{Cd} = E^0_{Cd} + \frac{RT}{nF}\log\frac{[NH_3]^4}{[Cd(NH_3)_4]^{2+}}$$

$$E_{Cd} = -0.67 - \frac{0.0591}{2}\log\frac{(1)}{[Cd(NH_3)_4]^{2+}}$$

$$\log[Cd(NH_3)_4^{2+}] = \frac{2-(0.88+0.61)}{(0.059)} = -9.15$$

anti log of $(-9.15) = 7.1\times10^{-10}$ M.

$$E_{Zn} = E^0_{Zn} + \frac{RT}{nF}\log\frac{[NH_3]^4}{[Zn(NH_3)_4]^{2+}}$$

$$E_{Zn} = -1.09 - \frac{0.0591}{2}\log\frac{(1)^4}{[Zn(NH_3)_4]^{2+}}$$

if $[Zn(NH_3)_4]^{2+} = 2\times10^{-6}$ M.

$$E_{Zn} = -1.09 - \frac{0.0591}{2}\log\frac{(1)^4}{(2\times10^{-6})} = 1.18 \text{ V against NHE.}$$

48.14 Find out the potential when 0.100 A was drawn from the cell consisting Cd/Cd^{2+}(1 M) || Cu (Cell with resistance = 4 ohms)?

Ans: $E=\left(E^0_{\text{Cu}}-E^0_{\text{Cd}}\right)$ as Cd = 1M = Cu equimolar solutions.

E_{Cu} = [0.337 – (–0.403)] = 0.740 Volts if 4 = ohmic resistance e = 0.1

E_{Cell} = (0.740 – 1R) = (0.740) – (0.1 × 4) = 0.340 V.

48.15 Calculate the required potential to produce a current of 0.100 A in the reverse direction in the cell consisting of $\text{Cd}|\text{Cd}^{2+}(1\text{M})||\text{Cu}^{2+}+(1\text{M})|\text{Cu}$? $\left(\text{Given } E^0_{\text{Cu}}=0.337,\ E^0_{\text{Cd}}=0.403\right)$

Ans: $E=\left(E^0_{\text{Cd}}-E^0_{\text{Cu}}\right)=(-0.403)-0.337=-0.740\text{ V}$

E_{Cell} = (–0.740) – (0.1 × 4) = –1.14 V.

48.16 If 30 ml iodide solution was analysed by coulometic titration, Ag^+ which was deposited at the anode with current of 40.2 μ a for passage of current for 283 seconds. Calculate the concentration of iodide ion the solution.

Ans: No. of Coulombs $=\dfrac{40.2\times 283}{1000}=11.38$ $\left(\text{If } I=40.2,\ t=283500\right)$

Nos. of far Cd = No. of $I=\dfrac{11.38}{96493}=1.179\times 10^{-4}$

$$\left[I^-\right]=\frac{1.79\times 10^{-4}}{30}\times\frac{1000}{\text{liter}}\times\frac{1\text{ mole}}{\text{eq. wt.}}=3.93\times 10^{-3}\text{ moles/L.}$$

48.17 Coulometric reduction of iron was carried out at Pt cathode when 39.3 ml of O_2 and H_2 was liberated at 765 mm and 23°C Calculate the amount of ferric salt consumed?

Ans: If we convert gas to standard conditions, we have

$$39.3\text{ ml}\times\frac{765}{760}\times\frac{273}{296}=36.48\text{ ml.}$$

If $4H^+ + 4e^- = 2H_2(g),\ 2H_2O \rightarrow O_2(g) + 4H^+ + 4e^-$

With 3 mole of gas with 4 mole of electrons 0.750 mol of gas is generated.

$$\text{Nos. of Faradays}=\frac{36.48}{22.400}\times\frac{1}{0.750}=2.17\times 10^{-3}$$

$$\text{Wt. of ferric salt}=\frac{2.171\times 10^{-3}\times 1\text{ eq}}{F}\times\frac{400\text{ g ferric salt}}{\text{mole}}\times\frac{1}{2\text{ equi/mol ferric sulphate}}$$

4.34 × 10^{-2} of salt.

48.18 Find out the magnitude of the overvoltage if during electrolysis at cathode copper was deposited which was 10^{-4} M but copper in bulk solution was 10^{-3} M?

Ans: The electrode potential was $E_1=E^0_{\text{Cu}}+\dfrac{RT}{nF}\log\left[\text{Cu}^{2+}\right]$

$$E_1=0.345+\frac{0.0591}{2}\log\left(10^{-2}\right)$$

$E_1 = 0.286$ V for copper

Because of concentrations over potential the real potential was

$$E_2 = 0.345 + \frac{0.0591}{2}\log\left(10^{-4}\right) = 0.227\text{ V in bulk of solution}$$

$(E_2 - E_1) = (0.227 - 0.286) = 0.59$ V is over potential.

48.19 What was the starting potential as well as the reversible cathode potential required for copper for plating from 0.1 M copper sulphate solution? $\left(E^0_{Cu} = 0.350\right)$.

Ans: As per Nernst equation

$$E = E^0_{Cu} + \frac{RT}{nF}\log\left(Cu^{+2}\right) = 0.350 - \frac{0.0591}{2}\log(0.1) = 0.35 - 0.03 = +0.03 = +0.32\text{V}$$

This is minimum potential to start deposition. Now reversible potential is

$$E = E^0_{Cu} + \frac{RT}{nF}\log\left(10^{-4}\right) = 0.350 - \frac{0.0591}{2}\log(0.0001) = 0.23\text{ V}.$$

48.20 What potential was needed to liberate oxygen at anode if the reaction proceeds as:

$$\frac{1}{2}O_2 + 2H^+ + 2e^- = H_2O \left(\text{Given } E^0_H = 1.229,\ H^+ = 1\text{M}\right),\ \text{pH} = 7.0?$$

Ans: The reaction was

$$\frac{1}{2}O_2 + 2H^+ + 2e^- = H_2O \qquad E^0_H = 1.23, \qquad [H^+] = 1\text{M}$$

By Nernst equation

$$E = E^0_H + \frac{RT}{nF}\log\left(H^+\right) \qquad \text{if } E^0_H = 1.229,\ [H^+] = 1.0,\ n = 1,\ RT/F = 0.0591$$

$$E = 1.229 - \frac{0.0591}{1}\log\left(H^+\right) \qquad \text{at pH} = 7.0 [H^+] = 10^{-7} \text{ as } n = 1$$

$$E = 1.229 - \frac{0.0591}{1}\log\left(10^{-7}\right)$$

$$E = 1.229 - \frac{0.0591}{1}\log\left(10^{-7}\right) = (1.229 - 0.413) = 0.81\text{ V}.$$

PROBLEMS FOR PRACTICE

48.21 What was the period needed for deposition of 0.4 gm of copper from acid solution of blue vitrol with current of 4 amps with only 50% current efficiency? **[Ans: 607 seconds]**

48.22 Find out the potential needed to pass a current of 0.05 A with resistance of 6.50 ohm in cell of Pt|Fe^{2+}(0.180) Fe^{3+}(0.0100 F)||Pd^2(0.09) M? **[Ans: –0.066 V]**

48.23 Ascertain the current efficiency for the following reaction when 0.0764 gm of lead dioxide was electrolysed with 0.240 amperes current for 19.6 minutes? The reaction was $Pb^{2+} + 2H_2O \rightarrow PbO_2(s) + 4H^+ + 2e^-$? **[Ans: 21.9%]**

48.24 The hydrochloric acid was titrated coulometrically with alkali hydroxide at constant current of 20.34 μA for 645.3 seconds. What was the amount of hydrochloric acid in the solution? **[Ans:** 0.1360]

48.25 The picric acid (0.914 gm) was dissolved in 0.1 M of hydrochloric acid and it was coulometrically analysed at –0.65v vs SCE with 65.7 coulombs of electricity. Estimate the number of electrons transferred in the process of reduction? **[Ans:** 17.1 electrons two mols reduced]

48.26 Find out the concentration of the unknown organic compound if 10 ml of the sample at –08 V was reacted. The area under current time curve was 33.7 mA? **[Ans:** 1.05×10^{-3} M]

48.27 Ascertain the number of coulombs of electricity required to reduce 56.783 grams of silver to elemental silver? **[Ans:** 50783 Coulombs]

48.28 How much of cadmium was deposited if 4.251 coulombs of electricity passed through cadmium chloride solution? **[Ans:** 1.40 mg cd]

48.29 Find out the concentration of chloride when 0.2 ml of sample was titrated when 2.70 mg of silver was needed to reach end point? **[Ans:** 0.1254 mg/L]

48.30 Estimate the period required to dissociate 2.456 gm of potassium chloride with current of 5.3 A? **[Ans:** 10.93 mt]

48.31 Calculate the potential of electrode of zinc in solution at the concentration of 1.0×10^{-3} gm liter? **[Ans:** 0.847 V]

48.32 Calculate the normality of the solution in titration of potassium permanganate with ferrous iron when it took 33 mt to reduce 25 ml solution at current of 250 ma?**[Ans:** W = 0.2052 gm/eq]

48.33 What was change in the potential during electrolysis of stannous chloride which in its concentration was varied from 0.02 gm to 0.001 mg? **[Ans:** E = 0.107 V]

48.34 Estimate a time required in seconds for a current of 0.5 amp to plate out 0.59 gm of silver with 80% current efficiency? **[Ans:** 111 seconds]

48.35 If solution of 1×10^{-3} m of silver contained gold ions, calculate the concentration of gold (III) ion when silver and gold were deposited simultaneously? **[Ans:** 2×10^{-4} M each]

48.36 Find out the normality of the acid solution when the calibration factor of an integrating motor was 0.00267 micro equi. percount when 10 ml produced 40.72 counts during titration? **[Ans:** 0.01087 N]

48.37 A 0.5 M potassium sulphate liberated 22.33 ml of ($H_2 + O_2$) gas at 24°C and 740 mm pressure if the partial pressure of water over electrolyte was 22 mm. Calculate the coulombs actually used? **[Ans:** 111.7 Q]

48.38 A 100 ml solution contained 0.05 M of lead chloride and 0.02 M tin chloride at pH = 3.0 when one ampere current was passed for 1500 seconds, what is the rise in mass of platinum electrode? **[Ans:** 1.094 g]

48.39 Predict over voltage of hydrogen needed to electrolyse nickel from the 0.01 M nickel chloride solution with 0.10 M of acid? **[Ans:** 0.250 V]

48.40 Calculate how may gms of potassium dichromate was equivalent to 1 μA per second in a typical coulometric titration? **[Ans:** 5.08×10^{-10} gm]

REFERENCES

48.1 S.M. Khopkar, *Basic Concepts of Analytical Chemistry,* 3rd edn., New Age International Publisher (P) Ltd. (2007).

48.2 D.J. Pietrzyke and C.W. Frank, *Analytical Chemistry: An Introduction,* Academic Press, pp. 600-1, 605; 608-9 (1979).

48.3 R.D. Braun, *Introduction to Instrumental Analysis,* McGraw-Hill and Co., pp. 807-19 (1987).

48.4 D.G. Peters, J.M. Hayes and G.M. Heiftze, *Chemical Separations and Measurements*, Saunders Golden Series, pp. 409-12 (1974).

48.5 D.A. Skoog and D.M. West, *Analytical Chemistry*, 3rd edn., Holt Reinhart and Winston, pp. 390-98, 406-26 (1979).

48.6 J.S. Frietz and G.H. Schenk, *Quantitative Analytical Chemistry,* Allyn and Beacon Inc., pp. 301-06 (1979).

48.7 Y. Lyalikov, *Physico-Chemical Analysis,* Mir Publishing House, pp. 337 (1964).

48.8 H.H. Willard, L.L. Merritt and J.A. Dean, *Instrumental Methods of Analysis,* D. Van Nostrand and Co., pp. 660, 669 (1980).

48.9 G.W. Ewing, *Instrumental Methods of Chemical Analysis,* 5th edn., McGraw-Hill and Co. (1978).

48.10 R.L. Peasok and L.D. Shield, *Modern Methods of Chemical Analysis,* Wiley-Interscience (1968).

48.11 J.J. Lingane, *Electroanalytical Chemistry,* 2nd edn., Interscience (1958).

48.12 D.G. Peters, J.M. Hayes and G.M. Hieftze, *A Brief Introduction to Modern Chemical Analysis,* Saunders and Co. (1980).

Chapter 49

Chemical Sensors and Biosensors

INTRODUCTION

The term detector, transducer and sensor are used in loose sense synonymously. However, they very much differ substantially from each another. For instance, the detectors refers to mechanical, electrical or chemical devices to identify or indicate a change in one of the variables like pressure, temperature, charge, magnetic field or nuclear radiation of molecules.

The term transducer refers specifically to device which provides the information in the domain of electrical area. Photodiode, photomultiplier tubes are transducers, because they provide information on current being voltage proportional to radiant power of elctromagnetic radiation. Finally term sensor is used to explain the analytical devices capable of monitoring specific chemical species on continued basis e.g., glass or ion selective eletrodes involving measurement of potential or fiber optic sensors to measure absorbance or transmittance. Such sensor has transducer juxtaposed with a phase which has chemically sensitive coating, e.g., optrodes consisting of transducer coupled with fiber optics coated with a substance which can provide signal when it is coated with specifically to fixed physical/chemical property e.g., QCM *i.e.,* quartz crystal microbalance is one containing piezoelectric device. Such sensors are used to analyse benzene, hydrogen sulphide, formaldehyde and hydrochloric acid gas. Phosgene and mustard gas can be similary analysed.

Now we consider difference between sensitivity and detection limit. Sensitivity is measure of capacity to know difference between two small differences in analysed concentration. It is obtainable from slope of calibration plot. While detection limit is that limit which represents minimum concentration or the mass of analyte which can be detected at known confidence level while selectivity refers to degree to which analytical method is free from interferences by other species in sample. However absence of interference is rare in chemical analysis.

The discovery of biosensor is of recent origin. When a sample is brought in proximity of immobilized enzyme, it shows the catalytic reaction to yield new species e.g., NH_3, CO_2, H_2 or H_2O_2. Most of biosensors are based upon membrane electrodes to facilitate analysis of complex organic molecules. Further biocatalyst facilitates mild condition reaction with no interferences. Only limitation is cost of enzyme used which is prohibitive. Therefore to reduce the cost, immobilized enzymes are used. Also they are used to conserve the enzymes.

In comparison to enzyme Fiber Optic Sensors or Optrodes are less expensive. They consist of reagent phase immobilized on the end of fiber. The reaction of analyt sample with reagent leads to change in the value of absorbance, reflectance, fluoroscence or luminescence which is feeded to the one of the transducer like photocell. Such sensors can be minaturised and they are simple to handle. The voltammetric sensors are also called electrodes which are commonly used. The classical example is that of oxygen sensor for DO analysis. It consist of Clark oxygen sensor, cell with platinum cathode and silver anode immersed in buffer solution of potassium chloride. The membrane is used as partition for diffusion of oxygen leading to change in the potential between two electrodes. The drop or rise in potential is proportional to the concentration. Alternatively one can also use enzyme based electrode in DO meter or glucose analysis. These are the few examples of voltammetric sensors. The non-dispersive infrared analyser also involves use of membrane sensitive, sensor at fixed wavelength. The change in capacitance is sensed by amplifier. Such change is due to unequal heating in vicinity of membrane electrodes which in turn leads to capacitance change.

Finally let us consider what is meant by sensitised? An atom gets excited by absorption of electrons and transfers its energy to an orbit of different elements. The different atom emits excited radiation while returning to ground state. Such atomic fluorescence is termed as "sensitised fluorescence" which has nothing to do with any kind of sensors or sensitive devices.

A change in mass can also be detected by piezoelectric sensor involving use of quartz. Fluoroscence is prefered for measuring microbial activity with biosensor or for glucose analysis by biocatalytic sensor. Tissue sensor involves the use of say mushrooms for analysis of alcohol mycotoxins.

The field of sensor is new and promising. Many more effective sensors may be devised in future for their applications in nanotechnology. However for all these sensors analytical chemistry forms the foundation of knowledge in development of such sensors.

PROBLEMS

Multiple Choice Questions

Tick mark the correct answer from following multiple choices:

49.1 The first true chemical sensor discovered was the classical
(*a*) Litmus paper (*b*) Filter paper
(*c*) Ion exchange paper (*d*) Chromatographic paper

49.2 The first sensor developed for quantitative analysis of anion is
(*a*) Glass electrode (*b*) Ion-selective electrode
(*c*) Reference electrode (*d*) Solid state electrode

49.3 In comparison with measurement by sensor, instrumental analysis is still superior as
(*a*) It is cheap (*b*) It is dependable
(*c*) It is accurate, rapid and reproducible (*d*) Easy to operate.

49.4 Good sensor should satisfy one of the following requirement
(*a*) Specificity and sensitivity (*b*) low cost of fabrication
(*c*) Multiple applications (*d*) Better shelf life.

49.5 Sensor involving monitoring of absorbance/transmittancy is called as
(*a*) Thermal sensor (*b*) Optical sensor
(*c*) Ampero or potentiometric sensor (*d*) Piezoelectric sensor

49.6 Ion selective electrode belongs to the one of the following technique of analysis as:
(*a*) Optical method (*b*) Thermal method
(*c*) Electroanalytical technique (*d*) Piezoelectric device

49.7 In thermal sensor a significant change of one of the following property is mainly measured.
(*a*) Temperature (*b*) Heat contents
(*c*) Entropy (*d*) Enthalpy

49.8 Dissolved oxygen (DO) meter utilises one of the principles of analysis.
(*a*) Conductimetry (*b*) Potentiometry
(*c*) Amperometry (*d*) Coulometry

49.9 The most dependable physical property for measurement in optical sensor is:
(*a*) absorbance (*b*) reflection
(*c*) scattering (*d*) reaction

49.10 Most reliable mode of deposition of reagent in optical sensor is:
(*a*) entrapment of sample (*b*) surface adsorption
(*c*) electrostatic extraction (*d*) chemical immobilisation

49.11 Piezoelectric sensor largely involves measurement of
(*a*) Mass (*b*) Density
(*c*) Sp. gravity (*d*) Viscosity

49.12 The kind of crystals used in piezoelectric sensor is made up of
(*a*) silica (*b*) alumina
(*c*) quartz (*d*) sodium oxide.

49.13 Quartz crystal microbalance (QCM) works well with following
(*a*) Gas (*b*) Solid
(*c*) Liquid (*d*) Vacuum

49.14 Surface accoustic wave sensors do not work well in liquids
(*a*) as process is expensive (*b*) as needs more signal
(*c*) wave is damped (*d*) surface is spoiled

49.15 Microorganism based sensors are example of
(*a*) Biosensors (*b*) Tissue sensors
(*c*) Transducer sensors (*d*) Live sensors

49.16 Biosensors involve the measurement of change in
(*a*) metabolic activity (*b*) mass
(*c*) color (*d*) temperature

49.17 On account of lack of specificity biosensors are used only for
(*a*) broad spectrum analysis (*b*) rapid analysis
(*c*) economical processing effect (*d*) measurement

49.18 The microbial activity is monitored by measuring
(*a*) fluoresence of radiation (*b*) scattering of radiation
(*c*) phosphorescence phenomena (*d*) reflection effect

49.19 Biocatalyst and its immobilisation is used in sensor as
(*a*) amperimetry (*b*) oxygen
(*c*) bios (*d*) piezoelectric

49.20 Biosensors are extensively used for analysis of
(*a*) Sucrose (*b*) Amino acid
(*c*) Proteins (*d*) Glucose

49.21 Biosensors are used for measuring BOD_5 of water as
(*a*) it is cheap (*b*) is reproducible
(*c*) sensitive (*d*) selective

49.22 Tissue sensors largely involves the use of the
(*a*) tissue of plant and animals (*b*) microorganisms
(*c*) biocatalyst (*d*) animal organs

49.23 For the analysis of alcohol tissue sensor is made up of
(*a*) Mushrooms (*b*) Potato
(*c*) Cucumber (*d*) Cabbage

49.24 Mushrooms are used as tissue sensor in analysis of
(*a*) Ascorbic acid (*b*) Glycolic acid
(*c*) Alcohol (*d*) Urea

49.25 Animal tissues are used as tissue sensor for analysis of
(*a*) Cyatochrome (*b*) Carbohydrate
(*c*) Alkaloids (*d*) Fatty acids

ANSWERS

49.1 a	49.2 b	49.3 c	49.4 a	49.5 b	49.6 c	49.7 b
49.8 b	49.9 a	49.10 d	49.11 a	49.12 c	49.13 c	49.14 c
49.15 a	49.16 a	49.17 a	49.18 a	49.19 c	49.20 d	49.21 b
49.22 a	49.23 a	49.24 c	49.25 a.			

REFERENCES

49.1 P. Warsfold, A. Townsend, C. Poole, *Encyclopedia of Separation Sciences*, 2nd edn., Elsevier Pub. House p. 224 (2004).

49.2 A. Townsend and P. Worsfold, *Encyclopedia of Separation Science*, 1st edn., Academic Press, London (1994).

49.3 D.A. Skoog, E.J. Holler and T.A. Nieman, *Principles of Instrumental Analysis,* Harcourt Brace College Publication, 5th edn., pp. 9, 10, 609, 179 and 400 (1997).

49.4 S.M. Khopkar, *Basic Concepts of Analytical Chemistry*, 3rd edn., New Age International Publishers (P) Ltd. (2007).

Chapter 50

On-line Analysis—Automation

INTRODUCTION

Analytical chemistry covers large number of techniques pertaining to the quantitative chemical analysis such as spectral methods for characterisation and determination, electroanalytical methods for the quantification, radioanalytical technique which are well suited for study of radioactive isotopes, and thermal methods for characterisation of compounds which are thermally stable. These spectral methods are most popular with chemist due to the simplicity, ease of operation, and selectivity in analysis.

However, when question arises as to how to analyse large number of samples for multi-element analysis many methods in spite of being selective and sensitive have to be discarded. This is because we require rapid responses to be recorded for large number of samples. There are limited physical properties which can be easily measured. Such properties include refractive index, magnetic moment, change in potential or conductance, absorbance produced with chromogenic ligands, development of the diffusion current with reducible ion. Although all properties are suitable depending upon level of constituents and background of sample, one can resort to any technique of analysis.

In continuous analysis, when multiple samples are encountered and when multi-element analysis is necessary, techniques like flow injection analysis is most suitable. For instance pollutants like H_2S; NO-NOx, have been analysed by making use of flow colorimetry. A change in conductance is also used to measure SO_2 from the atmosphere. In all such operations, we need rapid chemical reaction which would work with instantaneous response. Indefinite time lag in chemical reaction is useless for continous measurement.

Based upon the rapid variation in physical property, number of on-line methods have been developed. The primary criteria in the selection of such method is response time. When the other factor outweighs other consideration of the manpower needed. Those methods which need no personal supervision or observation are best for automated analysis. The method should be at the same time simple but rapid.

Apart from response time or manpower requirements some other considerations include selectivity of the method. The proposed method should be selective and specific for a particular component. The other criteria is to judge sensitivity of the method. A sensitivity is of prime importance in selection when we have to choose one method from number of available options.

In this chapter large number of multiple choice questions are presented which go into breadth and depth of methods useful for automated methods.

PROBLEMS

Multiples Choice Questions

Tick mark correct answer from following multiple choices:

50.1 On-line analysis is popular because it basically
(*a*) saves time and manpower (*b*) saves electricity
(*c*) controls quality (*d*) facilitates difficult separations.

50.2 Most common property monitored in on-line analysis is related to:
(*a*) optical (*b*) electrical
(*c*) magnetic (*d*) radiometric field.

50.3 Automated methods of analysis are
(*a*) rapid needing no manpower (*b*) need trained manpower
(*c*) are accurate and precise (*d*) permit limited analysis.

50.4 Simple instruments demonstrating principle of continuous analysis is:
(*a*) fraction collector (*b*) digital analyser
(*c*) chromatograph (*d*) polarograph.

50.5 The correct sequence in on-line analysis is:
(*a*) sample, transducer, recorder (*b*) recorder, transducer, sample
(*c*) sample, recorder, transducer (*d*) Transducer, sample, recorder

50.6 On-line analysis of coloured complex can use
(*a*) IR spectrophotometry (*b*) flame photometry
(*c*) flow colorimetry (*d*) coulometry.

50.7 Non-dispersive infrared technique is used for analysis of
(*a*) H_2S (*b*) $CO–CO_2$
(*c*) $NO - NO_2$ (*d*) SO_2.

50.8 Automated H_2S analyser involves measurement with
(*a*) colorimeter (*b*) gas chromatograph
(*c*) nephelometer (*d*) conductometer.

50.9 Automated SO_2 analyser measures:
(*a*) conductance (*b*) refraction
(*c*) scattering (*d*) magnetic moment.

50.10 On-line analysis is primarily useful as it
(*a*) controls quality on large scale (*b*) is economical
(*c*) is simple (*d*) needs small sample.

50.11 On-line analysis is
(*a*) cheap (*b*) permitts analysis of large number of samples
(*c*) depends on trained manpower (*d*) is useful at microgram level.

50.12 Most commonly used instrument in on-line analysis of colored complex is:
(*a*) polarograph (*b*) spectrophotometer
(*c*) thermal analyser (*d*) fluorimeter.

50.13 Karl Fischer titration for moisture analysis involves titration
(*a*) photometry (*b*) thermometry
(*c*) amperometry (*d*) potentometry.

50.14 Monitoring of SO_2 in autoanalysis involves measurement of
(*a*) potential (*b*) current
(*c*) resistance (*d*) conductance.

50.15 Paramagenetic property can be made use of for the measurement of
(*a*) SO_2 (*b*) NO
(*c*) O_3 (*d*) CO_2.

50.16 Chemiluminescence measurement is used to analyse
(*a*) CO_2 (*b*) CO
(*c*) $NO - NO_2$ (*d*) O_2.

50.17 The automated parameter for measurement of water analysis is
(*a*) BOD (*b*) COD
(*c*) DO (*d*) SPM.

50.18 The flow injection analysis involves
(*a*) mixing of liquid to produce colour (*b*) use of pump
(*c*) control of flow rate of liquid (*d*) raising temperature of sample.

50.19 In autoanalysis of fluoride by flow colorimetry of complex of zirconium with alizarine the intensity of the colour
(*a*) increases (*b*) decreases
(*c*) remains same (*d*) is bleached.

50.20 When mercury is analysed by flow colorimetry with ammonium thiocyanate the intensity of coloured complex
(*a*) decreases (*b*) increases
(*c*) no change (*d*) changes to other wavelength.

50.21 Flow injection analysis involving conductance measurement needs
(*a*) long time to attain steady state (*b*) no time for attaining equilibrium
(*c*) coloured samples (*d*) rise in temperature of sample.

50.22 Dispersion coefficient in flow injection analysis represents:
(*a*) geometry of channel (*b*) flow viscosity of liquid
(*c*) shape of dispresive zone variation (*d*) concentration changes.

50.23 The dispersion coefficient indicates volume of sample is very:
(*a*) limited (*b*) unlimited
(*c*) large (*d*) small.

50.24 The concentration gradient can be generated if
(*a*) zone of dispersion is in reaction column
(*b*) rate of flow of liquid varies
(*c*) pressure is dropped
(*d*) temperature is raised.

50.25 Peak height in flow injection is not dependent upon
(*a*) sample size (*b*) channel
(*c*) flow rate (*d*) volume of flowing solution.

50.26 The dispersion of zone in sample depends upon
(*a*) sq. root of travelled distance in conduit tube
(*b*) change of temperature
(*c*) length of channel
(*d*) diameter of channel.

50.27 The flow injection analysis in terms of 50% represents
(*a*) volume of dispersion (*b*) interacting liquids
(*c*) reaching complete steady state (*d*) attaining half steady state.

50.28 The flow injection analysis prefers straight tube to coiled tube as it
(*a*) facilitates fast flow (*b*) facilitates radial mixing
(*c*) better reaction promotion (*d*) facilitates axial dispersion.

50.29 Ideal length for channel for good dispersion is
(*a*) 100 mm (*b*) 50 mm
(*c*) 10 mm (*d*) 1 mm.

50.30 SO_2 analyser by flow colorimetry uses
(*a*) p-roseaniline lecuo base (*b*) N-napthyldiamine hydrochloride
(*c*) Nessler's reagent (*d*) potassium dichromate.

50.31 Flow injection colorimetry for $NO - NO_2$ uses
(*a*) N-napthylethylenediamine chloride (*b*) alizarine
(*c*) quinanzarine (*d*) rose bengal extra.

50.32 Magnitude of *D* with channel of length of 20 cm, flow rate of 1.5 ml/mt,with diameter of 0.3 mm is
(*a*) 10.0 (*b*) 5.0
(*c*) 20.0 (*d*) 1.0.

50.33 The diffusion coefficient represents ratio of
(*a*) Lo/L (*b*) id/ie
(*c*) Ro/R (*d*) Co/C.

50.34 Better dispersion is possible with
(*a*) long channel (*b*) short channel
(*c*) curved channel (*d*) linear channel.

50.35 The analysis of fluoride in presence of anions in water sample is done with
(*a*) ion analyser (*b*) fluorimeter
(*c*) nephelometer (*d*) turbidimeter.

50.36 In FIA of coloured molydenum complex is obtained by interactions of ammonium molydate with
(*a*) acid (*b*) alkali
(*c*) phosphate (*d*) sulphate.

50.37 Bromothymol blue in FIA gives strong colour due to action of
(*a*) $Na_2Cr_2O_7$ (*b*) $NaBO_3$
(*c*) $Na_2B_4O_7$ (*d*) H_3BO_3

50.38 The anion responsible for liberating free thiocyanate in alkali thiocyanate solution is

(*a*) chloride (*b*) sulphate
(*c*) phosphate (*d*) nitrate.

50.39 The chemiluminescence property facilitates the analysis of

(*a*) $CO - CO_2$ (*b*) H_2S
(*c*) NO – NOx (*d*) hydrocarbons.

ANSWERS

50.1 a	50.2 a	50.3 a	50.4 a	50.5 a	50.6 c	50.7 b
50.8 a	50.9 a	50.10 a	50.11 b	50.12 b	50.13 d	50.14 d
50.15 b	50.16 c	50.17 c	50.18 a	50.19 b	50.20 b	50.21 a
50.22 c	50.23 a	50.24 a	50.25 d	50.26 a	50.27 d	50.28 b
50.29 d	50.30 a	50.31 a	50.32 d	50.33 d	50.34 a	50.35 a
50.36 c	50.37 c	50.38 a	50.39 c.			

REFERENCES

50.1 S.M. Khopkar, *Basic Concepts of Analytical Chemistry,* 3rd edn., New Age International Publishers (P) Ltd. (2007).

50.2 G.D. Christen, *Analytical Chemistry,* 5th edn., John Wiley and Co. (1994).

50.3 G.D. Nichols, *On-line Process Analysers,* John Wiley and Co. (1988).

50.4 V. Cerda and G. Ramis, *An Introduction to Laboratory Automation,* John Wiley (1990).

50.5 S. Siggia, *Continuous Process Analysis,* Addison Wesley (1958).

50.6 J. Ruzicka and E.H. Hansen, *Flow Injection Analysis,* 2nd edn., John Wiley (1987).

50.7 S.M. Khopkar, *Environmental Pollution Analysis,* 2nd edn., New Age International Publishers (P) Ltd. (2009).

50.8 S.M. Khopkar, *The Design and Development of Gas Detectors – Research and Industry,* (CSIR) **29** 86(1984).

50.9 A.K. De, S.M. Khopkar and R.A. Chalmers, *Solvent Extraction of Metals,* Van Nostrand Reinhold and Co. (1970).

50.10 S.M. Khopkar, *Analytical Chemistry of Macrocyclic and Supramoleculor Compounds,* 2nd edn., Springer-Verlag (2006).

50.11 S.M. Khopkar, *Solvent Extraction Separation of Elements with Liquid Ion Exchangers,* New Age International Publishers (P) Ltd. (2007).

50.12 S.M. Khopkar, *Environmental Pollution Monitoring and Control,* New Age International Publishers (P) Ltd. (2006).

Appendix-1

International Atomic Weights

Symbol	*Mass*	*Symbol*	*Mass*	*Symbol*	*Mass*
Ac	227	Hf	178.49	Ra	226
Ag	107.85	He	4.00	Rn	222
Al	26.98	H	1.00	Re	186.20
Am	243	In	114.81	Rh	102.90
Ar	39.76	Ir	192.21	Rb	83.46
As	74.92	K	39.09	Ru	101.07
At	210	La	138.90	Sb	121.76
Au	196.9	Li	6.94	Sc	44.95
Ba	137.32	Lu	174.96	Se	78.96
Be	9.01	Mg	24.30	Si	28.08
Bi	208.98	Mn	54.93	Sn	118.7
B	10.81	Hg	200.59	Sr	87.62
Br	79.90	Mo	95.94	S	32.06
Cd	112.41	Na	22.99	Ta	180.94
Ca	40.07	Ne	20.17	Tc	98
C	12.01	Ni	58.69	Te	127.60
Ce	140.12	Nb	92.90	Tl	204.38
Cs	132.90	N	14.00	Th	232.03
Cl	35.45	O	15.99	U	238.02
Cr	51.99	Os	190.23	V	50.94
Co	58.93	Pb	207.3	W	183.85
Cu	63.54	Pd	106.42	Xe	131.29
F	18.99	P	30.97	Yt	88.95
Fe	55.84	Pt	195.07	Zn	65.39
Ga	69.72	Pa	231.03	Zr	91.22

Ref.: Pure and Applied Chem-68, 2339 (1996).

Appendix-2

Standard Electrode Potential

Ion	Potential	Ion	Potential
Al^{3+}	–1.662v	Ni^{2+}	–0.250
Ag^{+}	+0.799	N^{4+}	–0.230
As^{3+}	+0.559	Na^{+}	–2.714
Ba^{2+}	–2.906	O^{2-}	+1.776
Bi^{3+}	+0.200	O_2 (g)	+1.229
Br^{-}	+1.065	Pa^{3+}	+0
C^{4+}	+0.699	Pb^{2+}	–0.126
Cd^{2+}	–0.403	Pd^{2+}	+0.987
Ca^{2+}	–2.866	Pt^{2+}	+0.73
Ce^{4+}	+1.70	Pt^{4+}	+0.68
Cl^{-}	+1.359	Sb^{3+}	0.581
Cr^{3+}	–0.744	Se^{2+}	+1.15
Cr^{6+}	+1.33	Se^{4+}	+0.740
Co^{2+}	–0.277	Sn^{2+}	–0.136
Co^{3+}	+1.808	Sn^{4+}	+0.154
Cu^{2+}	+0.337	S^{2-}	+0.141
Fe^{3+}	+0.771	Tl^{+}	–0.336
Fe^{2+}	–0.440	Tl^{3+}	+1.25
H^{+}	0	Ti^{2+}	–0.369
Hg^{2+}	+0.788	Ti^{4+}	+0.099
In^{3+}	+0.5335	UO^{2+}	+0.334
K^{+}	–.985	V^{5+}	+1.00
Ll^{+}	–3.045	V^{3+}	–0.256
Mg^{2+}	–2.363	Zn^{2+}	–0.763
Mn^{2+}	–1.180		

Ref.: D.A. Skoog, F.J. Holler, T.A. Newman Principles of Instrumental analysis, 5th edn., Saunders Harcourt Co. (1998) pg. 30.

Appendix-3

Important Physical Constants

Speed of light (c)	–	2.99792×10^{3} ms^{-1}
Planck constant (h)	–	6.62608×10^{-34} js
Avagadros number (N)	–	6.022137×10^{23} particles mol^{-}
Faraday's Constant (f)	–	96,485.31 V mol^{-}
Gas Constant (R)	–	8.31 451 × JK^{-1} mol^{-}
	–	0.08205 Latmle^{-1} mol^{-}
Boltzmans constant (K)	–	1.38066×10^{-23} JK^{-1}
Mass of an electron (e)	–	9.109080×10^{-31} kg
Electric charge	–	1.6021×10^{-19}C

$$\left(\frac{RT}{F}\right) = 0.0591$$

Appendix-4

Values of 't' for various levels of probability

Degrees of Freedom	*Factor for confidence interval*			
	80%	*90%*	*95%*	*99%*
1.	3.08	6.31	12.7	63.7
2.	1.89	2.92	9.92	31.6
3.	1.64	2.35	5.84	12.9
4.	1.53	2.13	4.60	8.6
5.	1.48	2.02	4.03	6.86
6.	1.44	1.94	3.71	5.96
7.	1.42	1.90	3.50	5.40
8.	1.40	1.86	3.36	5.0
9.	1.38	1.83	3.25	4.78
10.	1.37	1.81	3.17	4.59
11.	1.36	1.80	3.11	4.44
12.	1.36	1.78	3.06	4.32
13.	1.35	1.77	3.01	4.22
14.	1.34	1.76	2.98	4.14
∞	1.29	1.96	2.58	3.29

Ref.: 1. F.J. Holler, Principles of Instrumental analysis, Harcourt Brace (1992) p. 144.

2. D.A. Skoog, D.M. West – Analytical Chemistry Holt, Reinhart, Winston p. 71 (1979).

Appendix-5

Values of Q for rejection of quotient at various confidence limits

Number of Observations	*90%*	*95%*	*99%*
3	0.941	0.98	0.994
4	0.765	0.85	0.936
5	0.642	0.73	0.821
6	0.560	0.64	0.740
7	0.507	0.59	0.680
8	0.478	0.54	0.634
9	0.437	0.51	0.598
10	0.412	0.48	0.568

Ref.: 1. D.A. Skoog, D.M. West, F.J. Holler, 6th edn., *Fundamental of analytical chemistry* - Harcourt Brace (1992) p. 46.

2. J.S. Fritz, G.H. Schenk—Quantitative Analytical Chemistry, 4th edn., p. 38 Allyn & Bacon p. 38 (1979).

Appendix-6

Values of F at 95% confidence level

V_1 / V_2	V_1					
	2	*3*	*4*	*5*	*6*	∝
2	19.00	19.16	19.25	19.30	19.33	19.50
3	9.55	9.28	9.12	9.01	8.94	8.53
4	6.94	6.59	6.39	6.26	6.16	5.63
5	5.79	5.41	5.19	5.05	4.95	4.36
6	5.14	4.76	4.53	4.39	4.28	3.67
∞	3.00	2.60	2.37	2.10	2.10	1.00

Ref.: H.A. Lattinen. W.E. Harris, Chemical Analysis, p. 545, McGraw Hill & Co. (1978).

Appendix-7

Chi square distribution

Degree of Freedom	*Probability (P)*						
	0.99	*0.95*	*0.90*	*0.50*	*0.10*	*0.01*	*0.005*
1	0.00016	0.0039	0.0158	0.455	2.71	2.71	7.78
2	0.01	0.1026	0.211	1.39	4.61	9.21	10.6
3	0.071	0.352	0.584	2.37	6.25	11.3	12.8
4	0.207	0.711	1.064	3.36	7.78	13.3	14.9
5	0.412	1.15	1.61	4.35	9.24	15.1	16.7
6	0.676	1.64	2.20	5.35	10.6	16.8	18.5
7	0.99	2.17	2.83	6.35	12.0	18.5	20.3
8	1.34	2.73	3.49	7.34	13.4	20.1	22.0
9	1.73	3.33	4.17	8.34	14.7	21.7	23.6
10	2.56	3.94	4.87	9.34	16.0	23.2	25.2
20	7.43	7.26	8.55	14.3	22.3	30.6	32.8
30	13.79	18.49	20.60	29.3	40.3	50.9	53.7

Ref.: H.A. Lattinen, W.E. Harris—Chemical Analysis, p. 546, Second Edition, McGraw Hill & Co. (1976).

Appendix-8

Confidence levels for different values of Z

Confidence level (%)	*Z*
50	± 0.67
68	± 1.0
80	± 1.29
90	± 1.64
95	± 1.96
96	± 2.0
99	± 2.55
99.9	± 3.29

Confidence limit for mean = $x \pm z\sigma$, if σ = standard deviation

Confidence limit for mean (μ) = $x \pm z\sigma/VN$ (N = nos. of measurements)

Ref.: 1. D.A. Skoog, F.J. Holler, T.A. Niemon, Principles of Instrumental analysis, 5th Ed. Harcourt, Brace (1998).

2. D.A. Skoog, D.M. West – Analytical Chemistry, 3rd Ed. p. 68, Holt, Reinhart, Winston (1979).

Appendix-9

The deviation for the determination of standard deviation from range ($s = K \times$ range)

N	*K deviation factor*
2	0.89
3	0.59
4	0.49
5	0.43
6	0.40
7	0.37
8	0.35
9	0.34
10	0.33

Ref.: J.S. Fritz, G.H. Schenk—Quantitative Analytical Chemistry, 4th edn., p. 35 Allyn & Bacon in (1979).

Appendix-10

Values of C_v/C_sV_s against varying R in ion selective electrode

R	C_V/C_sV_s	*R*	C_V/C_sV_s
1.28	0.113	1.56	0.870
1.30	0.140	1.58	0.973
1.32	0.170	1.60	1.086
1.34	0.203	1.62	1.213
1.36	0.240	1.64	1.353
1.38	0.280	1.66	1.510
1.40	0.325	1.68	1.691
1.42	0.373	1.70	1.894
1.44	0.425	1.72	2.126
1.46	0.485	1.74	2.397
1.48	0.548	1.76	2.711
1.50	0.618	1.78	3.088
1.52	0.694	1.80	3.536
1.54	0.778	1.82	3.8015

Ref.: H. Brewster, Solving problems in Analytical Chemistry, John Wiley & Co. (1980), p. 336, 346.

Index

G

H

I

N

O

P

Q

R

T

U

V

W

X

Y

Z